T0324165

TERRESTRIAL DEPOSITIONAL SYSTEMS

TERRESTRIAL DEPOSITIONAL SYSTEMS

Deciphering Complexities through Multiple Stratigraphic Methods

Edited by

KATE E. ZEIGLER
Zeigler Geologic Consulting
Albuquerque, NM
United States

WILLIAM G. PARKER
Petrified Forest National Park
Petrified Forest, AZ
United States

ELSEVIER

Elsevier
Radarweg 29, PO Box 211, 1000 AE Amsterdam, Netherlands
The Boulevard, Langford Lane, Kidlington, Oxford OX5 1GB, United Kingdom
50 Hampshire Street, 5th Floor, Cambridge, MA 02139, United States

Notices
Knowledge and best practice in this field are constantly changing. As new research and experience broaden
our understanding, changes in research methods, professional practices, or medical treatment may
become necessary.

Practitioners and researchers must always rely on their own experience and knowledge in evaluating and
using any information, methods, compounds, or experiments described herein. In using such information
or methods they should be mindful of their own safety and the safety of others, including parties for
whom they have a professional responsibility.

To the fullest extent of the law, neither the Publisher nor the authors, contributors, or editors, assume any
liability for any injury and/or damage to persons or property as a matter of products liability, negligence
or otherwise, or from any use or operation of any methods, products, instructions, or ideas contained in
the material herein.

Library of Congress Cataloging-in-Publication Data
A catalog record for this book is available from the Library of Congress

British Library Cataloguing-in-Publication Data
A catalogue record for this book is available from the British Library

ISBN: 978-0-12-803243-5

For information on all Elsevier publications visit our
website at https://www.elsevier.com/books-and-journals

Working together
to grow libraries in
developing countries

www.elsevier.com • www.bookaid.org

Publisher: Candice Janco
Acquisition Editor: Amy Shapiro
Editorial Project Manager: Tasha Frank
Production Project Manager: Anitha Sivaraj
Cover Designer: Matthew Limbert

Typeset by SPi Global, India

IN MEMORIAM

This volume is dedicated to William R. Dickinson (1931–2015):
   Professor Emeritus, University of Arizona, and legendary Colorado Plateau geologist
   "I am here tracing the History of the Earth itself, from its own Monuments."
                                                                    Jean André Deluc, 1794

(in "Geological Letters Addressed to Professor Blumenbach, Letter 3", *The British Critic*, p. 598.)

CONTENTS

Magnetostratigraphy of the Upper Jurassic Morrison Formation at Dinosaur National Monument, Utah, and Prospects for Using Magnetostratigraphy as a Correlative Tool in the Morrison Formation   279

S.C.R. Maidment, D. Balikova, A.R. Muxworthy

Terrestrial Carbon Isotope Chemostratigraphy in the Yellow Cat Member of the Cedar Mountain Formation: Complications and Pitfalls   303

M.B. Suarez, C.A. Suarez, A.H. Al-Suwaidi, G. Hatzell, J.I. Kirkland, J. Salazar-Verdin, G.A. Ludvigson, R.M. Joeckel

CONTRIBUTORS

A.H. Al-Suwaidi
Petroleum Institute University & Research Centre, Abu Dhabi, United Arabic Emirates

D. Balikova
Imperial College London, London, United Kingdom

A. Bercovici
Smithsonian Institution, Washington, DC, United States

G. Hatzell
University of Arkansas Fayettville, Fayettville, AR, United States

R.M. Joeckel
University of Nebraska, Lincoln, NE, United States

J.I. Kirkland
Utah Geological Survey, Salt Lake City, UT, United States

K.P. Kodama
Lehigh University, Bethlehem, PA, United States

G.A. Ludvigson
Kansas Geological Survey, Lawrence, KS, United States

S.C.R. Maidment
University of Brighton, Brighton, United Kingdom

J.W. Martz
University of Houston-Downtown, Houston, TX, United States

L.A. Michel
Southern Methodist University, Dallas, TX, United States
Tennessee Technological University, Cookeville, TN, United States

A.R. Muxworthy
Imperial College London, London, United Kingdom

T.S. Myers
Southern Methodist University, Dallas, TX, United States

W.G. Parker
Petrified Forest National Park, Petrified Forest, AZ, United States

J. Salazar-Verdin
University of Texas San Antonio, San Antonio, TX, United States

M.B. Suarez
University of Texas San Antonio, San Antonio, TX, United States

C.A. Suarez
University of Arkansas Fayettville, Fayettville, AR, United States

N.J. Tabor
Southern Methodist University, Dallas, TX, United States

J. Vellekoop
KU Leuven, Leuven, Belgium

K.E. Zeigler
Zeigler Geologic Consulting, Albuquerque, NM, United States

PREFACE

Stratigraphic correlation provides the framework for understanding ancient Earth systems. If stratigraphic resolution is poor, so will be the understanding of the system.
Olsen and Kent (2000)

The stratigraphic record of terrestrial deposits reflects a complicated interplay among many different depositional environments that includes both the depositional record and also the erosional forces that erase part of the story recorded by sedimentary rocks. Each outcrop of terrestrial strata holds a part of the history of a region and if multiple outcrops can be linked together, the bigger picture of a basin's development begins to emerge. It can be difficult to understand the depositional history of even a single outcrop. Additionally, correlations among outcrops near and far can be very difficult to determine correctly.

The study of complex depositional systems often relies on more than one stratigraphic technique to truly understand the sequence and timing of key events. For example, lithostratigraphy alone may not clarify complicated relationships in a rock sequence, or the farther one travels from the basin of interest, other techniques such as magnetostratigraphy and chronostratigraphy are key to testing, or even substituting for, long-distance biostratigraphic correlations. This book compiles detailed methodologies for techniques such as biostratigraphy, magnetostratigraphy, and chemostratigraphy and provides case studies that showcase the use of these techniques. We believe that readers will find the detailed information on these techniques and the associated case studies to be useful when planning and implementing large-scale studies of complex depositional systems.

By combining chapters that are oriented toward explanations of specific analytical techniques, experiments, sampling methods, etc., with examples of how these different methods are used, we hope to provide a useful "manual" for tackling these types of rock sequences. The overall style of this book was formulated to be approachable and straightforward for readers of all different levels of experience. Technique chapters provide detailed explanations of different analyses, data collection methods, and sampling techniques for use in beginning these types of research projects. Included chapters discuss collection of appropriate field data for stratigraphic and biostratigraphic studies, a guide for recognizing and characterizing paleosols, and methods for magnetostratigraphic studies of terrestrial rocks. Case studies show the full application of individual stratigraphic methods as well as how interdisciplinary projects can yield significant results including paleomagnetic and isotope chemostratigraphic studies of Mesozoic rocks of the western

United States, as well as formulation of long-range biostratigraphic units in Triassic rocks. Several of the chapters are coauthored by teams who have been working together on a rock sequence, but from different angles. This allows readers to see how different techniques can be interwoven to approach a better understanding of a complex rock sequence.

Terrestrial Depositional Systems: Deciphering Complexities through Multiple Stratigraphic Methods will be of great interest to graduate students and early-level professionals, as well as anyone looking for multidisciplinary discussions of depositional basin analysis.

Building Local Biostratigraphic Models for the Upper Triassic of Western North America: Methods and Considerations

W.G. Parker*, J.W. Martz†
*Petrified Forest National Park, Petrified Forest, AZ, United States
†University of Houston-Downtown, Houston, TX, United States

Contents

INTRODUCTION

Small-Scale Studies as the Foundation of Natural Science

Recently there has been a concerted effort to improve the scientific rigor of studies on Late Triassic biotic and environmental change in western North America. These studies have emphasized the need for thorough documentation of the evidence used to construct hypotheses. This documentation allows other researchers to fully comprehend and assess the evidence for presented arguments, and also to reexamine the evidence in order to confirm that it has been represented accurately. This principle has been applied to hypotheses involving vertebrate systematics (e.g., Nesbitt, 2011), the identification of vertebrate fossils using apomorphies and specified voucher specimens (e.g., Nesbitt et al., 2007; Irmis et al., 2007; Nesbitt and Stocker, 2008; Parker and Martz, 2011; Martz et al., 2013; Parker, 2013), and stratigraphic models (e.g., Martz and Parker, 2010; Ramezani et al., 2011; Irmis et al., 2011; Parker and Martz, 2011; Martz et al.,

Terrestrial Depositional Systems
http://dx.doi.org/10.1016/B978-0-12-803243-5.00001-7

2012, 2014; Atchley et al., 2013; Nordt et al., 2015). These well-documented studies essentially provide a detailed roadmap to their own potential falsification (Popper, 1962).

Although developing a large-scale synthesis of the history of life is a major goal of paleontology and geology, a synthesis is only as accurate as the smaller-scale studies on which it is based (Fig. 1). As noted by Holland (1989, p. 263), "It is easier to play with secondary data than to collect primary data." Understanding environmental and biotic change begins with local lithostratigraphic, biostratigraphic, and chronostratigraphic studies, as these make it possible to determine the order of events in Earth history. The development of stratigraphic models for particular study areas that are not only detailed and accurate, but also scientifically testable, requires detailed and scrupulously collected field data. Biostratigraphic and chronostratigraphic models are constructed by first developing detailed lithostratigraphic models using measured sections, onto which the stratigraphic level for fossil localities and sampling sites for geochemical, radioisotopic, and magnetostratigraphic samples may be plotted. The accuracy of a stratigraphic model therefore depends on the reliability of data collected from particular sites; a biostratigraphic model cannot be accurate if the superpositional relationships of fossil localities are not accurately and precisely known (e.g., Parker and Martz, 2011; Martz et al., 2013). Moreover, a

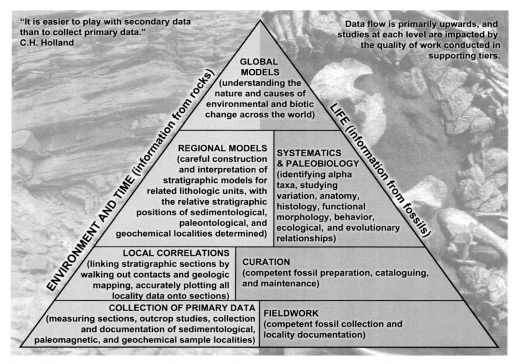

Fig. 1 The hierarchy of study scales, showing the reliance of large-scale synthesis on a large number of detailed local studies supported by thorough documentation.

biostratigraphic model is only as reliable as the taxonomic identification of the fossils used to construct it, so fossil identification requires great care. The absolute ages of environmental and biotic events, and where they fall on the geologic timescale, are likewise unlikely to be accurately known if the superpositional relationships of radioisotopic and magnetostratigraphic samples are not accurately documented. As local studies are combined into global models of environmental and biotic change, any errors that occur in the construction of stratigraphic models at small geographic scales will negatively impact the validity of global models derived from them (Fig. 1). Therefore, high-quality data collection and documentation at the local level is critical.

This paper is concerned with the methodology for constructing detailed, rigorous, and testable stratigraphic models on small geographic scales. These methods will be illustrated using local studies from a particular regional study area: the Upper Triassic continental sedimentary rocks of western North America assigned to the Chinle Formation and Dockum Group (Stewart et al., 1972; Blakey and Gubitosa, 1983; Lehman, 1994; Martz and Parker, 2010; Parker and Martz, 2011). This paper will not provide complete instructions on the basics of geologic mapping, measuring stratigraphic sections, and fossil collection. However, it will provide specific recommendations for applying these methods in the construction of detailed, accurate, well-documented, and testable lithostratigraphic, biostratigraphic, and chronostratigraphic models. To students, we offer this as a guide to important considerations that may not be covered in a basic field geology course. To many professional readers, this paper may seem to be reinventing the wheel by presenting concepts and methods with which they are already well acquainted. However, the fact that these basic methods of documentation are not applied consistently in many studies makes clear that they need to be reemphasized. Moreover, in the course of our research we have encountered particular problems and solutions to those problems that may not be immediately obvious, even to the experienced researcher.

The Historical Synthesis of Stratigraphy

The development of methods for reconstructing geologic time has been a growing synthesis of ideas, starting with the realization that stratigraphy is the key to building Earth history. Increasing scrutiny of the stratigraphic record has been accompanied by the discrimination of different concepts relating to that record. In the process, paleontologists and geologists have gained not only a more detailed understanding of environmental and biotic change over time, but also the ability to extend the correlation of age-equivalent strata over increasingly broad geographic areas, even into distinct sedimentary basins (Fig. 2A) and with greater precision and accuracy (Fig. 2B and C).

The invention and practical application of the foundations of modern historical geology (lithostratigraphy, biostratigraphy, and geologic mapping) began with British geologist William Smith in the early 19th century (e.g., Arkell, 1933; Hancock, 1977;

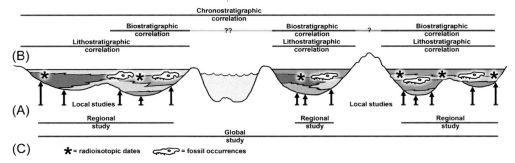

Fig. 2 The changing geographic scope and chronostratigraphic resolution of stratigraphic studies. (A) Hypothetical sedimentary basins separated by geographic barriers (mountains and oceans). Some sedimentary basins contain fossils that may or may not be the same taxa as in other basins (indicated by the skulls) and radioisotopic dates may also be available (indicated by the asterisks). (B) The geographic scopes at which different stratigraphic studies are capable of correlating; lithostratigraphic correlation is generally only possible within a basin or adjacent basins, while biostratigraphic correlation may permit correlation between basins or, more rarely, continents, and chronostratigraphic correlation using radioisotopic dating and other nonbiostratigraphic methods may permit correlations between continents. (C) The geographic scales of stratigraphic studies. Local studies are performed at particular locations within sedimentary basins, while regional studies occur up to the same scale at which lithostratigraphic correlation is possible. Global studies occur up to the maximum possible scale of chronostratigraphic correlation.

Winchester, 2009). The scale, detail, accuracy, and practical economic value of Smith's work was staggering (Winchester, 2009). However, Smith himself did not emphasize the potential of stratigraphy for reconstructing geologic history (Hancock, 1977; McGowran, 2005). It fell to other geologists in continental Europe and Great Britain to recognize that the vertical order of strata was equivalent to their relative ages, and that similar sequences of fossils could be identified even where the lithology of the rocks containing the fossils differed. The recognition that fossils could be used to correlate strata across a much larger geographic area compared to lithologic units led to the separation of lithostratigraphy and biostratigraphy (Fig. 2A). This insight, combined with the realization that the global stratigraphic record could be finely divided using different assemblages of fossils, was used to construct the global geologic timescale during the 19th century (Arkell, 1933; Hancock, 1977; Rudwick, 1985).

By the late 20th century, it became clear that distinguishing between stratigraphic units based on lithology (lithostratigraphy) and those based on fossil content (biostratigraphy) was insufficient. It was also necessary to distinguish these stratigraphic units from those based on age (chronostratigraphy) (Hedberg, 1965, 1976; Salvador, 1994; NACSN, 1983, 2005), as it was becoming clear that biostratigraphic units (biozones) could easily be diachronous if a fossil taxon living in different areas did not necessarily appear everywhere at the same time (e.g., Hancock, 1977, p. 19; Woodburne, 1989; McKenna and Lillegraven, 2005). However, when Hedberg (1965, 1976) first advocated chronostratigraphic units, the application of nonbiostratigraphic methods for determining strata of equivalent age (especially radioisotopic dates and magnetostratigraphy) was not yet widespread, and biostratigraphic correlation was still the most common means of establishing "time-stratigraphic" units of equivalent age in different areas (e.g., Berry, 1966, pp. 1493–1496; Tedford, 1970, p. 700; but see Evernden and James, 1964; Evernden et al., 1964). This essentially required treating chronostratigraphic and biostratigraphic units as synonymous, as they effectively had been since the 19th century, defeating the purpose of distinguishing them in the first place (Hancock, 1977). Moreover, Hedberg (1965, 1976) did not advocate creating new chronostratigraphic units, but rather adapting globally recognized biostratigraphic units such as the system and stage (e.g., Arkell, 1933, pp. 9–14). Hancock (1977, p. 19) disparaged this practice, noting that considering a biostratigraphic unit "to be chronostratigraphic is to debase a practical stratigraphic unit, as well as to deny the biological characteristics of the origin, dispersal, and extinction of species populations—none of which is likely to be isochronous."

Fortunately, important strides were made over subsequent decades in resolving this problem. One solution has been the identification of Global Stratotype Section and Points, organized by the International Commission on Stratigraphy, to establish chronostratigraphic-type sections and boundaries for series (and by extension, stages, and systems) that thus become conceptually independent of biozones. Consequently, systems and stages ceased to be defined by the fossils originally used to characterize them. Chronostratigraphic-type section boundaries approximate, but are no longer tied to, the biostratigraphic transitions originally used to recognize them (e.g., Walsh, 2004). The second solution has been the increasingly widespread availability of radioisotopic dates and magnetostratigraphy, which allows chronostratigraphic correlation without the use of fossils (e.g., MacDougal, 2008; Gradstein et al., 2012; Irmis et al., 2010, 2011; Olsen et al., 2011).

Evernden et al. (1964) and Evernden and James (1964) pioneered the use of Potassium-Argon dating of volcaniclastic sediments to test the established North American Land Mammal Ages (NALMAs) as well as associated floras. They were able to determine empirically that the proposed NALMAs were indeed superpositionally distributed and that assigned age assessments were more correct for the mammal fauna rather than those of the related floras. However, they did not thoroughly test whether these faunal assemblages were diachronous across North America.

Two critical consequences of the separation of chronostratigraphy and biostratigraphy are worth mentioning here, as they have major significance for vertebrate stratigraphy and geochronology, including for Late Triassic vertebrates (Woodburne, 1989; Irmis et al., 2010, 2011; Olsen et al., 2011, pp. 220–222; Martz and Parker, this volume):

1. Lithostratigraphic and chronostratigraphic methods of correlation allow the isochronous nature of biostratigraphic units to be fully tested rather than merely assumed. With these independent means of providing relative and absolute ages for strata, it

is now possible to identify diachronous faunal and floral distributions, and thereby track the movement of groups of organisms to different geographic regions and environments over time.

2. Historically, the separation of lithostratigraphy, biostratigraphy, and chronostratigraphy has allowed the correlation of age-equivalent strata to be extended over wider geographic areas with greater precision (Fig. 2A and B). During the 19th century, the recognition that similar sequences of fossils could be identified in geographic regions with different lithostratigraphic units allowed the relative ages of those strata to be at least approximated for the first time, and more detailed scrutiny of those fossil sequences allowed smaller and smaller divisions of time to be hypothesized, and over a wider area. With the application of nonbiostratigraphic dating methods, chronostratigraphy allows strata of the same age to be identified even when lacking the same fossils or any fossils at all, extending the range of chronostratigraphic correlations even further. Moreover, as radioisotopic dates are based on decay rates that do not vary geographically, the use of these dates to determine age-equivalent strata is potentially far more accurate and precise than using fossil distributions that might be diachronous.

If radioisotopic dates are scarce and the size of the study area is sufficiently confined, lithostratigraphy and biostratigraphy might still be used as proxies for chronostratigraphy (e.g., Desojo et al., 2013; Martz and Parker, this volume). Over short geographic distances, a particular bed or the distribution of a fossil taxon are less likely to be diachronous; in other words, it is unlikely that a particular bed was deposited, or that a fossil taxon lived, at radically different ages in areas a few kilometers apart. As a result, the presence of a particular bed or the true range of a fossil taxon is likely to be the same age everywhere they are observed for very small-scale studies. As the geographic scope of a study increases, this assumption becomes increasingly questionable. It is therefore essential to distinguish stratigraphic studies at varying scales.

As biostratigraphic methods developed, geologists also made important distinctions between strata of certain inferred age, and the time spans over which they formed (Arkell, 1933; Hancock, 1977; Walsh, 1998, 2000, 2004), as well as realizing that the incomplete nature of the fossil record distorts the apparent stratigraphic and geochronologic ranges of taxa (e.g., Foote and Raup, 1996; Marshall, 1995; Walsh, 1998; Barry et al., 2002; Behrensmeyer and Barry, 2005). The difference between biostratigraphy and biochronology, as well as between actual historical events unrecognizable without omniscience, and empirically determined but imperfectly perceived units of strata and time, are given much more detailed treatment by Walsh (1998, 2000) and Martz and Parker (this volume).

Scales of Stratigraphic Models

There is an intimate and inexorable relationship between stratigraphy and geography, something recognized since the time of William Smith (Winchester, 2009). Here we define three scales of stratigraphic models distinguished largely by the geographic scale they encompass (Fig. 2C). It is important to recognize that lithostratigraphic and biostratigraphic correlations tend to become coarser and less applicable, and nonbiostratigraphic methods of relative and absolute age determination tend to become more essential, as the geographic scope of the model increases. Each scale encompasses smaller scales; multiple local models are combined into regional models, and multiple regional and/or local models are combined into global models (Fig. 2C).

Local Stratigraphic Models

Local stratigraphic studies are the primary focus of this paper. A local stratigraphic model covers a geographic area across which the stratigraphic relationships of particular beds or localities can be determined with a high degree of precision (Fig. 3). Such a study area has an area that is large enough to encompass multiple lithostratigraphic measured sections (Fig. 3A and B) and fossil localities, but small enough that bed-level lithostratigraphic correlations are possible by thoroughly walking out contacts, or by identifying in closely associated outcrops a pattern of facies and beds that is so detailed and similar that bed-level correlations can be made with confidence (Fig. 3B). These bed-level correlations allow the relative stratigraphic positions of fossil localities and radioisotopic date samples in different parts of the study area to be determined within several meters of accuracy, permitting the construction of a composite local lithostratigraphic model (Fig. 3C). As such, local models provide the best possible balance between sample size (the geographic area is large enough to allow multiple fossil localities, radioisotopic date samples, and/or magnetostratigraphic samples to be incorporated) and precision (the geographic area is small enough that the relative stratigraphic positions of these localities and samples may be determined with high confidence). Careful documentation of stratigraphic sections, fossil localities, and geochemical samples is essential in the development of local studies so that the stratigraphic and geochronologic models can be verified by revisiting specific locations where data were gathered (e.g., McKenna and Lillegraven, 2005; Martz and Parker, 2010, this volume; Parker and Martz, 2011).

The maximum geographic area for which a stratigraphic model may be considered "local" is determined by the physical continuity and accessibility of exposed outcrop, and the amount of time the geologist is willing to expend completely walking out individual beds in order to precisely link localities. The area across which individual beds may be traced in fluvial facies could plausibly extend from a few square kilometers to a few hundred square kilometers (e.g., Kirkland, 2006; Martz, 2008; Martz and Parker, 2010; Martz et al., 2012, 2014; Hartman et al., 2014). Bed-level correlations for marine facies may be extended across larger areas (e.g., Brett, 2000). Compared to regional and global models, facies and particular beds within local study areas are less likely to be diachronous over local geographic scales. Moreover, endemism and diachronous geographic distributions of fossil taxa are also less likely to be significant at the local scale compared to regional (e.g., Martz et al., 2013, pp. 357–359) and global studies (e.g., Irmis et al., 2010, pp. 46–48; Olsen et al., 2011, pp. 220–222).

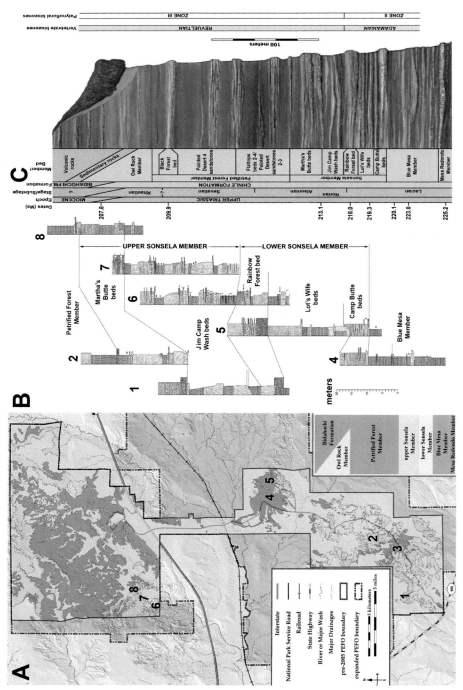

Fig. 3 Local lithostratigraphic model using the work of Martz and Parker (2010) and unpublished data for the Upper Triassic Petrified Forest National Park (see Parker and Martz, 2011, and Martz and Parker, this volume, for an example of local biostratigraphic model building for the same study area). (A) Simplifed geologic map of the Chinle Formation in Petrified Forest National Park. (B) Correlated lithostratigraphic sections identifying the same beds or facies packages within the Sonsela Member throughout the park; the precise stratigraphic level of fossil localities and radioisotopic date samples has been determined. (C) The local lithostratigraphic model for PEFO, a composite of correlated lithostratigraphic sections throughout the park calibrated by radioisotopic dates (Ramezani et al., 2011; Atchley et al., 2013).

The geographic scope of local models is determined by lithostratigraphy, and this is true even of local biostratigraphic models. Determining the precise relative ages of fossil localities in a local study area will be limited by the degree to which lithostratigraphic correlations are possible, as lithostratigraphy provides the scaffolding on which biostratigraphic models are built (see Martz and Parker, this volume). Hypothetically, if absolute age information is sufficiently detailed, it might be possible to build a local biostratigraphic model (i.e., one in which the superpositional relationships of fossil localities could be determined with high precision) that is not dependent on lithostratigraphy. For example, if two sections that shared the same fossil taxa were extremely well calibrated by radioisotopic dates, it would be possible to determine the precise relative superpositional relationships of fossil localities without having to correlate beds between sections. However, in practice this is not possible for Upper Triassic strata in western North America owing to the rarity of absolute ages except for a handful of local study areas (e.g., Ramezani et al., 2011; Irmis et al., 2011; Atchley et al., 2013). Determining the superpositional relationships of fossil localities with enough precision to construct a local-scale stratigraphic model requires detailed and accurate lithostratigraphy.

We may also draw an informal distinction between "hard local models" and "soft local models." In a hard local model, the lithostratigraphic correlation of beds between stratigraphic sections has been verified to the highest possible degree by actually walking out the units (e.g., Martz, 2008; Martz and Parker, 2010). In a soft local model, the stratigraphic section has been examined and measured in particular locations, and beds and facies packages have been correlated based on sedimentological similarities and equivalent stratigraphic position, but not by actually tracing these beds visually between sections (e.g., Kirkland et al., 2014; Martz et al., 2014). Constructing soft models rather than hard models may be a necessity when the time frame of the study does not permit walking out individual beds, but bed-level correlations in soft models must be considered inherently less reliable than for hard models.

There are only a handful of examples of extremely detailed local stratigraphic studies for Upper Triassic strata in western North America. However, the lithostratigraphy, biostratigraphy, and chronostratigraphy of the Chinle Formation in Petrified Forest National Park (PEFO) in northern Arizona are a notable exception (Fig. 3). At PEFO, an extremely detailed and thoroughly documented lithostratigraphic model for the Chinle Formation has been combined with thorough sedimentological descriptions and interpretations of depositional environments and paleoclimate, as well as exhaustive and precise documentation of localities for fossils and geochemical data (Ash, 1970; Long and Ballew, 1985; Long and Padian, 1986; Litwin et al., 1991; Therrien and Fastovsky, 2000; Heckert and Lucas, 2002; Woody, 2003, 2006; Parker, 2006; Parker and Martz, 2011; Olsen et al., 2011; Ramezani et al., 2011, 2014; Martz et al., 2012; Loughney et al., 2011; Reichgelt et al., 2013; Atchley et al., 2013; Nordt et al., 2015; Martz and Parker, this volume). The result is one of the most detailed stratigraphic and geochronologic models of Triassic environmental and biotic change in the world (Nordt et al., 2015).

Regional Stratigraphic Models

A regional stratigraphic model (Fig. 4) is defined as one occurring across a geographic area small enough to allow lithostratigraphic correlation of distinct facies packages between multiple local study areas (Fig. 4B and C), including those classified as formations or members (e.g., Blakey, 1990), and for the same stratigraphic sequence of fossil taxa to be identified in different local study areas. However, the size of a regional study area is too large for individual beds within facies packages to be reliably correlated across the entire region, or for the precise relative superpositional relationships of fossil localities in different local study areas to be determined. The geographic area of a regional study may extend from a few hundred square kilometers to an entire sedimentary basin or closely associated basins (Fig. 4A).

The Upper Triassic strata of the Chinle Formation, Dockum Group, and related strata of the western United States (Fig. 4) are predominantly fluvial in origin with some lacustrine, paludal, and eolian deposits (e.g., Dubiel, 1987; Blakey and Gubitosa, 1983; Lehman and Chatterjee, 2005; Dubiel and Hasiotis, 2011; Howell and Blakey, 2013). There have been numerous regional-scale lithostratigraphic studies of these units that correlate members, and even a few laterally extensive beds or facies packages within a member (e.g., Branson, 1927; Poole and Stewart, 1964; O'Sullivan, 1970; Stewart et al., 1959, 1972; McGowen et al., 1979, 1983; Blakey and Gubitosa, 1983, 1984; Lucas, 1993; Dubiel, 1987, 1994; Lehman, 1994; Lucas et al., 1994, 1997; Lehman and Chatterjee, 2005; Dubiel and Hasiotis, 2011; Martz et al., 2013; Kirkland et al., 2014, pp. 29–31). Moreover, the same fossil taxa can also be identified in Upper Triassic strata across the western United States, allowing biozones to be widely correlated (e.g., Colbert and Gregory, 1957; Litwin et al., 1991; Lucas, 1993, 1998, 2010; Martz et al., 2013, pp. 357–359; Martz et al., 2014; Martz and Parker, Chapter 2 of this volume). The Upper Triassic strata of the western United States, and more restricted areas within it, can therefore be treated as regional study areas.

The use of lithostratigraphy and biostratigraphy as proxies for chronostratigraphy begins to lose reliability at the regional scale. Laterally changing facies relationships, diachronous lithostratigraphic units (Figs. 4B and C), and localized habitat preferences or endemism of fossil taxa that may have been trivial (or at least undetectable) at the local scale may become an issue at regional and global scales (e.g., Behrensmeyer, 1978; Bown and Beard, 1990; McKenna and Lillegraven, 2005; also see the later discussion). Several studies have identified complex, and in some cases potentially diachronous, lateral facies changes within the Chinle Formation and Dockum Group (e.g., O'Sullivan, 1970; Stewart et al., 1972; Blakey and Gubitosa, 1983; Lehman, 1994; Martz, 2008, pp. 106–107; Martz et al., 2014, pp. 422–428) that must be considered when trying

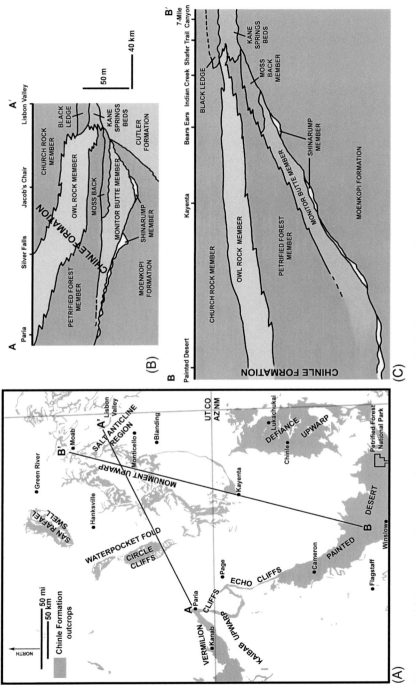

Fig. 4 Regional lithostratigraphic model of Blakey and Gubitosa (1983, 1984). (A) Map of Chinle Formation outcrops in southeastern Utah showing the transects of the A-A' and B-B' cross sections, modified from Blakey and Gubitosa (1984, Fig. 1). Chinle Formation outcrops based on Stewart et al. (1972, pl. 1). (B) A-A' cross section of Blakey and Gubitosa (1984, Fig. 6A). (C) B-B' cross section of Blakey and Gubitosa (1984, Fig. 6B). Note that the B end of the B-B' cross section is very close to Petrified Forest National Park (see Fig. 2), and is approximately equivalent to that local study area stratigraphically. In the B-B' cross section, the Monitor Butte Member probably correlates approximately with the Blue Mesa Member in PEFO, while the Petrified Forest Member correlates with the Sonsela and Petrified Forest Member of PEFO, and the Church Rock Member is not preserved in PEFO.

to determine the relative ages of strata in different local study areas. Individual channel sandstones in the Chinle Formation and Dockum Group are generally laterally discontinuous even within local study areas (e.g., Therrien and Fastovsky, 2000, p. 211; Martz, 2008, pp. 31–69; Martz and Parker, 2010, pp. 9–10, 17), and formations and members can vary greatly in thickness across a region and laterally interfinger with other units (Fig. 4B and C; e.g., Stewart et al., 1972; Blakey and Gubitosa, 1983). One should therefore avoid composite stratigraphic columns for regional models that present the precise relative superpositional relationships of fossil localities in widely separated local study areas, as such precise correlations are not actually possible (e.g., Parker and Martz, 2011, pp. 253–254); note that the biozone correlations presented by Martz and Parker (Chapter 2 of this volume) correlate vertebrate fossil occurrences with extreme caution. Moreover, possible local endemism has been recognized for phytosaurs in the Chinle Formation and Dockum Group; the taxa *Smilosuchus* and *Leptosuchus* are closely related, but seem to occur within distinct clades, with the former being known only from Arizona and the latter only from Texas (Stocker, 2010). Whether or not this apparent endemism is real or an artifact of sampling or preservation requires additional fossil collection and stratigraphic work.

Unfortunately, detailed radioisotopic dates of Upper Triassic strata in the western United States are mainly confined to the Chinle Formation of Petrified Forest National Park (Fig. 3C; Ramezani et al., 2011, 2014; Atchley et al., 2013; Nordt et al., 2015). Only a few particular beds have been dated elsewhere in the Chinle Formation (e.g., Irmis et al., 2011; Ramezani et al., 2014), and high-resolution U-Pb zircon dates are not yet available for the Dockum Group, Dolores Formation, or Popo Agie Formation. This means that regional chronostratigraphic and geochronologic models for the Late Triassic of the western United States must tie together particular local study areas within that region primarily using lithostratigraphic and biostratigraphic correlations relying on similar facies, palynofloras, plant megafossils, and vertebrate fossils as crude proxies for chronostratigraphic correlations (e.g., Branson and Mehl, 1929, pp. 17–18; Gregory, 1957; Ash, 1980; Litwin et al., 1991; Lucas and Hunt, 1993; Cornet, 1993; Lucas, 1993, 1998; Dubiel, 1994; Ash and Hasiotis, 2013; Martz et al., 2013; Martz and Parker, Chapter 2 of this volume), in spite of their relative unreliability.

Global Stratigraphic Models

A global stratigraphic model (Fig. 5) is one constructed on a scale at which lithostratigraphic correlation is no longer possible because facies packages do not extend between different regional study areas, and biostratigraphic correlations are problematic (especially as chronostratigraphic proxies) because of diachronous faunal distributions and/or endemism. Biostratigraphy was, of course, the original basis for the globally applied geologic timescale (Rudwick, 1985), and continues to be used as a chronostratigraphic proxy even

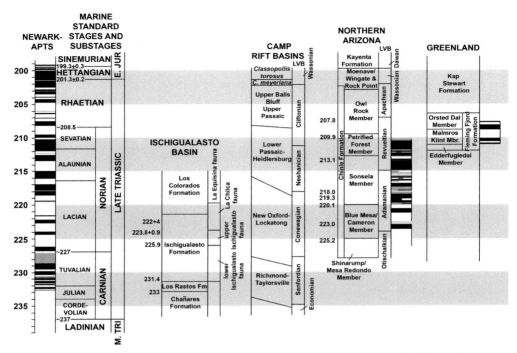

Fig. 5 Global chronostratigraphic correlations for Upper Triassic strata modified from Olsen et al. (2011, Fig. 16). Radioisotopic dates and magnestostratigraphy for northern Arizona from Ramezani et al. (2011, 2014), Atchley et al. (2013), and Zeigler et al. (Chapter 6 of this volume). Horizontal gray and white bars represent 5 million year increments.

today. However, as the use of radioisotopic dating and magnetostratigraphy becomes more widespread, global correlations tend to become chronostratigraphic in nature, tying together local and/or regional stratigraphic and geochronologic models from different parts of the world using nonbiostratigraphic methods (Fig. 5; e.g., Irmis et al., 2010, 2011; Olsen et al., 2011).

By the middle of the 20th century, when biostratigraphy was virtually the only method of intercontinental correlation, it had become accepted that the Chinle Formation, Dockum Group, and Popo Agie Formation were correlative with the German Keuper based on similarities between vertebrate faunas (e.g., Colbert, 1960). Some studies recognized finer correlations; basal phytosaurs were identified as occurring in the Popo Agie Formation, lower part of the Dockum Group, and the Blasensandstein, while more derived phytosaurs occurred in the upper Chinle Formation and Dockum Group and the Stubensandstein (e.g., Camp, 1930, pp. 4–6; Gregory, 1957, pp. 7–10; Colbert and Gregory, 1957; Colbert, 1960; Cooper, 1982). Palynomorph-based correlations to Upper Triassic marine strata in Europe were used to correlate the Chinle Formation and

Dockum Group to the Carnian and Norian stages of the Upper Triassic (Dunay and Fisher, 1974; Cornet, 1993; Litwin et al., 1991). Schultz (2005, pp. 126–140) provided a historical summary of global stratigraphic and geochronologic models for the Triassic Period, with an emphasis on Gondwana.

After the formulation of the Late Triassic land vertebrate "faunachrons" (Lucas and Hunt, 1993), an ambitious attempt was made to use Upper Triassic palynomorphs and vertebrates to precisely tie all four of Gregory's (1957, 1972) vertebrate faunas in the western United States to the geochronologic timescale, and to other Upper Triassic lithostratigraphic units throughout the world (e.g., Lucas, 1993, 2010; Lucas et al., 2007). Unfortunately, the use of vertebrates and palynofloras for global correlation has severe limitations (e.g., Schultz, 2005; Langer, 2005; Rayfield et al., 2005, 2009; Irmis et al., 2010; Olsen et al., 2011). Taxonomic endemism is common at global scales, even during the Late Triassic when the continents were joined (Ezcurra, 2010; Irmis et al., 2011). Higher taxa, and even alpha taxa, may have diachronous stratigraphic ranges over large geographic scales (Woodburne, 1989, pp. 221–223; Woodburne, 1996; Irmis et al., 2007, 2010, 2011).

Closely related alpha taxa on different continents may be lumped into a single alpha taxon to streamline intercontinental biostratigraphic and biochronologic correlations, but this increases the likelihood of diachronous distributions for the composite "alpha taxon" (Woodburne, 1989, pp. 221–223; Angielczyk and Kurkin, 2003; Rayfield et al., 2005, 2009; Irmis et al., 2010). For example, Heckert and Lucas (2000, 2002) considered the Late Triassic aetosaurs *Calyptosuchus wellesi* (Case, 1932; Long and Ballew, 1985) from the Chinle Formation of the western United States and *Aetosauroides scagliai* (Casamiquela, 1960; Desojo and Ezcurra, 2011) from the Ischigualasto Formation of Argentina and the Santa Maria Formation of Brazil to be congeneric with *Stagonolepis robertsoni* (Agassiz, 1844; Walker, 1961) from the Elgin Sandstone of Scotland. Consequently, they considered *Stagonolepis* to have a global distribution that could be used to correlate latest Carnian strata globally. However, these three genera are quite distinct (Desojo and Ezcurra, 2011; Parker, 2014), and radioisotopic calibration of the Chinle Formation and the Ischigualasto Formation indicates that *Calyptosuchus* appeared in North America in the early Norian, over 5 million years after the appearance of *Aetosauroides* in the late Carnian of South America (Desojo et al., 2013).

As with vertebrates, palynofloras have also been found to have globally diachronous distributions strongly dependent on localized climatological conditions (McKenna and Lillegraven, 2005; Whiteside et al., 2015). Although late Carnian palynofloras in Europe have been used to identify Upper Triassic strata in North America as being the same age (Dunay and Fisher, 1974; Cornet, 1993; Litwin et al., 1991), radioisotopic recalibration of the Late Triassic timescale and radioisotopic dates from western North America (Ramezani et al., 2011, 2014) indicate that similar palynofloras in the latter region are early Norian (e.g., Parker and Martz, 2011; Reichgelt et al., 2013). It has also been demonstrated that localized climatological fluctuations in the Late Triassic had a major impact on palynofloras, which probably explains why Late Triassic palynofloras had globally diachronous distributions (Whiteside et al., 2015).

As a result of these problems with using biostratigraphy as a chronostratigraphic proxy, other workers have preferred to base global Late Triassic chronostratigraphy and geochronology on radioisotopic dates and magnetostratigraphy (Fig. 5; Krystyn et al., 2002; Hounslow and Muttoni, 2010; Irmis et al., 2010, 2011; Kent et al., 2014). Biostratigraphic correlations may still be used as a crude proxy for chronostratigraphy on regional and even global scales when other data are unavailable (e.g., Desojo et al., 2013, pp. 208–215; Martz and Parker, Chapter 2 of this volume), but must be considered far less reliable due to the possibility of endemism and diachronous distributions (Irmis et al., 2010).

CONSTRUCTING LOCAL STRATIGRAPHIC MODELS
Choosing the Size of the Local Study Area

The transitions between local, regional, and global scales are somewhat subjective. As previously discussed, a local stratigraphic study occurs on a small enough geographic scale that bed-level correlations are possible by thoroughly walking out outcrops. For obvious reasons, such studies work best in areas with good continuous outcrop exposure, or if adjacent outcrops share such an unmistakable sequence of facies and beds that there are few doubts about how to correlate them. However, the geographic expanse of continuous outcrop and lithostratigraphic units are highly variable. As a result, a geographic area of a given size may contain extensive badlands where beds can be walked out for kilometers (suitable for a "hard" local study), or have outcrop distribution so patchy that individual beds are impossible to walk out (suitable for a "soft" local study if particular beds can be correlated between outcrops, or a regional study if they cannot).

The amount of time the researcher is willing to devote to the project will also place constraints on the geographic scope of the study area. For a local study area of tens to hundreds of square miles, a local study project may take years of concerted effort if it is being done with the proper attention and accuracy. The Upper Triassic outcrop incorporated into the Petrified Forest National Park local study area in northeast Arizona (Fig. 3; Martz and Parker, 2010; Parker and Martz, 2011; Martz et al., 2012) is over 150 square miles, while the local study area in southern Garza County, western Texas, studied by Martz (2008) is over 300 square miles; these studies, which included thorough geologic mapping took less than 2 years and about 5 years to complete, respectively. Both studies were facilitated by the fact that the researchers lived very close to the study area, allowing fieldwork to be frequent and relatively convenient.

Another consideration is the distribution of fossil localities of interest that might be tied into the stratigraphic section. An ideal local study area contains a large number of localities that occur in a small enough area that it is possible to tie them into the same lithostratigraphic model with high precision. Other spots that might be desirable to tie into the section might include lithostratigraphic type sections for formations, members, and/or beds; outcrop localities for depositional system or paleosol studies; and sampling sites for geochemical samples, including radioisotopic dates. In all these cases, incorporating sites into the stratigraphic model is of course dependent on being able to relocate them, which is only possible if adequate locality data was collected and made available (see the next section). Being able to plot all of these other sources of data onto individual stratigraphic sections (Fig. 3B) and onto the composite stratigraphic section for the local study area (Martz and Parker, Chapter 2 of this volume, Figs. 9–13), can provide invaluable information to augment the biostratigraphic model. With these multiple sources of data, it may be possible to precisely link biotic events to the geologic timescale, and demonstrate their relationship to environmental and climatic changes (e.g., Parker and Martz, 2011; Atchley et al., 2013).

Field Data Documentation
Photographic Documentation
Photographs are a tragically unappreciated but absolutely essential form of field documentation that we promote for both lithostratigraphic studies and fossil locality documentation (Figs. 6 and 7); as discussed here, photographs are more valuable than GPS coordinates or any other kind of field data. However, photograph information must be recorded so that the content of a photograph is understood, or the photo is useless (Fig. 6). Researchers who have had to make use of older field data know the consternation of looking through old photographs and trying to remember where, when, why, and of what the photo was taken; looking through the unlabeled photographs of another researcher is even worse, as there may be no way to determine where they were taken without relocating the site, and possibly no way at all to know exactly when it was taken or why. We suggest the following protocols for documenting photographs in the field:

1. Make certain the photograph shows what is required. This may seem obvious but in practice is often overlooked. Outcrop photographs used to document measured sections should show all units included in the sections (though not necessarily all in the same photo), and all locality photographs should include at least one showing the foreground and background clearly enough to precisely relocate a locality to within a square meter (Figs. 6 and 7A).

2. Field notebooks should be used to assign labels to all photographs as soon as they are taken (Fig. 6). The notebook entry should identify the year, month, day, and photograph number for the day, in that order (yymmdd_#). For example, the fifth photograph taken in the field on Aug. 12, 2013, would be labeled: 130812_5. When these labels are later applied to photograph images in the computer, it ensures that the

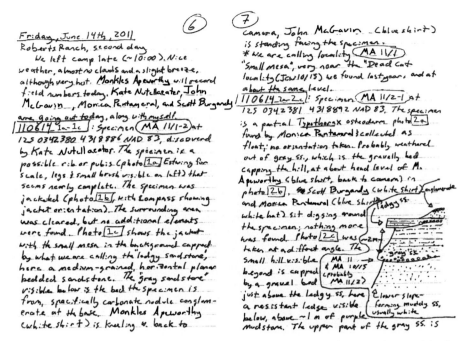

Fig. 6 Hypothetical field notes employing several of the reccommendations advocated in the text. Notes are organized by photograph numbers, which are enclosed in boxes and follow a year-month-day-photograph format advocated in the text; for example, 110614_1b is the second photograph taken on Jun. 14, 2011. Specimen numbers are encircled and follow a collector initials-year/site number-specimen number format; for example, field number MA 11/1-2 is the second specimen collected by Monkles Apeworthy from the first field locality he collected in 2011. Note that the GPS datum is given for all coordinates, and that identifying features of each photograph are listed, and that page numbers are also given.

images are automatically arranged in the chronological sequence in which they were taken (Fig. 7B), making them easy to find. While still in the field, it is recommended to regularly check the number of photographs taken on the digital camera to make sure that they match the number of labels you have assigned in your field notebook. If photos are taken for which labels were overlooked in the field, provide new labels and correct this numbering in your notes so that the labels remain in the same sequence as the photographs in your camera. For example, if it is realized that a photograph was taken between photographs labeled 130812_5 and 130812_6 in the field notes, there are two possible solutions. One is to renumber the latter photograph as 130812_7 in the field notes, renumber all subsequent photographs accordingly, and label the over-looked photo as 130812_6. An alternative that saves the trouble of renumbering pho-tographs is to simply number the new photograph 130812_5b, which will keep it in sequence within the computer (Fig. 7B).

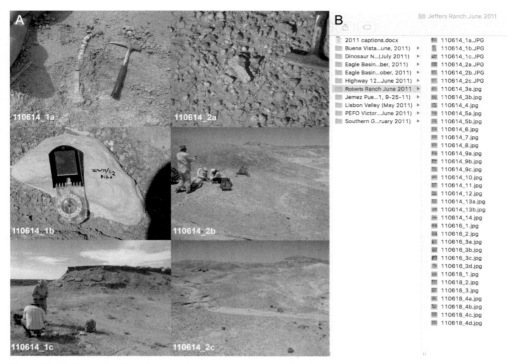

Fig. 7 Field photographs referenced in Fig. 6: (A) The field photographs, organized by number (note that the numbers are placed on the figure for convenience; it is not necessary to label the actual photographs as long as the photograph label gives the photo number). (B) Screen shot showing how the photographs are labeled sequentially when the photo numbers are used for the image name.

3. Each written photograph label in a field notebook should be followed by (at minimum) a clear description of what is in the photograph, why the photo was taken, and GPS coordinates (including the geodetic datum referenced) (Fig. 6). Field numbers for specimen localities shown in the photograph should be clearly identified, and ideally circled to make them easy to find when scanning through notes. Other helpful details might include who is in the picture, what they are wearing, and what they are doing. If there is any confusion about which label applies to which photograph, this additional information can help resolve it. Placing specific objects (rock hammers, hats, etc.) at specific spots of interest in the photo can help identify the exact spot from where something was collected, though make sure that the field notes identify what each object is marking.

4. As soon as possible after returning from the field (ideally on the same day) while photograph content is still fresh in your mind, transfer the photographs to a computer and with the field notebook in hand, change the photograph image names to the dated label assigned to it in the notebook (Fig. 7B). Ideally, field notes should also be transcribed into a word processing or spreadsheet file, with the photograph labels clearly identified and the photographs' content described. It is also recommended that specimen field numbers in the typed field notes be boldfaced to make them easy to find. This document can be kept in the same folder as the photograph files or if preferred the photos and captions can be inserted directly into the word processing document.

Collecting Paleontological Field Data

As noted by McKenna and Lillegraven (2005, p. 32), fossil locality data must be precisely given for a biostratigraphic hypothesis to be considered verifiable, and therefore scientific. Exact disclosure of locality data is controversial for several reasons, including that it opens the site and any remnant fossils to potential threat of poaching and/or vandalism. Accordingly, full disclosure of site locations is often omitted from published papers. Instead, authors will often state that locality data is on file at a specific institution for qualified researchers. Furthermore, as time passes and institutional knowledge is lost, so too may be these data, hindering future researchers from verifying the published data.

Many fossils are collected from federal lands, and it is often stated anecdotally that the publication of specific locality data is expressly forbidden by government agencies. However, this is not necessarily the case. Locality data are protected from disclosure by Section 6309 (2) of the Paleontological Resources Protection Act of 2009 (Public Law 111-11, Subtitle D), which states that "information concerning the nature and specific location of a paleontological resource shall be exempt from disclosure...unless the Secretary [of the Interior] determines that disclosure would not create risk of harm to or theft or destruction of the resource or site containing the resource." Thus if it is determined by the Secretary of the Interior (or their designated representative, which at the unit level includes land managers) that the disclosure of paleontological site information will cause no harm to the resource, it can be released. However, Section 6304 (c)(3) of the same law states that the Secretary or their designated representative must provide this permission in writing. We consider the repeatability of scientific work to be paramount to its long-term viability. As a result, only with written permission from landowners (federal, state, and private) should explicit locality data be provided given that it will not endanger further fossil resources. To do otherwise may be in violation of federal laws.

Regardless of whether the information is released publically, documentation of the collection of any fossil resources is absolutely essential to preserve the scientific data associated with all fossils. At a minimum this should include: (1) notes written on site while the work is being done documenting the date of the work being done (Fig. 6), (2) who is conducting the work, (3) the potential field identification of the fossils, (4) a photograph of the exact location with reference points noted as well as the direction of the shot (as described earlier), and (5) geographical and geologic data including spatial (e.g., GPS coordinates) and stratigraphic relation to specific beds (see the following discussion).

Any GPS data should include the geodetic datum used when recording coordinates (Behrensmeyer, 2011). Often disregarded, the utilized datum should be considered part of the written coordinates because, for example, the two most widely used datums in North America, NAD 27 and NAD 83 (=WGS84) are offset by approximately 90m. Relocation of sites should be unambiguous, especially since in typical badlands terrain, physical evidence of fossil collection disappears extremely rapidly.

All collected data should be recorded in the field notebook, which should serve as the basis for all associated information (Fig. 6; Behrensmeyer, 2011). Localities and collected specimens should be provided with unique numbers in the field. There are innumerable formats for field numbers, but the particular system utilized by the current authors begins with the initials of the field worker, the year, an informal locality number, and the number of the specimen recovered there. For example, if in 2014 the senior author discovered a new site and collected two specimens there, and this was the fifth locality he had personally documented that year, the field numbers would be WGP 14/5-1 and WGP 14/5-2. A new field number is assigned even if the locality has been previously visited and documented. This is to allow the field number to be used to find the notes pertaining to each discovery. For example, if in 2015 the senior author revisited the same locality described in the 2014 example and collected a new specimen there, a researcher examining the collections and seeing the field number associated with the specimen would immediately know to look in the 2015 notes rather than the 2014 notes for information on that particular specimen. The researcher would not know this if the original 2014 field number were used.

Field numbers should remain part of the official record for each locality and specimen as they are what tie specimens and data back to the field notebook. Notebooks should be photocopied regularly to prevent loss and/or damage (Behrensmeyer, 2011). An obvious but sometimes overlooked point is that it is essential to be able to determine which fossil in a museum collection corresponds to a unique field number and associated data. In addition to recording data in field notebooks, the same data must be kept with the specimen. A card or piece of paper with (at minimum) the field number should be bagged with the specimen, placed in the collection container, or written on the jacket, preferably with additional information such as GPS coordinates and photograph numbers. Field number information should never be discarded, even after a formal specimen number is assigned to the specimen. The original field number, and preferably the original specimen card, should always remain with the specimen, and field numbers written on the specimens should likewise never be removed. The advantages of this data retention can be enormous; not only can the number be used to locate field data in original field notebooks (which should also remain with the specimens), but also field numbers can be invaluable for reconstructing the distribution and association of material at a locality that is otherwise not available from any other source (Martz, 2008, pp. 15–23, Appendix; Parker, 2014, pp. 67–69, Fig. 4.1).

Of all of these means of documenting a locality, photographs are arguably the most important (Figs. 6 and 7A). In order for a future researcher to determine the exact stratigraphic position of a specimen, or to be able to collect more material for a previously collected specimen, it must be possible to confidently relocate the locality to about a meter. Plotting localities on a topographic map does not provide this level of precision. Neither necessarily do GPS coordinates, as coordinates have a margin of error; if several stratigraphically distinct beds occur within the margin of error of the coordinates, it may be impossible to know which bed produced the fossil specimen. However, if photographs of the locality were taken at the discovery and excavation of the specimen, it is possible to identify the exact spot from which it came (Fig. 7A). Even if all other locality information (including maps and GPS coordinates) are lost, a researcher may be able to relocate the locality precisely by matching up the backgrounds of the photos to a precision (within a square meter) that mapping and GPS coordinates could never provide.

Documenting the Geographic Distribution of Beds for Lithostratigraphy
Determining the Stratigraphic Sequence and Geographic Distribution of Beds
When beginning a lithostratigraphic study, particularly one employing geologic mapping (see the next paragraph), the first priority is to identify stratigraphic horizons or "marker beds" that can be traced out over long distances. Both the southern Garza County (Martz, 2008) and Petrified Forest National Park (Martz and Parker, 2010) studies relied primarily on fluvial channel sandstones, which are often erosionally resistant and identified visually (e.g., the Jasper Forest bed and Flattops One bed in Fig. 8A and B). This was an especially important consideration in the southern Garza County study, where the outcrop was concealed by broad areas that were heavily overgrown. Channel sandstones tend to cap cliffs and mesas above steep exposed slopes on which vegetation cannot as easily grow, making beds easy to trace. However, it is also possible to identify unconformities and paleosol horizons within fine-grained overbank sequences that may be traced on foot (e.g., Barry et al., 2002; Prochnow et al., 2006).

For a mapping project, the stratigrapher should have several resources at hand: (1) a field notebook; (2) topographic maps on which to draw field geologic maps; (3) a GPS unit; (4) a camera; and (5) information on important localities that are likely to be reachable on foot for that particular day including (most importantly) GPS coordinates and photographs of the sites showing the exact horizon where fossil specimens were recovered. It is helpful to mark these localities on your topographic map in advance. Consultation with existing, coarse-resolution geologic maps (i.e., maps that show formations and possibly members, but not many beds) may also be extremely helpful to get an idea as to the geographic distribution of outcrops, although they do not usually provide the detail of stratigraphic information required here. You do not require a Jacob's staff; measuring sections should be delayed as long as possible (see the following).

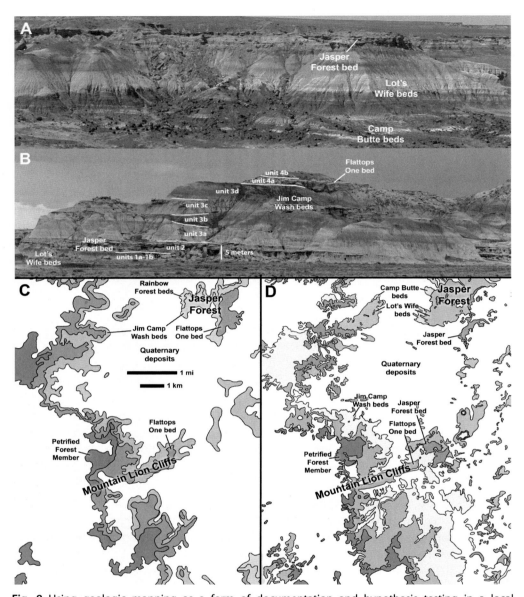

Fig. 8 Using geologic mapping as a form of documentation and hypothesis testing in a local lithostratigraphic study where the complete section is not present at any single locality: the Chinle Formation at Petrified Forest National Park (PEFO), northern Arizona. (A) Mesas at Jasper Forest capped by the bed identified by Woody (2006) as the Flattops One bed and by Martz and Parker (2010) as the Jasper Forest bed; the names Lot's Wife beds and Camp Butte beds are also from Martz and Parker (2010). (B) Mountain Lion Cliffs capped by the sandstone identified by Woody (2006) and Martz and Parker (2010) as the Flattops One bed. The identification of the lowermost beds in the section as the Jasper Forest bed and Lot's Wife beds is also from Martz and Parker (2010); this image is modified from Martz and Parker (2010, Fig. S8a) and identifies the units in the Mountain Lion Cliffs section. (C) Modified geologic map of the region of PEFO encompassing Jasper Forest and the Mountain Lion Cliffs, modified from Woody (2006, Fig. 9); note that if Woody's (2006) stratigraphic correlations are correct, the same sandstone capping the mesas at Jasper Forest in A should be traceable along the cliffs to Mountain Lion Cliffs. (D) Modified geologic map from the same area, modified from Martz and Parker (2010, Fig. S1); note that if Martz and Parker's (2010) stratigraphic correlations are correct, the sandstone capping the cliffs at Jasper Forest should be traceable to the base of the Mountain Lion Cliffs.

The choice of a starting point is subjective. The first outcrop examined will essentially provide the foundation for the day's work, so it is best to choose an outcrop that has at least one distinctive bed that might be traceable. This starting point may be at a particularly interesting fossil locality, type section, or simply an interesting outcrop containing potential marker beds. Having picked a starting point, pause for a time and simply examine the outcrop in front of you. At this early stage, do not interpret it (e.g., "This must be the Lithodendron Wash Bed, a fluvial sandstone"). An important part of doing proper geology (or science in general) is to avoid making assumptions and to build a model from scratch, assembling individual observations to see what picture they form without letting preconceptions prevent you from seeing what is directly in front of you. The most important tool for beginning to understand something you do not yet comprehend is to simply describe it as best you can. Begin by describing the outcrop to yourself while simply looking at it from a distance (e.g., "There is a purplish popcorn weathering slope, possibly mudstone, capped by a gray, resistant bed that seems to have distinct bedding in it, with a slope of interbedded white and reddish beds above that."). Having a thorough picture of what the outcrop looks like in your mind, now examine the individual layers in general while noting lithology, bedding, and fossils. Do not spend too long with this (the time for extremely detailed outcrop descriptions is when formally measuring sections, which should come later), but at least get an overall sense of the basic stratigraphy and the lithology of different beds.

Identify beds or contacts that might be traceable such as resistant sandstones and conglomerates, unconformities, easily recognized paleosols, or interesting packages of strata (Fig. 8). Give them informal names (e.g., the "a" bed, the "white ledgy beds," etc.), again avoiding any assumptions about what previously published units they might represent; the time for determining that is later. Identifying laterally continuous marker horizons is not an exact science. Obviously, it is impossible to be certain which horizons might be laterally continuous until they have actually been walked out. For example, you might find that a resistant sandstone bed can only be traced for a short distance before pinching out, or that two distinct sandstones occurring only a few meters apart are part of a package of discontinuous sandstone lenses that pinch in and out, and that it is more convenient and informative to map the whole package as a single unit. You have to start somewhere; make your best guess and modify your choices as needs be. Also, be clear to yourself how you are mapping the upper and lower contacts of your marker bed. In the case of fluvial sandstones, the lower boundary is usually a sharp unconformity that is easily identified, but the upper contact is likely to have a gradational contact with overlying fine-grained overbank deposits, so make a decision as to exactly where in that fining-upward sequence you want to mark the upper contact. Map marked beds on your topographic base map for the outcrop right in front of you. Identify the upper and lower boundaries for the beds, being careful not to map beyond what you can actually see. Take some time to draw a simple picture of the section in your field notebook, and mark GPS coordinates accordingly. If you take a photograph, be sure to document photograph information (see the previous description).

Now, start walking. Your path should be guided by: (1) the direction of fossil localities of interest, and (2) the directions of continuous outcrop extending from the section you just examined. As you walk, map the unit or units you identified. If an interesting facies change or new bed appears, perhaps take some time to examine it, or even draw another rough stratigraphic section identifying the important beds and facies changes. As you go, also keep an eye out for outcrops that might be good for measured sections. Photograph and document them as you go, as well as any other interesting sedimentary structures or fossils you encounter. Take lots of notes and photographs, and GPS coordinates, of where these things were observed. Once you are at the outcrop, data are relatively easy and inexpensive to collect, so verge on the side of overdocumentation during the fieldwork.

From time to time, you may encounter places where you have to decide whether to follow one informally named marker bed or another. This might be because the outcrop containing both beds splits into more than one (i.e., the exposures of these beds head in different directions), or because you can only keep one of the beds in sight (e.g., it is easiest to trace a bed capping a cliff by walking along the base of the cliff, but a second marker bed above may not be visible unless you climb to the top of the cliff to keep it in sight). Do not stress out over the decision; mapping will help you remember where you need to come back to in order to trace the second bed. By the time you are done, you will probably have crisscrossed the whole area, checking and rechecking your correlations between outcrops.

Using the GPS coordinates, map, and photographs, relocate important fossil localities as you come to them, determine where they fit into your informal stratigraphic scheme, and draw them into your informal stratigraphic sketches. At the end of the day, you should be able to sit down and draw a composite stratigraphic section (again, just an informal sketch) combining all the beds you have identified, and plotting all the localities you relocated. Your lithostratigraphic model and biostratigraphic model are starting to come together, both on paper and in your head, having been built by direct observation rather than assumption.

Geologic Mapping as Stratigraphic Documentation

When beginning the project, the first question to ask is whether or not geologic mapping will be helpful in documenting the stratigraphic model. If it is possible to observe the entire section at one outcrop (Fig. 9A), mapping out individual beds is obviously not needed to determine their superpositional relationships. Moreover, if the exposures are on a steep cliffside, outcrop belts may be so narrow in map view that mapping individual members and beds is not possible (Fig. 9B).

However, mapping is useful in cases where (1) the stratigraphic section is thick, and (2) the entire section is not observable in one place, such that the precise superpositional

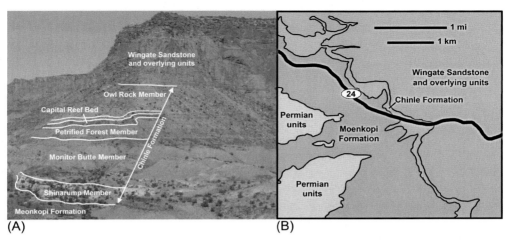

Fig. 9 A local study where geologic mapping would not be especially helpful for documenting or hypothesis testing in a local lithostratigraphic study: the Chinle Formation at Capitol Reef National Park, southern Utah. (A) The complete Chinle Formation section is exposed in a single outcrop so that there is no ambiguity about the details of the lithstratigraphic model or the superpositional relationships of beds. (B) Geologic map for the same area, modified from Morris et al. (2010), demonstrating that, unlike in Figs. 8C and D, Chinle Formation outcrops are too thin in map view to practically distinguish different members, much less individual beds.

relationships of beds scattered across the local study area are not immediately obvious (Figs. 8C and D). Both these criteria were met in southern Garza County and Petrified Forest National Park. In both areas, it was determined that lithostratigraphic miscorrelations of particular beds had resulted in important biostratigraphic miscorrelations (Parker and Martz, 2011, pp. 233–235, 245; Martz et al., 2013, p. 342).

In these cases, it may be essential to trace beds from outcrop to outcrop just to understand the overall stratigraphy of the area (Figs. 8C and D), much less to be able to put all localities into a single detailed and accurate stratigraphic model (Fig. 3). Mapping of these beds essentially provides a testable hypothesis regarding the lithostratigraphic model by claiming a certain relationship between particular beds and geography. If a lithostratigraphic model developed for an area and documented by mapping is accurate, then the geographic distribution of those beds should be as they appear on the maps (Figs. 8C and D). Consider the following example from a particularly critical stratigraphic interval at PEFO.

Jasper Forest is a large accumulation of fossil wood in the southern part of PEFO. The majority of the wood accumulation is in a small valley that opens to the northwest, surrounded by high bluffs. These bluffs are capped by a thick sandstone unit named the Jasper Forest bed (Fig. 8A; Raucci et al., 2006; Martz and Parker, 2010), which is the main petrified wood-bearing unit in the southern part of PEFO. This sandstone has been assigned to many bed-level units including the "Sonsela Sandstone bed" (Billingsley,

1985), Flattops One bed (Woody, 2003, 2006), and Agate Bridge bed (Heckert and Lucas, 2002). The latter two correlations considered the Jasper Forest bed to be continuous with the Flattops One bed (sensu Billingsley, 1985) in the southern area of PEFO, which caps the Mountain Lion Cliffs south of Jasper Forest (Figs. 8B and C). Initial geologic mapping (Woody, 2006) illustrated this proposed correlation by showing the same sandstone bed capping Jasper Forest and Mountain Lion Cliffs (Fig. 8C). However, later geological mapping determined that this lithostratigraphic correlation was erroneous and that the sandstone at Jasper Forest was lower stratigraphically than the Flattops One bed (Fig. 8D; Raucci et al., 2006; Martz and Parker, 2010). Because the type section of the Agate Bridge bed was located in the southern end of the park away from the geographic feature named Agate Bridge, and actually represented the Flattops One bed (Heckert and Lucas, 2002), this name could not be retained for the bed at Jasper Forest and therefore the name for this bed given by Raucci et al. (2006), the Jasper Forest bed, was utilized by Martz and Parker (2010) (Fig. 8).

The exact relationships between these two prominent sandstone beds were carefully worked out by walking out each exposed bed between Jasper Forest, where the bluffs are capped by the Jasper Forest bed, and Mountain Lion Mesa, to the southwest, which is capped by the Flattops One bed. At the Mountain Lion Cliffs (Fig. 8B), there are several sandstone marker beds in the cliff-forming exposures of the Jim Camp Wash beds (sensu Martz and Parker, 2010). Which of these is equivalent to the Jasper Forest bed has been contested (Cooley, 1957; Roadifer, 1966; Ash, 1970; Herrick et al., 1999; Heckert and Lucas, 2002). Detailed mapping of the area and repeatedly walking out the beds until a consensus was reached between the authors (Martz and Parker, 2010; Martz et al., 2012), resulted in the determination that the bed identified by Herrick et al. (1999) as the Jasper Forest bed, which lies well below Flattops One bed, was indeed the Jasper Forest bed (Fig. 8B and D).

This correlation was aided by recognition that the strata above the Jasper Forest bed, the Jim Camp Wash beds, Martha's Butte beds, and the "monotonous purple beds" (Martz and Parker, 2010) were continuous and maintained the same superpositional relationships throughout the study area (Fig. 3; Martz et al., 2012). This recognition also led to the determination that the allegedly distinct Rainbow Forest bed (e.g., Cooley, 1957; Ash, 1970) is actually equivalent to the Jasper Forest bed (Martz and Parker, 2010). These revised correlations greatly facilitated completion of the park map (Martz et al., 2012) and greatly clarified biostratigraphic relations in the southern end of the park. Indeed, they lead to a considerably revised understanding of the nature and timing of biotic change within PEFO (Parker and Martz, 2011).

Other Methods of Lithostratigraphic Documentation

In constructing lithostratigraphic models, the important consideration is clear, repeatable documentation of the geographic distribution of marker beds that other geologists can repeat. There are other ways to accomplish this without detailed geologic mapping.

For example, Kirkland (2006) and Hartman et al. (2014) constructed detailed lithostratigraphic models of marker beds by presenting bed-level lithostratigraphic correlations between outcrops documented by outcrop photographs on which the marker horizons were identified, and the precise geographic position of the outcrops was identified (these photographs should be included in projects utilizing mapping anyway to augment measured sections, as discussed in the next section). These clearly labeled and located outcrop photographs serve the same essential purpose as mapping, as they allow the reader to retrace the steps of the stratigrapher, walking the beds out from outcrop to outcrop to determine if the stratigrapher's correlations were accurate (Behrensmeyer, 2011).

Measuring Sections

Walking out beds and mapping them in is the ideal starting point for a project, before measuring stratigraphic sections, a lesson learned by the authors the hard way. The best-measured sections are those that (1) can be measured at known fossil localities (or other localities of interest), allowing relevant stratigraphic horizons to be plugged into the section, or a nearby outcrop with better exposure for which the precise stratigraphic level of the locality of interest can be determined; and (2) contain more than one traceable marker horizon so that stratigraphic sections can be tied together (Fig. 3B).

Which horizons will turn out to be most useful for correlation between sections might not be clear in advance. The best outcrops for measuring sections that meet one or both of these criteria might not be immediately obvious and might only be recognized after some time has already been spent exploring the outcrops. It is desirable to not waste time measuring redundant sections if the significant lithostratigraphic information is better represented in other outcrops. In some cases, one may spend time measuring a poorly exposed section at a fossil locality or a well-exposed section with one marker bed, to later discover a short distance away a much better outcrop with multiple marker beds, and that the level of a nearby fossil locality can be easily identified. Measuring sections may also go more quickly if the stratigrapher has already spent time becoming familiar with the stratigraphic units and geographic distribution of exposures. Therefore, put off formally measuring sections as long as possible.

When you finally have a good handle on the stratigraphy and how it relates to the topography, you are ready to identify outcrops at which to measure sections. Choose outcrops onto which fossil locality data can be precisely placed, and sections containing more than one marker bed, permitting the sections to be correlated. Always remember that the point of measuring sections is to be able to correlate lithostratigraphic sections with fossil locality data precisely plotted in order to generate a precise and detailed composite section (Fig. 3B and C). Do not waste time measuring sections that do not facilitate this goal in some way.

Lithostratigraphic type sections are useful to include as well, as they allow you to finally give formal names to your informally named marker beds. If these type sections have already been well described, you may simply refer to them in your correlations, or you might choose to redocument them if they have been poorly described. The same applies to all previously published sections; if you can examine the outcrop with section description in hand and identify the units described with relative ease, it will certainly save time to use the published sections. Measure your sections far enough apart that you do not have a lot of redundant sections from the same area showing exactly the same beds, but close enough together that you can easily correlate marker beds between them.

As with mapping (or using labeled outcrop photographs) for correlation, the key concept is repeatability. You want future researchers to not only be able to relocate the outcrop you measured, but to be able to identify your units. Stratigraphers often overlook this critical point and create extremely detailed but poorly described units that make it extremely difficult for anyone else to identify their units when visiting the same outcrop. Your reader should be able to stand in front of the same outcrop with your published stratigraphic section in hand and visually identify each unit you describe. If only a few units are described, this might be easy to do with only a written section description. However, if many units are identified, and especially if there are lateral changes in thickness or facies over a short area, a section description, even when accompanied by a drawn stratigraphic section, may be virtually useless.

The best tool for solving this problem is the use of labeled photographs (Fig. 8B). Ideally, on the day you measure the section, you should carry a copy of the photograph of the outcrop that you took on a previous day. This allows you to label unit contacts directly on the photograph to be redrawn onto a digitized photo later. Hartman et al. (2014) used this method for the lectostratotype section of the Hell Creek Formation, as we did for the Chinle Formation at PEFO (Fig. 8B; Martz and Parker, 2010).

There is considerable subjectivity in deciding how to subdivide a measured section into units and how to number them (e.g., giving every bed its own number, grouping them subjectively with individual beds having subunit numbers, whether or not to use the same numbers for the same beds in different outcrops, etc.). However, at the very least, important marker beds used to correlate throughout the local area should be clearly delineated (Fig. 3B), and the precise stratigraphic horizon of important fossil localities, geochemical samples, or any other data of interest should be very clearly identified in the unit description in terms of its distance from the unit boundaries (i.e., "fossil locality x is located in unit y about z meters above the base of this unit, in a thin lens of intrabasinal conglomerate").

Building Local Biostratigraphic Models
Taxonomic Identification of Paleontological Specimens

As stated earlier, robust biostratigraphic hypotheses are contingent on correct determination of the taxonomic assignments of voucher specimens. Historically, fossil specimens have been assigned to taxa based on the expert opinion of the paleontologist making the identification centered on phenetic characteristics present in the fossil specimen. However, the adoption of phylogenetic systematics has resulted in the realization that many of these characteristics used historically to make taxonomic assignments can actually be plesiomorphic for a clade, and thus not actually diagnostic to a specific taxon (e.g., Nesbitt and Stocker, 2008). Phylogenetic-based methods allow for determination of plesiomorphic and homoplastic characters and allow referrals of specimens to taxa based solely on synapomorphy (e.g., Bell et al., 2010). Furthermore, stratophenetic approaches (identifying fossils as distinct taxa based on their relative stratigraphic positions) are also to be avoided as inherently circular; fossils need to be identified using independent morphologic criteria and have the stratigraphic ranges determined afterwards (e.g., Bell et al., 2004; McKenna and Lillegraven, 2005). All fossils used for biostratigraphic work should be identified utilizing discrete apomorphies in the context of a phylogenetic analysis (Bell et al., 2004, 2010; Bever, 2005; Nesbitt et al., 2007; Nesbitt and Stocker, 2008). Fossils should not be identified on the basis of overall similarity or homoplasy. For example, in Triassic rocks of the American Southwest, hollow vertebrae were once considered to allow referral of material to the theropod dinosaurs (e.g., Colbert, 1989; Hunt et al., 1998). However, the presence of hollow vertebrae is more widespread throughout archosauriformes, especially in the Shuvosauridae, which are crocodile-line archosaurs very distantly related to dinosaurs (Nesbitt, 2007, 2011). Thus, the presence of hollow vertebrae now only allows assignment to the level of Archosauria.

Limitations of this method are that for some groups, phylogenetic relationships are not well resolved, restricting the ability to assign specimens to specific clades. Also, poorly preserved or fragmentary specimens often will only be able to be assigned at best to a least inclusive clade (e.g., Archosauromorpha). Nonetheless, identifications at coarser taxonomic levels still provide useful information regarding the presence of certain clades in faunal assemblages (Nesbitt and Stocker, 2008).

There is a notable difference in the effects on workflow if the fossils used for voucher specimens are immediately identifiable in the field or if investment needs to be made in the lab before the identification can be made. In the former scenario when attempting to identify units from a fresh perspective without using preexisting data, direct identification of a particular index fossil in the field allows one to readily hypothesize the appropriate biozone and probably the stratigraphic unit and immediate incorporate these data into the field study. However, if the fossils can only be identified after preparation, their identity may not be established until after the field exercise is completed. This may result in the changing of hypotheses and possibly the need to return to the field site to continue work. Recognition of this problem and its potential effects on fieldwork is an important factor to consider in the initial planning of a new study.

Establishing a Testable Biozonation

If a lithostratigraphic model is sufficiently detailed, and fossil locality data are robust, plotting fossil occurrences on a lithostratigraphic column is relatively simple. It is best to plot the data on individual measured sections, but plotting them on a composite section for the local study area is an overall goal for any study (see Martz and Parker, Chapter 2 of this volume, Figs. 9–12). Generally, alpha taxa should be utilized; however, an apomorphy-based approach may not allow all taxa to be resolved at that precise a systematic level, especially if phylogenetic relationships within the group are uncertain. In those cases, decisions will have to be made about which occurrences to identify as alpha taxa, and which as higher taxa (i.e., more inclusive clades).

It is important to remember that the lowest- (LO_k) and highest-known occurrences (HO_k) (sensu Walsh, 2000) almost certainly do not represent the actual range of a taxon and are artifacts of preservation and sampling, subject to change as sampling increases (Walsh, 1998). When possible, confidence intervals should be determined to provide baseline data on the incompleteness of the fossil record for each taxon (Strauss and Sadler, 1987; Marshall, 1990). This is especially true in cases where a multitaxon extinction event is hypothesized (e.g., Marshall, 1995). Unfortunately, these techniques have not been applied yet to the Chinle Formation where a potential connection between faunal change and environmental change has been proposed but not established (Parker and Martz, 2011, p. 252; Atchley et al., 2013). Establishing a more exact calibration of locality ages can also clarify and support or falsify these relationships between environmental and biotic change. Such work is ongoing for the Chinle Formation combining biostratigraphic information with radioisotopic dates and sequence stratigraphy (Atchley et al., 2013; Ramezani et al., 2014; Nordt et al., 2015). Incorporation of magnetostratigraphic data is also underway (Zeigler et al., Chapter 6 of this volume). A particular case study for using local biostratigraphic models to establish a regional biostratigraphic model is given in much more detail by Martz and Parker (Chapter 2 of this volume).

Publication

When it is time to publish, remember the emphasis on repeatability. Describe all fossil localities in detail and in stratigraphic context and provide specific voucher specimens for all taxa listed. When compiling faunal lists for various fossil localities, specific voucher specimens must be provided for all taxa deemed present in the assemblage. These voucher designations need to be accompanied by a detailed rationale listing the apomorphies that allow assignment of the material to a specific clade (Nesbitt and Stocker, 2008). Faunal lists provided without these data are subjective, not testable, and should be avoided if possible. Another researcher should be able to find every fossil used to identify that taxon at a given stratigraphic level and check your identification. Therefore only cataloged fossils in permanent repositories should be used as voucher specimens. When possible the voucher specimens should be figured to aid with recognition.

Print publishing has historically been limited by manuscript size and the use of color, and often the resulting publication was only available to those associated with institutions that had a journal subscription. The recent advent of affordable open-access publishing options has eliminated many of these barriers. Open-access, digital publishing allows for much larger papers that can include the unlimited use of color and even irregular-sized figures such as maps at much cheaper cost than old methods using printed media. For detailed stratigraphic studies, a publication that permits both color and extensive supplemental material is invaluable. Both criteria are best served by online publications such as *PLOS ONE, PeerJ,* or those that at least make supplemental data available electronically. Publishers have become increasingly appreciative of the need to provide this data in recent years, and even journals that publish short articles usually allow extensive supplemental materials to be published online. Color photographs are invaluable for any publication dealing with geologic data, as rock colors are important characteristics (this is less likely to be important in the case of fossils, where color is usually a diagenetic artifact).

In the case of published measured sections, included information should comprise not only a detailed measured section, but photographs of the measured rock outcrop on which particular units are labeled and the contacts drawn in with lines. GPS coordinates, including the utilized datum, should also be included to ensure that your sections and units can be precisely replicated by interested researchers (Martz and Parker, 2010). This may present problems if a lithostratigraphic section was measured at a fossil locality, where it may be desired to keep precise locality data confidential. There are two possible solutions: Either indicate that the section coordinates are available to qualified researchers, use the locality section but do not state that fossil localities occurred at those exact coordinates, or measure a section nearby close enough that fossil data can be precisely placed. Again, color photographs of outcrops and extensive section descriptions are best placed in an online supplement or appendix. When presenting these stratigraphic columns, make sure that individual localities are plotted on individual sections, with the composite section only being presented to synthesize the data in these individual sections.

Creating your figures with care and providing key data with the emphasis on repeatability in your manuscript will enhance the importance and value of your publication, ensuring that future workers can follow your rationale at every step of the process. Furthermore, if done well your work will form a baseline that future workers can build on rather than having to redo all of your efforts because your final work was ambiguous and unclear.

Remember too that fossils are not the only data collected at the outcrop. Collections also include specimens for paleomagnetism, geochemistry, and geochronology. The collection of these data should be documented the same way as for fossils and deposited in an established museum collection. Nondestructive analyses such as LA-ICP-MS and some geochemistry can be repeated on the same specimens. When possible, raw data should be uploaded to online databases for preservation and future distribution.

Conclusions and Prospectus

In recent years there has been an increased emphasis on the need for thorough documentation of specimen identifications, measured sections, and published figures to improve the scientific rigor of published paleontological and biostratigraphic work. This emphasis on documentation allows for increased repeatability and critical analysis of published studies contributing to their greater utility. In this paper we have provided recommendations for applying paleontological and geological methods to construct accurate, well-documented, and testable lithostratigraphic, biostratigraphic, and chronostratigraphic models. These models are developed at the local, regional, and global scales recognizing the utility and limitations of each. Local studies form the foundation of detailed biostratigraphic models, based on detailed lithostratigraphic correlations, whereas regional and global studies are heavily reliant on chronostratigraphy for correlations of biozones.

Documentation is crucial to clarify hypotheses as well as to provide a repeatable framework for future studies. If errors are identified, this framework can be used to quickly identify where mistakes were made and to understand the rationale behind stated hypotheses. Final publication should be in a medium that allows large amounts of supplementary data, color figures, and irregular figures such as maps and measured sections to be included. These improvements require the addition of considerable supplemental data to the paper, and presenting data clearly enough that results can be replicated or falsified with ease.

ACKNOWLEDGMENTS

Thank you to the staff of Petrified Forest National Park and to the landowners of Garza County, Texas, where many of the methods presented here were refined. Discussions with Bill Mueller, Sterling Nesbitt, Randall Irmis, Michelle Stocker, Sarah Werning, Jonathan Weinbaum, and Matthew Brown helped clarify many of the ideas presented herein. Reviews by Eric Scott and Adam Marsh greatly improved the manuscript. This is Petrified Forest National Park Paleontological Contribution No. 42.

REFERENCES

Agassiz, L., 1844. Monographie des poisons fossiles du vieux grés rouge ou Système Dévonian (Old Red Sandstone) des Isles Britanniques ed de Russie. Jent et Gassman, Neuchâtel. 171 pp.

Angielczyk, K.D., Kurkin, A.A., 2003. Has the utility of *Dicynodon* for Late Permian terrestrial biostratigraphy been overstated? Geology 31, 363–366.

Arkell, W.J., 1933. The Jurassic System in Great Britain. Clarendon Press, Oxford.

Ash, S.R., 1970. *Pagiophyllum simpsonii*, a new conifer from the Chinle Formation (Upper Triassic) of Arizona. J. Paleontol. 44, 945–952.

Ash, S.R., 1980. Upper Triassic floral zones of North America. In: Dilcher, D.L., Taylor, T.M. (Eds.), Biostratigraphy of Fossil Plants. Dowden, Hutchinson & Ross, Stroudsburg.

Ash, S.R., Hasiotis, S.T., 2013. New occurrences of the controversial Late Triassic plant fossil *Sanmiguelia* Brown and associated ichnofossils in the Chinle Formation of Arizona and Utah, USA. Neues Jahrbuch für Geologie und Paläontologie Abhandlungen 268, 65–82.

Atchley, S.C., Nordt, L.C., Dworkin, S.I., Ramezani, J., Parker, W.G., Ash, S.R., Bowring, S.A., 2013. A linkage among Pangean tectonism, cyclic alluviation, climate change, and biologic turnover in the Late Triassic: the record from the chinle formation, southwestern United States. J. Sediment. Res. 83, 1147–1161.

Barry, J.C., Morgan, M.E., Flynn, L.J., Pilbeam, D., Behrensmeyer, A.K., Raza, S.M., Khan, I.A., Badgley, C., Hicks, J., Kelley, J., 2002. Faunal and environmental change in the Miocene Siwaliks of northern Pakistan. Paleobiology 28, 1–71.

Behrensmeyer, A.K., 1978. Taphonomic and ecologic information from bone weathering. Paleobiology 4, 150–162.

Behrensmeyer, A.K., 2011. Linking researchers across generations. In: Canfield, M.R. (Ed.), Field Notes on Science and Nature. Harvard University Press, Cambridge, MA.

Behrensmeyer, A.K., Barry, J.C., 2005. Biostratigraphic surveys in the Siwaliks of Pakistan: a method for standardized surface sampling of the vertebrate fossil record. Palaeontol. Electron. 8 (1). 24 p.

Bell, C.J., Head, J.J., Mead, J.I., 2004. Synopsis of the herpetofauna from Porcupine Cave. In: Barnosky, A.D. (Ed.), Biodiversity Response to Climate Change in the Middle Pleistocene: The Porcupine Cave fauna From Colorado. University of California Press, Berkeley.

Bell, C.J., Gauthier, J.A., Bever, G.S., 2010. Covert biases, circularity, and apomorphies: a critical look at the North American Quaternary herpetofaunal stability hypothesis. Quat. Int. 217, 30–36.

Berry, W.B.N., 1966. Zones and zones-with exemplification from the Ordovician. Bull. Am. Assoc. Pet. Geol. 50 (7), 1487–1500.

Bever, G.S., 2005. Variation in the ilium of North American *Bufo* (Lissamphibia: Anura) and its implications for species-level identification of fragmentary anuran fossils. J. Vertebr. Paleontol. 25, 548–560.

Billingsley, G., 1985. General stratigraphy of the Petrified Forest National Park, Arizona. Mus. North. Ariz. Bull. 54, 3–8.

Blakey, R.C., 1990. Stratigraphy and geologic history of Pennsylvanian and Permian rocks, Mogollon Rim region, central Arizona and vicinity. Geol. Soc. Am. Bull. 102 (9).

Blakey, R.C., Gubitosa, R., 1983. Late Triassic paleogeography and depositional history of the Chinle Formation, southern Utah and northern Arizona. In: Reynolds, M.W., Dolly, E.D. (Eds.), Mesozoic Paleogeography of West-Central United States. Society of Economic Paleontologists and Mineralogists, Rocky Mountain Section, Denver, CO.

Blakey, R.C., Gubitosa, R., 1984. Controls of sandstone body geometry and architecture in the Chinle Formation (Upper Triassic), Colorado Plateau. Sediment. Geol. 38, 51–86.

Bown, T.M., Beard, K.C., 1990. Systematic lateral variation in the distribution of fossil mammals in alluvial paleosols, lower Eocene Willwood Formation, Wyoming. Geol. Soc. Am. Spec. Pap. 243, 135–151.

Branson, E.B., 1927. Triassic-Jurassic "red beds" of the Rocky Mountain region. J. Geol. 35, 607–630.

Branson, E.B., Mehl, M.G., 1929. Triassic amphibians from the Rocky Mountain region. Univ. Mo. Stud. 4, 1–87.

Brett, C.E., 2000. A slice of the "layer cake": the paradox of "frosting continuity". Palaios 15, 1–3.

Camp, C.L., 1930. A study of the phytosaurs with description of new material from western North America. Mem. Univ. Calif. 10, 1–160.

Casamiquela, R.M., 1960. Noticia preliminar sobre dos nuevos estagonolepoideos Argentinos. Ameghiniana 2 (1), 3–9.

Case, E.C., 1932. A perfectly preserved segment of the armor of a phytosaur, with associated vertebrae. Mus. Paleontol. Univ. Michigan 4, 57–80.

Colbert, E.H., 1960. Triassic rocks and fossils. N. M. Geol. Soc. Guideb. 11, 55–62.

Colbert, E.H., 1989. The Triassic dinosaur *Coelophysis*. Mus. North. Ariz. Bull. 57, 1–160.

Colbert, E.H., Gregory, J.T., 1957. Correlation of continental Triassic sediments by vertebrate fossils. Geol. Soc. Am. Bull. 68, 1456–1467.

Cooley, M.E., 1957. Geology of the Chinle Formation in the Upper Little Colorado Drainage Area, Arizona and New Mexico. University of Arizona, Tucson, AZ. Masters Thesis.

Cooper, M.R., 1982. A Mid-Permian to Earliest Jurassic tetrapod biostratigraphy and its significance. Arnoldia Zimbabwe 9, 77–103.

Cornet, B., 1993. Applications and limitations of palynology in age, climatic, and paleoenvironmental analyzes of Triassic sequences in North America. Bull. New Mex. Mus. Nat. Hist. Sci. 3, 75–93.

Desojo, J.B., Ezcurra, M.D., 2011. A reappraisal of the taxonomic status of *Aetosauroides* (Archosauria, Aetosauria) specimens from the Late Triassic of South America and their proposed synonymy with *Stagonolepis*. J. Vertebr. Paleontol. 31 (3), 596–609.

Desojo, J.B., Heckert, A.B., Martz, J.W., Parker, W.G., Schoch, R.R., Small, B.J., Sulej, T., 2013. Aetosauria: a clade of armoured pseudosuchians from the Upper Triassic continental beds. In: Nesbitt, S.J., Desojo, J.B., Irmis, R.B. (Eds.), Anatomy, Phylogeny, and Paleobiology of Early Archosaurs and Their Kin. Special Publications of the Geological Society of London, London.

Dubiel, R.F., 1987. Sedimentology of the Upper Triassic Chinle Formation, southeastern Utah: paleoclimatic implications. J. Ariz. Nev. Acad. Sci. 22, 35–45.

Dubiel, R.F., 1994. Triassic deposystems, paleogeography, and paleoclimate of the Western Interior. In: Caputo, M.V., Peterson, J.A., Franczyk, K.J. (Eds.), Mesozoic Systems of the Rocky Mountain Region. Society of Economic Paleontologists and Mineralogists Rocky Mountain Section, Denver, CO.

Dubiel, R.F., Hasiotis, S.T., 2011. Deposystems, paleosols, and climatic variability in a continental system: the Upper Triassic Chinle Formation, Colorado Plateau, U.S.A. In: Davidson, S., North, C. (Eds.), From River to Rock Record: The Preservation of Fluvial Sediments and Their Subsequent Interpretation. SEPM, Tulsa, OK.

Dunay, R.E., Fisher, M.J., 1974. Late Triassic palynofloras of North America and their European correlatives. Rev. Palaeobot. Palynol. 17, 179–186.

Evernden, J.F., James, G.T., 1964. Potassium-Argon dates and the Tertiary floras of North America. Am. J. Sci. 262, 945–974.

Evernden, J.F., Savage, D.E., Curtis, G.H., James, G.T., 1964. Potassium-Argon dates and the Cenozoic mammalian chronology of North America. Am. J. Sci. 262, 145–198.

Ezcurra, M.D., 2010. Biogeography of Triassic tetrapods: evidence for provincialism and driven sympatric cladogenesis in the early evolution of modern tetrapod lineages. Proc. R. Soc. B 277, 2547–2552.

Foote, M., Raup, D.M., 1996. Fossil preservation and the stratigraphic ranges of taxa. Paleobiology 22 (02), 121–140.

Gradstein, F.M., Ogg, J.G., Schmitz, M., Ogg, G. (Eds.), 2012. The Geologic Time Scale 2012. 2 vols. Elsevier, Amsterdam.

Gregory, J.T., 1957. Significance of fossil vertebrates for correlation of Late Triassic continental deposits of North America. In: XX Congreso Geologico Internacional, Sección II—El Mesozoico del Hemisferio Occidental y sus Correlaciones Mundiales. International Geological Congress Mexico City.

Gregory, J.T., 1972. Vertebrate faunas of the Dockum Group, eastern New Mexico and West Texas. In: New Mexico Geological Society Guidebook. 23, New Mexico Geological Society, New Mexico, pp. 120–130.

Hancock, J.M., 1977. The historic development of concepts of biostratigraphic correlation. In: Kauffman, E.G., Hazel, J.E. (Eds.), Concepts and Methods of Biostratigraphy. Dowden, Hutchinson, and Ross, Stroudsberg, PA, pp. 3–22.

Hartman, J.H., Butler, R.D., Weiler, M.W., 2014. Context, naming, and formal designation of the Cretaceous Hell Creek Formation lectostratotype, Garfield County, Montana. In: Wilson, G.P., Clemens, W.A., Horner, J.R., Hartman, J.H. (Eds.), Through the End of the Cretaceous in the Type Locality of the Hell Creek Formation in Montana and Adjacent Areas. Geological Society of America, Boulder, CO.

Heckert, A.B., Lucas, S.G., 2000. Taxonomy, phylogeny, biostratigraphy, biochronology, paleobiogeography, and evolution of the Late Triassic Aetosauria (Archosauria: Crurotarsi). In: Zentralblatt für Geologie und Paläontologie Teil I, Heft 11–12.

Heckert, A.B., Lucas, S.G., 2002. Revised Upper Triassic stratigraphy of the Petrified Forest National Park, Arizona, U.S.A. Bull. New Mex. Mus. Nat. Hist. Sci. 21, 1–36.

Hedberg, H.D., 1965. Chronostratigraphy and biostratigraphy. Geol. Mag. 102 (5), 451–461.

Hedberg, H.D., 1976. International Stratigraphic Guide: A Guide to Stratigraphic Classification, Terminology, and Procedure, first ed. Wiley, New York.

Herrick, A.S., Fastovsky, D.E., Hoke, G.D., 1999. Occurrences of *Zamites powellii* in Oldest Norian Strat in Petrified Forest National Park, Arizona. Petrified Forest National Park, Navajo, AR. National park service technical report, NPS/NRGRD/GRDTR-99/03, 91–95.

Holland, C.H., 1989. Synchronology, taxonomy, and reality. Philos. Trans. R. Soc. Lond. B 325, 263–277.

Hounslow, M.W., Muttoni, G., 2010. The geomagnetic polarity timescale for the Triassic: linkage to stage boundary definitions. Geol. Soc. Lond. Spec. Publ. 334, 61–102.

Howell, E.R., Blakey, R.C., 2013. Sedimentological constraints on the evolution of the Cordilleran arc: new insights from the Sonsela Member, Upper Triassic Chinle Formation, Petrified Forest National Park (Arizona, USA). Geol. Soc. Am. Bull. 125, 1349–1368.

Hunt, A.P., Lucas, S.G., Heckert, A.B., Sullivan, R.M., Lockley, M.G., 1998. Late Triassic dinosaurs from the western United States. Geobios 31, 511–531.

Irmis, R.B., Parker, W.G., Nesbitt, S.J., Liu, J., 2007. Early ornithischian dinosaurs: the Triassic record. Hist. Biol. 19, 3–22.

Irmis, R.B., Martz, J.W., Parker, W.G., Nesbitt, S.J., 2010. Re-evaluating the correlation between Late Triassic terrestrial vertebrate biostratigraphy and the GSSP-defined marine stages. Albertiana 38, 40–52.

Irmis, R.B., Mundil, R., Martz, J.W., Parker, W.G., 2011. High-resolution U–Pb ages from the Upper Triassic Chinle Formation (New Mexico, USA) support a diachronous rise of dinosaurs. Earth Planet. Sci. Lett. 309, 258–267.

Kent, D.V., Santi Malnis, P., Columbi, C.E., Alcobar, O.A., Martinez, R.N., 2014. Age constraints on the dispersal of dinosaurs in the Late Triassic from magnetochronology of the Los Colorados Formation (Argentina). Proc. Acad. Natl. Sci. Phila. 111, 7958–7963.

Kirkland, J.I., 2006. Fruitland Paleontological area (Upper Jurassic, Morrison Formation), western Colorado: and example of terrestrial taphofacies analysis. Bull. New Mex. Mus. Nat. Hist. Sci. 36, 67–96.

Kirkland, J.I., Martz, J.W., Deblieux, D.D., Madsen, S.K., Santucci, V.L., Inkenbrandt, P., 2014. Paleontological Resources Inventory and Monitoring Chinle and Cedar Mountain Formations, Capitol Reef National Park, Utah. Capitol Reef National Park, Utah. Utah geological survey unpublished report.

Krystyn, L., Gallet, Y., Besse, J., Marcoux, J., 2002. Integrated upper Carnian to lower Norian biochronology and implications for the Upper Triassic magnetic polarity time scale. Earth Planet. Sci. Lett. 203, 343–351.

Langer, M.C., 2005. Studies on continental Late Triassic tetrapod biochronology. II. The Ischigualastian and a Carnian global correlation. J. S. Am. Earth Sci. 19, 219–239.

Lehman, T.M., 1994. The saga of the Dockum Group and the case of the Texas/New Mexico boundary fault. Bull. New Mex. Bur. Min. Mineral Resour. 150, 37–51.

Lehman, T.M., Chatterjee, S., 2005. The depositional setting and vertebrate biostratigraphy of the Triassic Dockum Group of Texas. J. Earth Syst. Sci. 114, 325–351.

Litwin, R.J., Traverse, A., Ash, S.R., 1991. Preliminary palynological zonation of the Chinle Formation, southwestern U.S.A., and its correlation to the Newark Supergroup (eastern U.S.A.). Rev. Palaeobot. Palynol. 68, 269–287.

Long, R.A., Ballew, K.L., 1985. Aetosaur dermal armor from the Late Triassic of Southwestern North America, with special reference to material from the Chinle Formation of Petrified Forest National Park. Mus. North. Ariz. Bull. 54, 45–68.

Long, R.A., Padian, K., 1986. Vertebrate biostratigraphy of the Late Triassic Chinle Formation, Petrified Forest National Park, Arizona: preliminary results. In: Padian, K. (Ed.), The Beginning of the Age of Dinosaurs: Faunal Changes Across the Triassic-Jurassic Boundary. Cambridge University Press, Cambridge.

Loughney, K.M., Fastovsky, D.E., Parker, W.G., 2011. Vertebrate fossil preservation in blue paleosols from the Petrified Forest National Park, Arizona, with implications for vertebrate biostratigraphy in the Chinle Formation. Palaios 26 (11), 700–719.

Lucas, S.G., 1993. The Chinle Group: revised stratigraphy and biochronology of Upper Triassic nonmarine strata in the western United States. Mus. North. Ariz. Bull. 59, 27–50.

Lucas, S.G., 1998. Global Triassic tetrapod biostratigraphy and biochronology. Palaeogeogr. Palaeoclimatol. Palaeoecol. 143, 347–384.

Lucas, S.G., 2010. The Triassic timescale based on nonmarine tetrapod biostratigraphy and biochronology. Geol. Soc. Lond. Spec. Publ. 334, 447–500.

Lucas, S.G., Hunt, A.P., 1993. Tetrapod biochronology of the Chinle Group (Upper Triassic), western United States. Bull. New Mex. Mus. Nat. Hist. Sci. 3, 327–329.

Lucas, S.G., Anderson, O.J., Hunt, A.P., 1994. Triassic stratigraphy and correlations, southern High Plains of New Mexico-Texas. Bull. New Mex. Bur. Min. Mineral Resour. 150, 105–126.

Lucas, S.G., Heckert, A.B., Estep, J.W., Anderson, O.J., 1997. Stratigraphy of the Upper Triassic Chinle Group, Four Corners Region. N. M. Geol. Soc. Guideb. 48, 81–108.

Lucas, S.G., Hunt, A.P., Heckert, A.B., Spielmann, J.A., 2007. Global Triassic tetrapod biostratigraphy and biochronology: 2007 status. Bull. New Mex. Mus. Nat. Hist. Sci. 41, 229–240.

MacDougal, D., 2008. Nature's Clocks: How Scientists Measure the Age of Almost Everything. University of California Press, Berkeley, CA. 288 p.

Marshall, C.R., 1990. Confidence intervals on stratigraphic ranges. Paleobiology 16, 1–10.

Marshall, C.R., 1995. Distinguishing between sudden and gradual extinctions in the fossil record: predicting the position of the Cretaceous-Tertiary iridium anomaly using the ammonite fossil record on Seymour Island, Antarctica. Geology 23, 731–734.

Martz, J.W., 2008. Lithostratigraphy, Chemostratigraphy, and Vertebrate Biostratigraphy Of the Dockum Group (Upper Triassic), of Southern Garza County, West Texas. Texas Tech University, Lubbock, TX. PhD Thesis.

Martz, J.W., Parker, W.G., 2010. Revised lithostratigraphy of the Sonsela Member (Chinle Formation, Upper Triassic) in the southern part of Petrified Forest National Park, Arizona. PLoS ONE 5 (e9329), 1–26.

Martz, J.W., Parker, W.G., Skinner, L., Raucci, J.J., Umhoefer, P., Blakey, R.C., 2012. Geologic Map of Petrified Forest National Park, 1:50,000. Arizona Geological Society, Arizona.

Martz, J.W., Mueller, B.D., Nesbitt, S.J., Stocker, M.R., Atanassov, M., Fraser, N.C., Weinbaum, J.C., Lehane, J., 2013. A taxonomic and biostratigraphic re-evaluation of the Post Quarry vertebrate assemblage from the Cooper Canyon Formation (Dockum Group, Upper Triassic) of southern Garza County, western Texas. Earth Environ. Sci. Trans. R. Soc. Edinb. 103, 339–364.

Martz, J.W., Irmis, R.B., Milner, A.R.C., 2014. Lithostratigraphy and biostratigraphy of the Chinle Formation (Upper Triassic) in southern Lisbon Valley, southeastern Utah. In: Maclean, J.S., Biek, R.F., Huntoon, J.E. (Eds.), Geology of Utah's Far South. Utah Geological Association, Salt Lake City, UT.

McGowen, J.H., Granata, G.E., Seni, S.J., 1979. Depositional framework of the lower Dockum Group (Triassic), Texas Panhandle. Tex. Bur. Econ. Geol. 97, 1–60.

McGowen, J.H., Granata, G.E., Seni, S.J., 1983. Depositional setting of the Triassic Dockum Group, Texas panhandle and eastern New Mexico. In: Reynolds, M.W., Dolly, E.D. (Eds.), Mesozoic Paleogeography of West-Central United States. Society of Economic Paleontologists and Mineralogists, Rocky Mountain Section, Denver, CO.

McGowran, B., 2005. Biostratigraphy: Microfossils and Geologic Time. Cambridge University Press, Cambridge. 480 p.

McKenna, M.C., Lillegraven, J.A., 2005. Problems with Paleocene palynozones in the Rockies: Hell's Half Acre revisited. J. Mamm. Evol. 12, 23–51.

Morris, T.H., Manning, V.W., Ritter, S.M., 2010. Geology of Capitol Reef National Park, Utah. In: Sprinkel, D.A., Chidsey Jr., T.C., Anderson, P.B. (Eds.), Geology of Utah's parks and monuments. Utah Geological Association, Salt Lake City, UT, pp. 85–107. Utah Geological Association Publication 28.

Nesbitt, S.J., 2007. The anatomy of *Effigia okeeffeae* (Archosauria, Suchia), theropod-like convergence, and the distribution of related taxa. Bull. Am. Mus. Nat. Hist. 302, 1–84.

Nesbitt, S.J., 2011. The early evolution of archosaurs: relationships and the origin of major clades. Bull. Am. Mus. Nat. Hist. 352, 1–292.

Nesbitt, S.J., Stocker, M.R., 2008. The vertebrate assemblage of the Late Triassic Canjilon Quarry (northern New Mexico, USA), and the importance of apomorphy-based assemblage comparisons. J. Vertebr. Paleontol. 28, 1063–1072.

Nesbitt, S.J., Irmis, R.B., Parker, W.G., 2007. A critical re-evaluation of the Late Triassic dinosaur taxa of North America. J. Syst. Palaeontol. 5, 209–243.

Nordt, L.C., Atchley, S.C., Dworkin, S.I., 2015. Collapse of the Late Triassic megamonsoon in western equatorial Pangea, present-day American Southwest. Geol. Soc. Am. Bull. 127, 1798–1815.

North American Commission on Stratigraphic Nomenclature (NACSN), 1983. North American stratigraphic code. Am. Assoc. Pet. Geol. Bull. 67, 841–875.

North American Commission on Stratigraphic Nomenclature (NACSN), 2005. North American stratigraphic code. Am. Assoc. Pet. Geol. Bull. 89, 1547–1591.

Olsen, P.E., Kent, D.V., Whiteside, J.H., 2011. Implications of the Newark Supergroup-based astrochronology and geomagnetic polarity time scale (Newark-APTS) for the tempo and mode of the early diversification of the Dinosauria. Earth Environ. Sci. Trans. R. Soc. Edinb. 101, 201–229.

O'Sullivan, R.B., 1970. The Upper Part of the Upper Triassic Chinle Formation and Related Rocks, Southeastern Utah and Adjacent Areas. U.S. Geological Survey Professional Paper, 22.

Parker, W.G., 2006. The stratigraphic distribution of major fossil localities in Petrified Forest National Park, Arizona. Mus. North. Ariz. Bull. 62, 46–61.

Parker, W.G., 2013. Redescription and taxonomic status of specimens of *Episcoposaurus* and *Typothorax*, the earliest known aetosaurs (Archosauria: Suchia) from the Upper Triassic of western North America, and the problem of proxy "holotypes". Earth Environ. Sci. Trans. R. Soc. Edinb. 103, 313–338.

Parker, W.G., 2014. Taxonomy and Phylogeny of the Aetosauria (Archosauria: Pseudosuchia) Including a New Species From the Upper Triassic of Arizona. The University of Texas at Austin, Austin, TX. PhD Dissertation, 437 p.

Parker, W.G., Martz, J.W., 2011. The Late Triassic (Norian) Adamanian-Revueltian tetrapod faunal transition in the Chinle Formation of Petrified Forest National Park, Arizona. Earth Environ. Sci. Trans. R. Soc. Edinb. 101, 231–260.

Poole, F.G., Stewart, J.H., 1964. Chinle Formation and Glen Canyon Sandstone in northeastern Utah and northwestern Colorado. Geol. Surv. Prof. Pap. 501-D, D30–D39.

Popper, K.R., 1962. Conjectures and Refutations: The Growth of Scientific Knowledge. Basic Books, New York.

Prochnow, S.J., Nordt, L.C., Atchley, S.C., Hudec, M.R., 2006. Multi-proxy paleosol evidence for middle and late Triassic climate trends in eastern Utah. Palaeogeogr. Palaeoclimatol. Palaeoecol. 232 (1), 53–72.

Ramezani, J., Hoke, G.D., Fastovsky, D.E., Bowring, S.A., Therrien, F., Dworkin, S.I., Atchley, S.C., Nordt, L.C., 2011. High-precision U-Pb zircon geochronology of the Late Triassic Chinle Formation, Petrified Forest National Park (Arizona, USA): temporal constraints on the early evolution of dinosaurs. Geol. Soc. Am. Bull. 123, 2142–2159.

Ramezani, J., Fastovsky, D.E., Bowring, S.A., 2014. Revised chronostratigraphy of the lower Chinle Formation strata in Arizona and New Mexico (USA): high-precision U-Pb geochronological constraints on the Late Triassic evolution of dinosaurs. Am. J. Sci. 314, 981–1008.

Raucci, J.J., Blakey, R.C., Umhoefer, P.J., 2006. A new geological map of Petrified Forest National Park with emphasis on members and key beds of the Chinle Formation. Mus. North. Ariz. Bull. 62, 157–159.

Rayfield, E.J., Barrett, P.M., McDonnell, R.A., Willis, K.J., 2005. A Geographical Information System (GIS) study of Triassic vertebrate biochronology. Geol. Mag. 142, 327–354.

Rayfield, E.J., Barrett, P.M., Milner, A.R., 2009. Utility and validity of Middle and Late Triassic 'land vertebrate faunachrons'. J. Vertebr. Paleontol. 29, 80–87.

Reichgelt, T., Parker, W.G., Van Konijnenburg-Van Cittert, J.H.A., Martz, J.W., Kuerschner, W.M., Conran, J.G., 2013. The palynology of the Sonsela Member (Late Triassic, Norian) at Petrified Forest National Park, Arizona, USA. Rev. Palaeobot. Palynol. 189, 18–28.

Roadifer, J.E., 1966. Stratigraphy of the Petrified Forest National Park, Arizona. University of Arizona, Tucson, AZ. Doctoral Thesis.

Rudwick, M.J.S., 1985. The Meaning of Fossils: Episodes in the History of Palaeontology. University of Chicago Press, Chicago, IL, ISBN: 0226731030. p. 24.

Salvador, A. (Ed.), 1994. International Stratigraphic Guide, second ed. International Union of Geological Sciences and the Geological Society of America, Boulder, Colorado. 214 p.

Schultz, C.L., 2005. Biostratigraphy of the non-marine Triassic: is a global correlation based on tetrapod faunas possible? In: Koutsoukos, E.A.M. (Ed.), Applied Stratigraphy. Springer, Dordrecht.

Stewart, J.H., Williams, G.A., Albee, H.F., Raup, O.B., 1959. Stratigraphy of Triassic and associated formations in part of the Colorado Plateau region. U.S. Geol. Surv. Bull. 1046-Q, 487–576.

Stewart, J.H., Poole, F.G., Wilson, R.F., 1972. Stratigraphy and origin of the Chinle Formation and related Upper Triassic strata in the Colorado Plateau region. U.S. Geol. Surv. Prof. Pap. 690, 1–336.

Stocker, M.R., 2010. A new taxon of phytosaur (Archosauria: Pseudosuchia) from the Late Triassic (Norian) Sonsela Member (Chinle Formation) in Arizona, and a critical reevaluation of *Leptosuchus* Case, 1922. Paleontology 53 (5), 997–1022.

Strauss, D., Sadler, P.M., 1987. Confidence Intervals for the Ends of Local Taxon Ranges. University of California, Riverside, CA. Department of statistics technical report.

Tedford, R.H., 1970. Principles and practices of mammalian geochronology in North America. Proc. North Am. Paleontol. Conv. Part F, 666–703.

Therrien, F., Fastovsky, D.E., 2000. Paleoenvironments of early theropods, Chinle Formation (Late Triassic), Petrified Forest National Park, Arizona. Palaios 15 (3), 194–211.

Walker, A.D., 1961. Triassic Reptiles from the Elgin Area: *Stagonolepis*, *Dasygnathus*, and their allies. Philos. Trans. R. Soc. Lond. 244, 103–204.

Walsh, S.L., 1998. Fossil datum and paleobiological event terms, paleontostratigraphy, chronostratigraphy, and the definition of Land Mammal "Age" boundaries. J. Vertebr. Paleontol. 18 (1), 150–179.

Walsh, S.L., 2000. Eubiostratigraphic units, quasibiostratigraphic units, and "assemblage zones." J. Vertebr. Paleontol. 20 (4), 761–775.

Walsh, S.L., 2004. Time and time-rock again: an essay on the (over)simplification of stratigraphy. Palaeontol. Newsl. 57, 19–25.

Whiteside, J.H., Lindström, S., Irmis, R.B., Glasspool, I.J., Schaller, M.F., Dunlavey, M., Nesbitt, S.J., Smith, N.D., Turner, A.H., 2015. Extreme ecosystem instability suppressed tropical dinosaur dominance for 30 million years. Proc. Natl. Acad. Sci. U.S.A. 112 (26), 7909–7913.

Winchester, S., 2009. The Map that Changed the World. Harper Collins, New York. 368 p.

Woodburne, M.O., 1989. Hipparion horses: a pattern of endemic evolution and intercontinental dispersal. In: Prothero, D.R., Schoch, R.M. (Eds.), The Evolution of Perissodactyls. Oxford University Press, New York.

Woodburne, M.O., 1996. Precision and resolution in mammalian chronostratigraphy: principles, practices, examples. J. Vertebr. Paleontol. 16, 531–555.

Woody, D.T., 2003. Geologic Reassessment of the Sonsela Member of the Chinle Formation, Petrified Forest National Park. Museum of Northern Arizona, Arizona, MS.

Woody, D.T., 2006. Revised stratigraphy of the lower Chinle Formation (Upper Triassic) of Petrified Forest National Park, Arizona. Mus. North. Ariz. Bull. 62, 17–45.

Revised Formulation of the Late Triassic Land Vertebrate "Faunachrons" of Western North America: Recommendations for Codifying Nascent Systems of Vertebrate Biochronology

J.W. Martz*, W.G. Parker[†]
*University of Houston-Downtown, Houston, TX, United States
[†]Petrified Forest National Park, Petrified Forest, AZ, United States

'Pretty good! This must have been what happened,' dreamed he.

(McKenna and Lillegraven, 2006, p. 6)

Contents

Terrestrial Depositional Systems
http://dx.doi.org/10.1016/B978-0-12-803243-5.00002-9

INTRODUCTION
The Late Triassic Land Vertebrate "Faunachrons"

The Triassic Period was one of the most critical intervals in Earth history, a time when the terrestrial ecosystems recovered from the Permian extinction with a spectacular adaptive radiation that included early dinosaurs, crocodylian-line archosaurs, and mammals (e.g., Sues and Fraser, 2010; Nesbitt, 2011). At the end of the Triassic, a globally variable array of terrestrial ecosystems in which different amniote groups enjoyed varying degrees of success (Ezcurra, 2010; Sues and Fraser, 2010) gave way to uniformly dinosaur-dominated ecosystems as a result of complex environmental changes that are only beginning to be understood (e.g., Rowe et al., 2010; Irmis, 2011; Olsen et al., 2011; Parker and Martz, 2011; Irmis et al., 2011). Late Triassic biostratigraphy and biochronology, which provide a framework for understanding these biotic changes, have therefore been subjects of considerable interest on both global and regional scales (e.g., Colbert and Gregory, 1957; Zawiskie, 1986; Lucas, 1993b, 1998; Schultz, 2005; Irmis et al., 2010). Notwithstanding the gradual ascendency of nonbiostratigraphic methods (especially radioisotopic dating and magnetostratigraphy) for reconstructing the Triassic chronology (e.g., Irmis et al., 2010, 2011; Olsen et al., 2011), biostratigraphy and its more abstract chronological derivative, biochronology (Tedford, 1970; Walsh, 1998, 2001), remain essential components for understanding the nature and causes of biological changes when coupled with sedimentological and geochemical data firmly linked to the geologic timescale (e.g., Barry et al., 2002; Whiteside et al., 2011; Atchley et al., 2013; Eberth et al., 2013).

We will reevaluate Late Triassic vertebrate biochronology in western North America in detail, building on previously presented methodologies for developing detailed and accurate Upper Triassic lithostratigraphic and biostratigraphic models (e.g., Parker, 2006; Martz, 2008; Martz and Parker, 2010; Irmis et al., 2010, 2011; Parker and Martz, 2011; Martz et al., 2013, 2014; Parker and Martz, Chapter 1 of this volume). We will first discuss various biostratigraphic and biochronologic definitions and concepts, then review the historical development of the Late Triassic land vertebrate "faunachrons" in western North America (e.g., Camp, 1930; Colbert and Gregory, 1957; Gregory, 1957, 1972; Long and Ballew, 1985; Long and Padian, 1986; Lucas and Hunt, 1993; Lucas, 1993b, 1998; Hunt et al., 2005; Irmis et al., 2010; Parker and Martz, 2011), making comparisons between their development and those of other biostratigraphic and biochronologic models (e.g., Arkell, 1933; Tedford, 1970; Hancock, 1977; Savage, 1977; Woodburne, 1977, 1987, 1989, 1996, 2004; Flynn et al., 1984; Walsh, 1998, 2001; Barry et al., 2002; McGowran, 2005). These comparisons will then be used to make recommendations for how the Late Triassic land vertebrate "faunachrons" should be redefined and further developed, incorporating recent systematic revisions of North American phytosaur taxa and considering the current limits of resolution on stratigraphic and geochronologic data. Finally, this paper will apply this revised methodology to Upper Triassic continental strata in western North America to provide the best possible estimates for the Late Triassic land vertebrate "faunachrons."

Parker and Martz (Chapter 1 of this volume) introduce terms for stratigraphic studies at varying geographic scales. *Local* stratigraphic studies cover geographic areas small enough that it is possible to make bed-level correlations within formations and members, allowing the relative stratigraphic positions of particular fossil and geochemical sampling localities to be determined with a high degree of precision. *Regional* stratigraphic studies cover geographic areas small enough (usually a sedimentary basin or several closely associated sedimentary basins) to correlate thicker facies packages (formations, members, and potentially facies variations within members), and to identify similar biostratigraphic patterns that usually involve the same alpha taxa. However, regional models are large enough that precise bed-level correlations are impossible, and the relative positions of particular stratigraphic horizons can only be crudely approximated. *Global* stratigraphic studies cover geographic areas large enough that lithostratigraphic correlations are impossible, and biostratigraphic correlations are hampered by regional endemism, making chronostratigraphy based on radioisotopic age dating the most reliable means of correlation (Irmis et al., 2010). This paper will confine its discussion of Upper Triassic strata to a particular regional study area and to local study areas within it: the western United States, where lithostratigraphic and biostratigraphic models for the Chinle Formation and the Dockum Group can be correlated across western Texas and the Four Corners states (e.g., Gregory, 1957; Lucas, 1993a,b, 1998; Dubiel, 1994; Martz et al., 2013).

Institutional Abbreviations

AMNH FR: American Museum of Natural History, New York, NY; *CM*: Carnegie Museum of Natural History, Pittsburgh, PA; *GR*: Ruth Hall Museum at Ghost Ranch, Abiquiu, NM; *MNA*: Museum of Northern Arizona, Flagstaff, AZ; *MOTT*: Museum of Texas Tech University, Lubbock, TX (localities); *NMMNHS*: New Mexico Museum of Natural History and Science, Albuquerque, NM; *PEFO*: Petrified Forest National Park, AZ; *TMM*: Texas Vertebrate Paleontology Collections, Jackson School Museum of Earth History, Austin, TX; *TTU P*: Texas Tech University, Lubbock, TX (fossil specimens); *UCMP*: University of California Museum of Paleontology, Berkeley, CA; *YPM*: Yale Peabody Museum, New Haven, CT.

THE PATH FROM BIOSTRATIGRAPHY TO BIOCHRONOLOGY
Biozones, Biochronozones, and Biochrons

The difference between stratigraphic and chronologic (temporal) terms and concepts (Figs. 1 and 2; Table 1) is often blurred in the literature, as is the precise meaning of

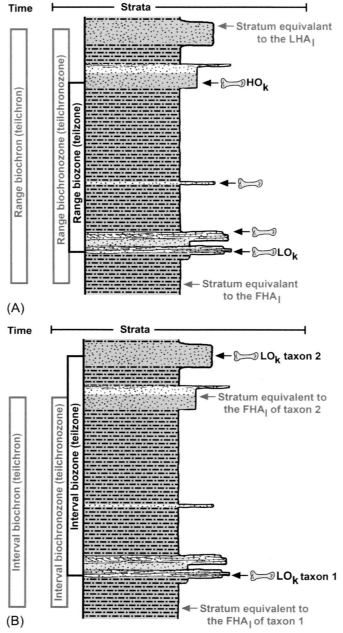

Fig. 1 Eubiostratigraphic teilzones applied to a particular local stratigraphic section, incorporating epistemological (observable) biostratigraphic data in black, and ontological (unknowable) stratigraphic horizons and units in gray. (A) A range teilzone bounded by the lowest and highest known occurrences of a single taxon. (B) An interval teilzone bounded by the lowest occurrences of two taxa. Note that the known (epistemological) fossil record does not give a completely accurate representation of the true (ontological) ranges of taxa. FHA_l, local first historical appearance (ontological); HO_k, highest known occurrence (epistemological); LHA_l, local last historical appearance (ontological); LO_k, lowest known occurrence (epistemological).

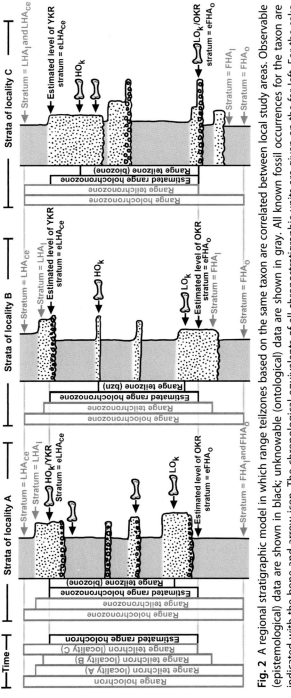

Fig. 2 A regional stratigraphic model in which range teilzones based on the same taxon are correlated between local study areas. Observable (epistemological) data are shown in black; unknowable (ontological) data are shown in gray. All known fossil occurrences for the taxon are indicated with the bone and arrow icon. The chronological equivalents of all chronostratigraphic units are given on the far left. For the sake of convenience, sedimentation rates are assumed to have been precisely equivalent for all three local study areas, so that the intervals between isochronous stratigraphic levels are the same thickness. Note that the OKR and YKR are equivalent to the local LO_k and HO_k respectively, and that both the range holochronozone and estimated range holochronozone are correlated between all three sections. Note that "stratum =" indicates the stratigraphic level that was being deposited during a paleobiological *event*. $eFHA_o$, estimated oldest first historical appearance (epistemological); $eLHA_{ce}$, estimated time of complete extinction (epistemological); FHA_o, oldest first historical appearance (ontological); LHA_{ce}, time of complete extinction (ontological); OKR, oldest known record (epistemological); YKR, youngest known record (epistemological); other abbreviations as in Fig. 1.

Table 1 Epistemological (operational) biostratigraphic, chronostratigraphic, and geochronologic terms discussed in the text

Biostratigraphic terms	Chronostratigraphic terms	Geochronologic (time) terms
Fossil occurrences	Chronostratigraphic terms	Estimated paleobiological events
LO$_k$: Lowest known stratigraphic occurrence. The stratigraphically lowest occurrence of a taxon at a particular local study area (Figs. 1 and 2)	*OKR*: Oldest known record. Of all the known LO$_k$s for a taxon at different local study areas, the one thought to be the oldest (Fig. 2)	*eFHA$_o$*: Estimated oldest first historical appearance. The best estimate for the time when a species originally appeared through evolution, or immigrated into a specified regional study area (Fig. 2). Time equivalent of the OKR. Note that in a local study area, the stratigraphic position of the *local* first historical appearance (*FHA$_l$*) would be estimated as the LO$_k$, making it redundant to define an estimated local first historical appearance (*eFHA$_l$*)
HO$_k$: Highest known stratigraphic occurrence. The stratigraphically highest occurrence of a taxon at a particular local study area (Figs. 1 and 2)	*YKR*: Youngest known record. Of all the known HO$_k$s for a taxon at different local study areas, the one thought to be the youngest (Fig. 2)	*eLHAce*: Estimated time of complete extinction. The best estimate for the time when a species went extinct, or emigrated from a specified regional study area. Time equivalent of the YKR

Row labels: "Lowest/oldest" (first data row), "Highest/youngest" (second data row)

Biozone	Estimated biochronozones	Estimated biochrons
Eubiozone: A body of strata bounded by the LO$_k$ and/or HO$_k$ of one or more specified taxa	*Estimated eubiochronozone*: The best estimate of the body of strata deposited between two paleobiological events	*Estimated eubiochron*: The best estimate of the time that elapsed between two paleobiological events

Row label: "Bounded by known fossil occurrences/estimates of paleobiological events…"

	Eubiozones (including teilzones)	Estimated Eubiochronozones (including teilchronozones and holochronozones)	Estimated Eubiochrons (including teilchrons and holochrons)
...for a local study area	Teilzone: A eubiozone at a particular local study area. As boundaries are based on known fossil occurrences, the teilzone may or may not be good approximation of the teilchronozone (Fig. 2)	Estimated teilchronozone: The best estimate for total interval of strata that was deposited between two paleobiological events (the local immigration and/or emigration of the taxa used) at a local study area (Fig. 2). As this would effectively be identical to the teilzone, it is not used here	Estimated teilchron: The best estimate of the time that elapsed between two paleobiological events (the evolution, immigration, emigration, and/or extinction of the taxa used) in a local study area (Fig. 2). As this would effectively be identical to the estimated holochron if local absolute age dates are unavailable, it is not used here
...for a global or a specified regional study area	No specified term; for this study, simply referred to as the eubiozone	Estimated holochronozone: The best estimate for total interval of strata that was deposited between two paleobiological events globally or in a specified regional study area (Fig. 2)	Estimated holochron: The best estimate for the time that elapsed between two paleobiological events globally or in a specified regional study area (Fig. 2)
Bounded vaguely or by lithologic changes or their time equivalents	Quasibiozone: A body of strata characterized by its known fossil content but bounded vaguely or by lithologic changes	Quasibiochronozone: A body of strata deposited when a specified set of taxa existed (which may have exceeded the ranges of known fossil occurrences) but bounded vaguely or by lithologic changes	Quasibiochron: The interval of time equivalent to a quasibiochronozone, and not known to correspond precisely to the actual time ranges of taxa
Number of taxa used	**Eubiozones (including teilzones)**	**Estimated Eubiochronozones (including teilchronozones and holochronozones)**	**Estimated Eubiochrons (including teilchrons and holochrons)**
One taxon	Range eubiozone: The stratigraphic interval between the LO_k and HO_k of a specified taxon (see Figs. 1A and 2)	Estimated range eubiochronozone: The stratigraphic interval between the OKR and YKR of a specified taxon (Fig. 2 as estimated range holochronozone)	Estimated range eubiochron: The time that elapsed between the $eFHA_o$ and $eLHA_{ce}$ of a single taxon (Fig. 2 as estimated range holochron)
Two taxa...	Interval eubiozone: The stratigraphic interval between the LO_ks or HO_ks of two specified taxa (Fig. 1B)	Estimated interval eubiochronozone: The stratigraphic interval between the OKRs or YKRs of two specified taxa	Estimated interval eubiochron: The time that elapsed between the $eFHA_o$s or $eLHA_{ce}$s of two specified taxa

Continued

Table 1 Epistemological (operational) biostratigraphic, chronostratigraphic, and geochronologic terms discussed in the text—cont'd

Number of taxa used	Eubiozones (including teilzones)	Estimated Eubiochronozones (including teilchronozones and holochronozones)	Estimated Eubiochrons (including teilchrons and holochrons)
...using lowest/oldest	LO_k-LO_k interval eubiozone: The stratigraphic interval between the LO_ks of two specified taxa (see Figs. 1B and 4)	OKR-OKR interval eubiochronozone: The stratigraphic interval between the OKRs of two specified taxa	$eFHA_o$-$eFHA_o$ interval eubiochron: The stratigraphic interval between the $eFHA_o$s of two specified taxa
...using highest/youngest	HO_k-HO_k interval eubiozone: The stratigraphic interval between the HO_ks of two specified taxa	YKR-YKR interval eubiochronozone: The stratigraphic interval between the YKRs of two specified taxa	$eLHA_{ce}$-$eLHA_{ce}$ interval eubiochron: The stratigraphic interval between the $eLHA_{ce}$s of two specified taxa
One set of multiple taxa...	Assemblage eubiozone (sensu Walsh, 2001): A body of strata defined by one set of multiple taxa with the boundaries being a LO_k and HO_k of one or two of the taxa (see Fig. 3A)	Assemblage eubiochronozone (sensu Walsh, 2001): A body of strata defined by one set of multiple taxa with the boundaries being a OKR and YKR one or two of the taxa	Assemblage eubiochron (sensu Walsh, 2001): A body of strata defined by one set of multiple taxa with the boundaries being the $eFHA_o$ and $eLHA_{ce}$ of one or two of the taxa
Two sets of multiple taxa...	Multiple-taxon interval biozone (sensu Walsh, 2001): A body of strata defined by two set of multiple taxa with the boundaries being LO_ks and/or HO_ks from one taxon in each group (Fig. 3B)	Multiple-taxon interval eubiochronozone (sensu Walsh, 2001): A body of strata defined by two set of multiple taxa with the boundaries being OKRs and/or YKRs from one taxon from each group	Multiple-taxon interval eubiochron (sensu Walsh, 2001): A body of strata defined by two set of multiple taxa with the boundaries being the $eFHA_o$ and/or $eLHA_{ce}$ of one taxon from each group
...using lowest/oldest	Multiple-taxon LO_k-LO_k interval eubiozone: A body of strata defined by two set of multiple taxa with the boundaries being LO_ks from one taxon in each group. (see Fig. 3B)	Multiple-taxon OKR-OKR interval eubiochronozone (sensu Walsh, 2001): A body of strata defined by two set of multiple taxa with the boundaries being OKRs of one taxon from each group (Fig. 14)	Multiple-taxon $eFHA_o$-$eFHA_o$ interval eubiochronozone (sensu Walsh, 2001): A body of strata defined by two set of multiple taxa with the boundaries being $eFHA_o$s of one taxon from each group (Fig. 14)
...using highest/youngest	Multiple-taxon HO_k-HO_k interval eubiozone: A body of strata defined by two set of multiple taxa with the boundaries being HO_ks from one taxon in each group	Multiple-taxon YKR-YKR interval eubiochronozone (sensu Walsh, 2001): A body of strata defined by two set of multiple taxa with the boundaries being YKRs of one taxon from each group	Multiple-taxon $eLHA_{ce}$-$eLHA_{ce}$ interval eubiochronozone (sensu Walsh, 2001): A body of strata defined by two set of multiple taxa with the boundaries being $eLHA_{ce}$s of one taxon from each group

Definitions are applicable to the current study unless otherwise noted.

particular terms, even when their stratigraphic or chronologic nature is specified (e.g., Walsh, 1998). This confusion is largely because of the varied usage of many terms that have been constantly defined and redefined in various ways, or simply applied without an explicit explanation. Two important sources of ambiguity must be addressed here before considering their application to Late Triassic biochronology:

1. Whether or not a term refers to a stratigraphic concept based on fossil datums, or a chronologic concept based on biological events (e.g., Arkell, 1933; Berry, 1966; Ludvigson et al., 1986; Walsh, 1998).

2. Whether or not a term is actually applicable using empirically determined data. This particular source of confusion stems from the incompleteness of the fossil record. Because of the rarity of fossil preservation and the capriciousness of fossil discovery (e.g., Behrensmeyer, 1982), the known stratigraphic ranges for fossil taxa, and the time that elapsed during the deposition of those stratigraphic intervals, is almost certainly shorter than the total stratigraphic and temporal intervals during which taxa actually existed (e.g., Johnson, 1979; Woodburne, 1996; Marshall, 1998; Walsh, 1998; Patzkowsky and Holland, 2012). It is therefore important to distinguish between the actual stratigraphic and temporal ranges of taxa, and the shorter apparent ranges available to us through the imperfect fossil record. Walsh (1998, p. 151) distinguished between *ontological (or theoretical)* concepts relating to the real world that we could discern if we had unlimited data and infallible methodologies with which to reveal it, and *epistemological (or operational)* concepts relating to the world perceived through the practical limitations of empirical data and scientific methodology. Differing concepts of biostratigraphic and biochronologic terms often depend on if the concepts are ontological or epistemological in nature. It is important to recognize from the outset that all stratigraphic and temporal units actually identified and utilized by paleontologists *must* be epistemological in nature. This commonsense observation is not always reflected in the way biostratigraphic and biochronologic units based on vertebrates are defined and discussed.

Our following discussion of various stratigraphic and chronologic concepts will make much reference to the first and second editions of the International Stratigraphic Guide (hereafter Hedberg, 1976, and Salvador, 1994, respectively), and the 1983 and 2005 editions of the North American Stratigraphic Code (hereafter NACSN, 1983, and NACSN, 2005, respectively). We will also make particularly frequent reference to discussions by Arkell (1933), Walsh (1998, 2001), and Woodburne (2004). Walsh (1998, 2001) in particular discussed terms for a wide variety of ontological and epistemological datums, events, and intervals of strata and time. We will only discuss a handful of those concepts with particular relevance to the following discussion, and propose some practical modifications. The reader is invited to make frequent referral to Table 1, which provides definitions for the terminology used here.

Biozones

A biostratigraphic unit, or biozone (Figs. 1–4), is defined by NACSN (1983, 2005) as "a body of rock defined and characterized by its fossil content." The definitions of Hedberg (1976) and Salvador (1994) are consistent with the NACSN's. Historically, however, the meaning of the term has been more confused. Although the term *zone* originally had a clear stratigraphic intent (discussed by Arkell, 1933), "biozone" has also been used to refer

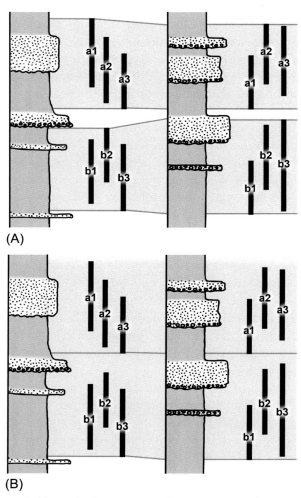

(A)

(B)

Fig. 3 Eubiozones bounded by multiple taxa, using the terminology of Walsh (1998, 2001). (A) Two disjunctive assemblage "fossilzones" (biozones), each bounded by the LO_ks and HO_ks of one set of taxa. (B) Two disjunctive base-disjunctive base multiple-taxon (assemblage) interval "fossilzones" (biozones) bounded by the LO_ks of two sets of taxa; notice that these boundaries eliminate the gaps between biozones occurring in the disjunctive assemblage biozones.

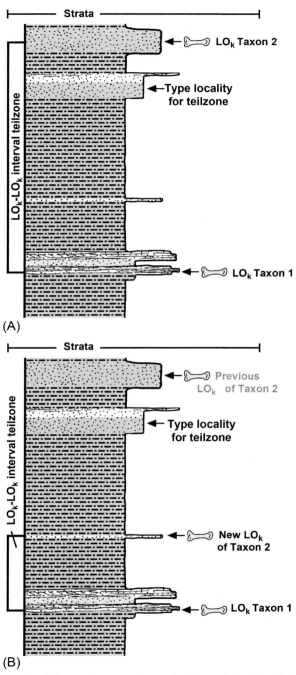

Fig. 4 The impact of new fossil discoveries on an interval teilzone defined by the LO_ks of two taxa that also have a type locality. (A) The biozone bounded by original LO_ks, showing the stratigraphic level of a type locality for the biozone, occurring within the teilzone as expected. (B) After the discovery of a stratigraphically lower LO_k for Taxon 2 (but not for Taxon 1), the stratigraphic range of the teilzone changes (in this case, shortened), and the type locality for the biozone now occurs outside of it.

to a unit of time (see discussion in Walsh, 1998, p. 155). Walsh (1998, 2001) therefore suggested substituting the term "fossizone" or "fossilzone" for "biozone." However, in light of the explicitly stratigraphic connotations of the term *biozone* used by most authors (e.g., Arkell, 1933; Hedberg, 1976; Salvador, 1994; NACSN, 1983, 2005; Woodburne, 2004), that term is preferred here. Similarly, the term *teilzone* (Arkell, 1933) is used to refer to a biozone from a particular local study area. Following Arkell (1933) and Walsh (1998), the suffix "-zone" applied to a unit has an explicitly stratigraphic rather than temporal connotation.

A *biostratigraphic unit* or *biozone* is a material thing, a body of sedimentary rock strata. A biozone can be seen, photographed, and walked on. A piece of a biozone can be broken off and looked at with a hand lens. Biostratigraphic units are made out of precisely the same materials as lithostratigraphic and chronostratigraphic units (*sensu* Hedberg, 1976; Salvador, 1994; NACSN, 1983, 2005): sedimentary rocks and whatever they contain, including fossils. Indeed, a particular stratum is part of more than one of these units (e.g., part of the biostratigraphic Revueltian biozone, the lithostratigraphic Chinle Formation, and the chronostratigraphic Triassic System). Biozones differ from lithostratigraphic and chronostratigraphic units in that they are characterized and often bounded by fossil datums, rather than by lithology or age (e.g., Hedberg, 1976; Salvador, 1994; NACSN, 1983, 2005). Precisely how a biozone should be bounded is debatable, but there is a clear trend in both invertebrate biostratigraphy (e.g., McGowran, 2005) and vertebrate biostratigraphy (e.g., Woodburne, 1977; Walsh, 1998; Parker and Martz, 2011) favoring explicit boundary definitions that use the highest or (more commonly) lowest stratigraphic occurrences of particular taxa (Figs. 1–4).

In the interest of stabilizing terminology in vertebrate biostratigraphy, we follow Walsh (2001) and Woodburne (2006, p. 229) in using the terms *lowest known occurrence* (LO_k) and *highest known occurrence* (HO_k) to describe the stratigraphically lowest and highest known occurrences of a fossil taxon (Figs. 1, 2, and 4; Table 1) in a particular local study area (*sensu* Parker and Martz, Chapter 1 of this volume). If multiple local study areas are sufficiently well correlated to determine which local LO_k is the oldest and which local HO_k is the youngest, these occurrences are referred to as the *oldest known record* (OKR) and *youngest known record* (YKR), respectively (Walsh, 1998; Fig. 2 and Table 1). All four of these terms refer to empirically determined (i.e., epistemological) pieces of data that can actually be observed, and are unlikely to represent the first and last members of the taxon to exist locally, regionally, or globally (Figs. 1 and 2).

It should be noted that a wide variety of other terms have also been applied to the lowest and highest known stratigraphic occurrences of fossils, most notably *first appearance datum* (FAD) and *last appearance datum* (LAD) (e.g., Berggren and Van Couvering, 1978, pp. 44–47; Salvador, 1994, p. 56; Schoch, 1989, p. 201; Lindsay and Tedford, 1990). However, as discussed by Walsh (1998), the terms *FAD* and *LAD* have also been used to describe events rather than fossil occurrences, including in Late Triassic vertebrate biochronology (e.g., Lucas, 1998), and therefore have a somewhat historically confused meaning. Elsewhere (Parker and Martz, 2011), we had adopted the terms *lowest known stratigraphic datum* and *highest known stratigraphic datum* (Opdyke et al., 1977; Lindsay and Tedford, 1990). These terms were advocated by Walsh (1998) but later abandoned (Walsh, 2001, p. 761) in favor of LO_k and HO_k, and we follow suit.

Walsh (2001) discussed the difference between what he called "eubiostratigraphic units," which are bounded by LO_ks and/or HO_ks (Figs. 1–5B), and "quasibiostratigraphic units," which are bounded by some other criterion, such as an unconformity or lithologic change (Figs. 5A and 8A), or may even have no formal boundaries at all. For the sake of brevity, we use the shorthand terms *quasibiozone* and *eubiozone* for these units (Table 1). Eubiozones and quasibiozones are both recognized and characterized by their fossil content, making both units biozones (Hedberg, 1976; NACSN, 1983, 2005; Salvador, 1994). Because they are empirically determined by known fossil occurrences, both also are explicitly epistemological units.

It should be noted that the stratigraphic limits of both types of biozones, especially eubiozones, are always tenuous and subject to revision (Fig. 4). As new fossil discoveries are made, and as identifications change for known fossil specimens, the specimens representing LO_ks and HO_ks (potentially including the OKR and YKR) will also change, moving the boundaries of the biozone (e.g., McKenna and Lillegraven, 2006). Because the boundaries of quasibiozones are not based on LO_ks and/or HO_ks, they are not as sensitive to new fossil discoveries as eubiozones. For example, Walsh (2001) noted that a lithostratigraphic unit with lithologic boundaries (such as unconformities) but still characterized by its fossil content would be a quasibiozone, specifically what he referred to as a "paleontologically distinct lithozone" (Figs. 5A and 8A). The use of such quasibiozones does not require that the biostratigraphic ranges of taxa be worked out in detail, only that it be known which broad lithostratigraphic unit they were derived from. This also means that slight range extensions of taxa within the "paleontologically distinct lithozone" (i.e., the discovery of lower LO_ks and/or higher HO_ks) are unlikely to change the boundaries of the unit unless they extend across the lithologic boundary.

"Assemblage zones" have been defined in a variety of ways in the literature, many of them vague. Salvador (1994, pp. 62–63) defined an assemblage zone as "a stratum or body of strata characterized by a distinctive assemblage or association of three or more fossil taxa that, taken together, distinguishes it in biostratigraphic character from adjacent strata." The definitions of Hedberg (1976, p. 50) and the NACSN (2005, Article 50d) are similar, although Hedberg's does not specify a membership of three or more taxa. According to both Hedberg (1976, p. 52) and Salvador (1994, p. 63), the boundaries of an assemblage zone "are drawn at surfaces (biohorizons) marking the limits of occurrence of the assemblage characteristic of the unit."

However, Hedberg (1976) stated that biohorizons are "*commonly* used as a biozone boundary" (italics ours), which implies that other boundaries may be used. Indeed

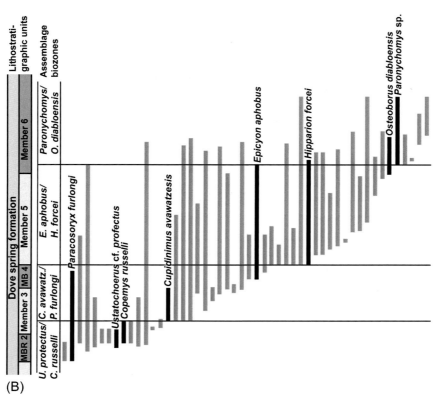

(A)

Hayden (1869)		Cope (1884)	Matthew (1924) "faunal zones"
Loup River beds	F	*Megalonyx* beds	Equus
		Equus beds	
	E	*Procamelus* beds	Hipparion
	D	*Ticholeptus* beds	Merychippus
White River Group			Porohippus
	C	Truckee Fm.	Miohippus
	B	White River Fm.	Mesohippus
	A		
		Unita Fm.	Epihippus
		Bridger Fm.	Orohippus
Wind River deposits		Green River Fm.	Eohippus
		Wasatch Fm.	
			Ptilodus
		Puerco Fm.	Polymastodon

(B)

Fig. 5 Changing boundary concepts for Cenozoic mammalian biozones in the western United States. (A) Quasibiostratigraphic units from the early days of mammalian biostratigraphy redrawn and simplified from Tedford (1970, Fig. 1); note that biostratigraphic subdivisions of Matthew (1924) began as lithostratigraphic units (Hayden, 1869; Cope, 1884). (B) Eubiostratigraphic units delineated by the precisely plotted LO$_k$s of particular taxa, independent of lithostratigraphic boundaries, redrawn and simplified from Woodburne (2006, Fig. 10). The ranges of taxa important to bounding and defining the biozones are in black; notice that neither taxon ranges nor biozone boundaries correspond with regularity to the boundaries of lithostratigraphic units.

Hedberg (1976, Fig. 4, p. 51) and NACSN (1983, Fig. 5A) indicated that at least some assemblage zones may not be bounded by the LO_ks and HO_ks of taxa at all, and are therefore quasibiozones. Woodburne (2004, p. 6) preferred to treat all assemblage zones as being imprecisely bounded quasibiozones, recognizing that because assemblage zones are characterized by multiple taxa, it is unlikely that the LO_ks of all taxa will occur at precisely the same stratigraphic level (Fig. 3).

However, assemblage zones can also be eubiozones (Fig. 3). Salvador (1994, Fig. 10, p. 62) and the NACSN (1983, Fig. 5B) show assemblage zones bounded by the LO_ks and HO_ks of *several* taxa, although which taxa are used varies locally as best fits the assemblage. More recently, the NACSN (2005, Article 51, Fig. 5A) seems to prefer assemblage zones that are bounded by the lowest and highest stratigraphic occurrences of particular taxa, but acknowledged (NACSN, 2005, Article 51a) that these boundaries might be "difficult to define precisely." Walsh (1998, 2001) presented several biozone ("fossilzone" in his terminology) concepts based on the LO_ks and/or HO_ks of multiple taxa, and these were also flexible with respect to which taxon marked the lower boundary of the biozone in a particular area (Fig. 3). This addressed the concern of Woodburne (2004, p. 6) about boundaries being imprecise when multiple taxa are used. Within Walsh's (2001) eubiostratigraphic assemblage zone concept, he distinguished between "assemblage biozones" (Fig. 3A) and "multiple-taxon interval biozones" (Fig. 3B; Walsh (1998) had previously identified the latter as "assemblage interval biozones"). Assemblage biozones (*sensu* Walsh, 2001) have lower and upper boundaries defined by the LO_ks and HO_ks from one specified group of taxa (essentially range zones using multiple taxa), whereas multiple-taxon interval biozones (*sensu* Walsh, 2001) are defined by the LO_ks and/or HO_ks of two different sets of taxa (essentially interval zones using multiple taxa).

Biochrons

A "biochronologic unit" or "biochron" is generally agreed to represent some interval of time relating to a fossil taxon or taxa, rather than a material entity (Hedberg, 1976; Salvador, 1994; Walsh, 1998; Woodburne, 2004). As such, the boundaries of a biochron are events, rather than observable markers such as fossils or lithologic boundaries. Following Arkell (1933) and Walsh (1998), the suffix "-chron" is used to designate a unit of time rather than of strata. However, the understanding of what sort of events are used to bound biochrons, and therefore what exactly the unit of elapsed time represents, tends to be far more contentious and confused than for biozones.

A biochron is identified by both editions of the International Stratigraphic Guide (Hedberg, 1976; Salvador, 1994) as a unit of time corresponding to the duration of a biozone. This would make at least some biochrons more or less epistemological in nature, direct temporal reflections of empirically determined biozones and radioisotopic dates, and their boundaries would represent the time of *burial* of the *known* fossil occurrences that form the biozone boundaries (at least in the case of eubiozones). In other words, by this definition a biochron is the estimated time that elapsed between the burial of the LO_ks and/or HO_ks used to bound a biozone, not the time that elapsed between the evolution and/or extinction of the bounding taxa.

However, Walsh (1998) used the term "fossilzone-chron" to refer to the time equivalent of a biozone ("fossil-zone" in his terminology), and used the term "biochron" in a very different way. According to Walsh (1998), a biochron is a strictly ontological concept, the temporal equivalent of a biochronozone (see the next section) rather than a biozone. The boundaries of biochrons represent paleobiological events (Figs. 1 and 2): the appearance of a taxon in an area through evolution or immigration, called the *first historical appearance* (FHA), or the disappearance of a taxon in an area through emigration or extinction, called the *last historical appearance* (LHA).

Walsh (1998) made several fine distinctions between different types of FHAs and LHAs. The paleobiological events representing the appearance and disappearance of a taxon globally or across an entire specified regional study area (Fig. 2) were called the *oldest first historical appearance* (FHA_o) and *time of complete extinction* (LHA_{ce}), whereas paleobiological events representing the appearance and disappearance of a taxon in a particular local area were called the *local first historical appearance* (FHA_l) and *local last historical appearance* (LHA_l). The FHA_o is also the oldest FHA_l anywhere for the taxon, and the LHA_{ce} is the also youngest LHA_l anywhere for the taxon. Biochrons based on these paleobiological events could therefore be local, regional, or global in nature (*sensu* Parker and Martz, Chapter 1 of this volume). For example, a biochron based on the FHAs of two taxa (Fig. 1B) could correspond to the total interval between the evolution of those taxa anywhere in the world (a "holochron" *sensu* Walsh, 1998; Fig. 2), or the interval between the times when the taxa appeared in a particular local or regional area (a "teilchron" *sensu* Walsh, 1998; Figs. 1 and 2). This distinction is less important to Late Triassic biochronology in western North America for reasons discussed later.

Just as Walsh (2001) found it useful to distinguish between biostratigraphic units bounded by specific fossil datums ("eubiostratigraphic units") and those with boundaries defined by some other criterion, or none at all ("quasibiostratigraphic units"), we find it useful to extend this terminology to biochronologic units. A eubiochronologic unit or eubiochron is defined here as an interval of time bounded by the FHAs and/or LHAs of specific taxa. A quasibiochronologic unit or quasibiochron is here defined as an interval of time during which a specified faunal assemblage existed, but for which the FHAs and/or LHAs of particular taxa do not precisely bound the time interval. For example, if a vertebrate fauna is not precisely delineated by FHAs and/or LHAs but is associated with a lithostratigraphic unit, the approximate time duration over which that unit was deposited could be considered a quasibiochron.

Walsh (1998, 2001) outlined several assemblage and multiple-taxon interval eubiochron concepts. As with assemblage and multiple-taxon interval biozones (Fig. 3), assemblage eubiochrons are bounded by the FHAs and LHAs on one group of taxa, and multiple-taxon interval biochrons are bounded by the FHAs and/or LHAs of two sets of taxa (Walsh, 2001).

Biochronozones

Walsh (1998) defined a biochronozone as a unit of strata bounded by the stratigraphic equivalents of paleobiological events, and as such is an ontological unit representing the stratigraphic equivalent of a biochron. Biochronozones are chronostratigraphic units or chronozones (NACSN, Articles 66, 75) rather than biozones, in that their boundaries are the stratigraphic equivalents of *events* (i.e., FHAs and/or LHAs) rather than fossil occurrences (LO$_k$s and/or HO$_k$s). The boundaries of a biochronozone are the particular layers of strata being deposited when a taxon evolved, immigrated, went extinct, and/or emigrated within a particular area of interest; biochronozones can correspond either to global holochrons (as holochronozones) or more local teilchrons (as teilchronozones) (Walsh, 1998). However, as with biochrons this distinction is less important to Late Triassic vertebrate biochronology (see the following).

Given the incomplete nature of the fossil record (e.g., Behrensmeyer, 1991; Walsh, 1998; Marshall, 1998), it is highly unlikely that the layers of strata being deposited when a taxon evolved or went extinct (FHAs and LHAs) correspond to OKRs or YKRs (i.e., it is highly unlikely that the very first and last individuals to ever exist will be preserved as fossils, much less discovered; Fig. 2). Therefore biochronozones (*sensu* Walsh, 1998) have two critical differences from biozones:

1. Paleontologists cannot actually identify biochronozones because they are ontological rather than epistemological units.
2. The stratigraphic bases of an interval biochronozone (one defined by two or more taxa) will almost certainly be lower than the base of a biozone defined by the same taxa (Figs. 1 and 2).

As with biozones and biochrons, we can distinguish different biochronozone concepts based on boundary criteria. A *eubiochronozone* (equivalent to Walsh's, 1998 biochronozone) is here defined as a stratigraphic interval bounded by layers of strata deposited at the same time that specified taxa evolved and/or went extinct (i.e., the stratigraphic equivalent of a eubiochron). A *quasibiochronozone* is defined here as an interval of strata deposited during the time interval that specified group of taxa existed, but for which there is no precise boundary definition based on the evolution and/or extinction of taxa (i.e., the temporal equivalent of a quasibiochron).

Estimated Biochronozones and Estimated Biochrons

Walsh's (1998, 2001) ontological biochron concept leaves us with a serious difficulty: If biochrons are truly bounded by ontological paleobiological events, then they are impossible to recognize! This means that no paleontologist has ever truly identified a biochron, and that the units of time referred to as biochrons in the literature are really only epistemological *estimates* of biochrons using empirically determined biostratigraphic datums and biozones (Tedford, 1970; Emry, 1973; Woodburne, 1977, 2006), and other data such as radioisotopic dates. By extension, the same is true of biochronozones. As acknowledged by Walsh (2001, p. 772), the boundaries of theoretical biochrons "can only be approached asymptotically with the collection of more and more data" from biostratigraphy. We introduce the terms *estimated biochronozone* and *estimated biochron* (Table 1) to refer to these epistemological approximations which, unlike biochronozones and biochrons (*sensu* Walsh, 1998, 2001), can actually be empirically determined.

Estimated holochronozones are estimated biochronozones bounded by the OKR and/or YKR for the taxa defining the unit (Fig. 2), which represent the best available approximation for the stratigraphic level of the FHA_o and/or time of complete extinction (LHA_{ce}). By extension, estimated biochrons are the best estimates for the time that elapsed between the temporal equivalents of the OKR and YKR, referred to here as the estimated oldest first historical appearance ($eFHA_o$) and estimated time of complete extinction ($eLHA_{ce}$). As the OKR and YKR occur within particular local study areas (and probably not the same one), it is necessary to approximate the equivalent stratigraphic horizons in other local study areas in order to correlate the estimated biochronozone regionally or globally (Fig. 2).

Estimated biochronozones have two major advantages over biozones (including teilzones):

1. Estimated biochronozones identify strata where fossil occurrences of the boundary-defining taxon may occur below known fossil occurrences within local study areas, allowing predictions of how the local biostratigraphic ranges of taxa will change with increased sample sizes.

2. Estimated biochronozones are chronostratigraphic rather than biostratigraphic units (NACSN, 2005, Article 75). Their boundaries are ideally not correlated using fossil occurrences, but using strata thought to be equivalent in *age* to the OKRs and/or YKRs that represent their boundaries. Consequently, unlike biozones, estimated biochronozones can also be correlated globally into areas where boundary-defining taxa are absent, as long as absolute age data allows chronostratigraphic correlation of strata to the OKRs of boundary-defining taxa. Unlike other chronostratigraphic units (such as series and systems), estimated biochronozone boundaries do not have a fixed global stratotype section and point (Walsh, 2004), but are free to move if different specimens are re-identified as the OKR.

Ideally, the correlation of biochronozones would be accomplished using radioisotopic dates and/or magnetostratigraphy, allowing isochronous stratigraphic horizons to be identified in different local study areas. Unfortunately, radioisotopic dates for Upper Triassic strata in western North America are scarce (Ramezani et al., 2011, 2014; Irmis et al., 2011; Atchley et al., 2013), and it is therefore necessary to use biostratigraphic and/or lithostratigraphic correlations as crude proxies for chronostratigraphic correlation (e.g., Lucas, 1993b; Desojo et al., 2013, pp. 208–215). Such correlations require the dubious assumption that lithostratigraphic and/or biostratigraphic boundaries are approximately the same age everywhere (i.e., that the depositional and biotic changes that initiated deposition of these stratigraphic units occurred everywhere at the same time), and the correlation of estimated biochronozones should therefore always be considered extremely tentative and subject to improved radioisotopic age data (Irmis et al., 2010).

The unavoidable crudeness of these correlations is the reason why no distinction is made here between local and regional or global biochrons (teilchrons and holochrons) or between local and regional or global biochronozones (teilchronozones and holochronozones). It would be impossible to distinguish local and global estimates for units at the current level of resolution available. To do so would require using the same local set of radioisotopic dates from Petrified Forest National Park (PEFO) and a few other local study areas (Ramezani et al., 2011, 2014; Irmis et al., 2011; Atchley et al., 2013), making the endeavor inherently circular. For example, the beginning of the Revueltian estimated holochron (defined below) is ~215 Ma based on radioisotopic data from PEFO (Dunlavey et al., 2009; Parker and Martz, 2011; Atchley et al., 2013). Technically, this date should only be the beginning of the Revueltian estimated teilchron (local biozone) for the PEFO local study area; in other words, the Revueltian began in PEFO at 215 Ma but may have begun at slightly different times in other local study areas. However, since there are no radioisotopic dates available for the base of the Revueltian estimated biochronozone anywhere else in western North America, the boundary ages for all local teilchrons would have to be estimated using the PEFO dates, and tied to these other local study areas using lithostratigraphy and/or biostratigraphy. It is clearly pointless and circular to talk about estimated teilchrons from different local study areas as separate time intervals when they are based on the same dates from a single local study area. It is therefore best to simply identify a single estimated eubiochronozone and a corresponding estimated eubiochron concept, which are specifically an estimated holochronozone and estimated holochron (Table 1). At some point in the future, radioisotopic calibration of Upper Triassic strata in the western United States may improve sufficiently that separate teilchron estimates can be made for different local study areas, making a distinction between estimated teilchrons and an estimated holochron meaningful.

"Faunachrons"

In formulating the North American Land Mammal "Ages," Wood et al. (1941) intended the term *age* to reflect the NALMA's temporal intent. However, an "age" is not a biochronologic unit, but a different kind of geochronologic unit: the temporal equivalent of a chronostratigraphic stage (NACSN, 2005, Article 81). As such, an age is linked to the boundary stratotype for the stage (Walsh, 2004). As discussed by Walsh (1998, pp. 169–170), establishing an immovable boundary stratotype (or "golden spike") for biozones (and by extension, estimated biochronozones and estimated biochrons) is inadvisable, because it makes the unit inflexible to future fossil finds outside its defined boundaries (Fig. 4; see the following).

Lucas (1993a) introduced the term *faunachron* as a biochronologic term intended as a replacement for the misleading use of the term *age* in vertebrate biochronology. Lucas (1992, p. 88) stated that a faunachron represents "the interval of geologic time that corresponds to the *duration of a taxon*" (italics ours), and also cited the claim by Berggren and Van Couvering (1978, p. 74) that biochrons are "created with inferred or abstract limits not inferred in the type section itself," which indicates that biochrons are hypothetical (i.e., ontological) temporal units. As is discussed in more detail in the following, Lucas and Hunt (1993) and Lucas (1993a) did not initially define "faunachrons" explicitly in terms of paleobiological events relating to specified taxa, treating them instead as quasibiochrons as defined here.

However, Lucas (1998) later used paleobiological events to define "faunachrons," specifically FHAs (which he referred to as first appearance datums, or FADs). This would make "faunachrons" a type of biochron *sensu* Walsh (1998, 2001), specifically a eubiochron as defined here. Walsh (2001) also stated that he considered a "faunachron" to be a type of "assemblage biochron," defined by paleobiological events. However, as Lucas (1998, 2010) defined "faunachrons" as being based on the first appearances of individual taxa, the term is somewhat misleading. As only two particular taxa bound each "faunachron" (i.e., the "faunachron" is not defined by a "fauna"), it does not meet the definition of either an assemblage biochron or multiple-taxon interval biochron *sensu* Walsh (2001). Instead, Lucas's (1998, revised for Lucas, 2010) "faunachron" definitions correspond to interval biochrons *sensu* Walsh (1998, p. 158). As the term "faunachron" is not only misleading (in not being defined by multiple taxa) but also redundant (in that other terms cover the same concept), we suggest abandoning it altogether.

Definition and Characterization of the NALMAs

Various authors (e.g., Tedford, 1970; Woodburne, 1977; Lucas, 1992) have noted that the North American Land Mammal "Ages" of Wood et al. (1941) are biochronologic rather the biostratigraphic units, intended to represent intervals of time rather than stratigraphic intervals characterized by their fossil content. Given the sporadic nature of the vertebrate fossil record (e.g., Barry et al., 2002; Behrensmeyer and Barry, 2005; Rogers and Kidwell, 2007), biochrons in the ontological sense (*sensu* Walsh, 1998) have often been deemed more attractive to mammal workers compared to biozones. As noted by Berggren and Van Couvering (1978, p. 7), "Mammalian biochrons…originate as local zones tied to reference sections and 'type faunas,' even though they are commonly liberated from such earthly bondage at birth." However, as discussed previously, it is of course impossible to extract unbiased, ontological, biochronologic information from an imperfect rock record (Emry, 1973; Woodburne, 1977).

Fortunately, the practical dependence of biochronology on biostratigraphy is not always neglected; as discussed in Woodburne (1987), both NALMAs and their subdivisions are often treated as true biostratigraphic units discussed in terms of the stratigraphic occurrence of fossils, even when the LO_ks and/or HO_ks of individual taxa aren't used in *defining* a NALMA boundary (e.g., Woodburne, 1996; Archibald et al., 1987). The following discussion will emphasize both the biochronologic nature and biostratigraphic basis of the NALMAs.

Much of the pioneering work on mammalian biostratigraphy in the American West (Fig. 5A) used units that were not bounded by lowest and highest fossil occurrences (Tedford, 1970), and were therefore quasibiostratigraphic rather than eubiostratigraphic units. Lithostratigraphic units were broadly characterized in terms of their fossil content (Fig. 5A), making them "paleontologically distinct lithozones" *sensu* Walsh (2001). The Wood Committee (Wood et al., 1941), in creating the first NALMAs, also sometimes employed such quasibiozones. For example, the Bridgerian age was "based on the Bridger formation of southwestern Wyoming…the time of deposition of Bridger A–D inclusive, with the enclosed faunas" (Wood et al., 1941, p. 10).

Although this approach of using vaguely bounded quasibiostratigraphic units is convenient for mammalian paleontologists dealing with low biostratigraphic resolution, it can also be a crutch for studies that employ poorly documented biostratigraphic data. Prothero and Emry (1996, p. 678) criticized the approach for "unacceptably mixing lithostratigraphy and biostratigraphy," and Prothero (1990, p. 240) observed, "Too often an index fossil is equated with the formation, and no attempt is made to document the actual range of the fossil within the formation. This results in a loss of resolution of the data. The stratigraphic range of the fossil is often reported to be the same as the total thickness of the formation, which may artificially extend the range" (Fig. 5A; e.g., Tedford, 1970, Fig. 6). Referring to quasibiozones that have no concrete boundaries at all (even lithologic ones) as "paleontologically distinct 'fuzzy' zones," and noted (Walsh, 2001, p. 767) that "…the concept of a 'fuzzy zone'…has an obvious temporal analog in the numerous mammal 'biochrons' whose boundaries are not rigorously defined by the evolution, immigration, and extinction of specified taxa, but whose contents are instead loosely characterized by a 'central core' of overall faunal aspect."

Quasibiozones and quasibiochrons continue to be used in mammalian biostratigraphy and biochronology. Cifelli et al. (2004) discussed pre-Campanian (Late Cretaceous) NALMAs as lithostratigraphic units characterized by their fossil content, but not given explicit taxon-based boundaries (e.g., "the Mussentuchit local fauna, collected from a restricted stratigraphic interval in the upper parts of the Cedar Mountain Formation"; Cifelli et al., 2004, p. 22). The Wasatchian through Duchesnean NALMAs have also been historically treated as quasibiozones (specifically, "paleontologically distinct lithozones") and/or quasibiochrons (e.g., Krishtalka et al., 1987; Robinson et al., 2004). For example, although Robinson et al. (2004) identified numerous first and last appearances for taxa within these biochrons, it is unlikely that all of these taxa appeared at precisely the same geologic instant, and Robinson et al. (2004) do not specify how to decide which of the particular taxon FHAs and LHAs represent the precise beginning and end of the biochron. As such, the precise boundaries for these "ages" must be considered somewhat vaguely defined (Woodburne, 2004), making them quasibiochrons as defined here.

Matthew (1924) may have been the first to define mammalian ranges independent of lithostratigraphy, although his concept was eubiochronologic rather than eubiostratigraphic: "Matthew did not directly state, but his usage indicates, that these 'faunal zones' are bounded by the temporal range of the horse genus in question, supplemented by the first appearance and limited occurrence of other mammalian genera within those time intervals" (Tedford, 1970, p. 672). In recent decades there has been an effort to incorporate detailed biostratigraphic information into mammalian biochronology (Fig. 5B), and to adapt the NALMAs into eubiozones and/or eubiochrons (e.g., Repenning, 1967; Tedford, 1970; Woodburne, 1977, 1987, 2004; Barry et al., 2002; Lofgren et al., 2004). Repenning (1967) and Woodburne (1977, 1987) suggested that NALMAs should be defined using the lowest known occurrences (LO$_k$s) of individual taxa; the same type of boundary definition was being advocated in microfossil biostratigraphy at about the same time (McGowran, 2005, pp. 28–29). As already discussed, such boundaries are (at least hypothetically) less arbitrary than those defined based on multiple taxa, and avoid the problem of potential gaps between eubiozones and eubiochrons because the top of each eubiozone or eubiochron is defined by the base of the next (Figs. 1, 3, and 4). NALMAs may also be subdivided using LO$_k$s or FHAs (e.g., Woodburne, 1996; Archibald et al., 1987; Lofgren et al., 2004). When eubiostratigraphic in nature, such units are "lowest occurrence interval zones" *sensu* Salvador (1994, p. 59), and "LO-LO single-taxon interval fossilzones" *sensu* Walsh (2001, p. 760).

However, for any eubiozone or eubiochron definition to be applied (regardless of whether it is based on a single taxon or multiple taxa), detailed biostratigraphic information is required. If the precise lowest and highest known stratigraphic positions of taxa are not known, it is meaningless to use them to define eubiozones, and by extension, estimated eubiochronozones and biochrons. This problem was noted by Emry et al. (1986, p. 128) in discussing the Chadronian, Orellian, and Whitneyan NALMAs: "Although we advocate the development of mammalian biochronology based on biostratigraphic units...such a biochronology cannot yet be realized...much additional work is needed before detailed information on most of the taxa will be available. In any new definitions, the limits of the ages should be based...[on] faunal breaks, and these cannot be recognized until we have detailed biostratigraphic information..." If such detailed data is lacking, then there may be no alternative than to treat NALMAs as quasibiozones or quasibiochrons, regardless of how they have been defined. For example, the absence of detailed biostratigraphic data made Matthew's (1924) zones difficult to separate from lithologic units (Fig. 5A), even if they were eubiochronologic in concept. Fortunately, detailed biostratigraphic data has been collected for several study areas that puts individual specimens in a precise stratigraphic context relative to other specimens (Fig. 5B), radioisotopic dates, and magnetostratigraphic data (e.g., Behrensmeyer and Barry, 2005). These data have then been used to develop remarkably detailed models of vertebrate faunal change and its relationship to environmental changes (Barry et al., 2002).

The Development of Late Triassic Terrestrial Vertebrate Biostratigraphy and Biochronology in the Western United States

Systematic and Biostratigraphic Foundations for the Late Triassic Land Vertebrate "Faunachrons"

Phytosaurs (Fig. 6) were large predatory archosauriforms superficially resembling extant crocodylians (e.g., Stocker and Butler, 2013). Phytosaurs are one of the most commonly encountered vertebrate fossils in Upper Triassic continental strata in western North America (e.g., Ballew, 1989; Long and Murry, 1995; Irmis, 2005; Parker and Martz, 2011; Stocker and Butler, 2013), and have played a key role in Upper Triassic biostratigraphy and biochronology in the region from its very beginnings. Alpha taxonomy within the group (Fig. 6) historically has been remarkably unstable (Table 2) and based almost exclusively on skull morphology (e.g., Long and Murry, 1995; Stocker and Butler, 2013; Hungerbühler et al., 2013).

It has been recognized since roughly the first half of the 20th century, even before the application of cladistic methods, that phytosaur skulls in western North America show gradational morphological changes that follow the same sequence as their stratigraphic position (Fig. 6; e.g., Huene, 1926; Camp, 1930; Colbert and Gregory, 1957; Gregory, 1957, 1972; Ballew, 1989; Long and Murry, 1995; Stocker and Butler, 2013). Several phytosaur taxa lack derived character states present in other phytosaurs. The phytosaur genera *Parasuchus* Lydekker, 1885 (*sensu* Kammerer et al., 2015; =*Paleorhinus* Williston 1904), *Promystriosuchus* Case 1922, and *Wannia* Stocker 2013, are identified as basal, plesiomorphic, or "primitive" because they possess external nares that are at least partially anterior to the antorbital fenestra, squamosals that are narrow bars in dorsal view, and supratemporal fenestrae that are level with the skull roof (Fig. 6A). The more derived genera *Angistorhinus* Mehl 1913 and *Brachysuchus* Case 1929 possess external nares that are posteriorly displaced over the antorbital fenestra, although they possess narrow squamosals and supratemporal fenestrae level with the skull roof as in *Parasuchus*, *Promystriosuchus*, and *Wannia*. These taxa are claimed to be older than most other phytosaur genera, even in cases where it cannot be clearly demonstrated that these taxa occur stratigraphically below other phytosaurs (Huene, 1926; Camp, 1930; Colbert and Gregory, 1957; Gregory, 1957, 1962, 1972; Gregory and Westphal, 1969;

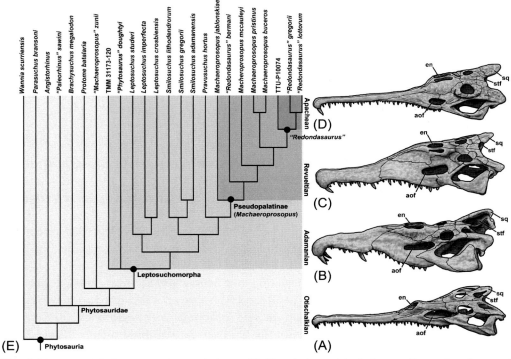

Fig. 6 Phytosaur phylogeny and morphology. (A) Nonphytosaurid phytosaur *Parasuchus bransoni* (based on TMM 31100-101). (B) Nonpseudopalatine leptosuchomorph *Smilosuchus gregorii* (based on UCMP A272/27200). (C) Pseudopalatine leptosuchomorph *Machaeroprosopus buceros* (based on UCMP V2816/34246). (D) Pseudopalatine leptosuchomorph *Machaeroprosopus* ("*Redondasaurus*") *gregorii* (based on holotype YPM 3294). (E) Phylogeny showing which nodes define the base of the Late Triassic land vertebrate teilzones, estimated holochronozones, and estimated holochrons; taxa that are not used to define unit boundaries are excluded. Most of phylogenetic tree after Kammerer et al. (2015), phylogeny of Pseudopalatinae after Hungerbühler et al. (2013). *aof*, antorbital fenestra; *en*, external nares; *sq*, squamosal; *stf*, supratemporal fenestra.

Chatterjee, 1978, 1986; Hunt and Lucas, 1991) although in some cases it can (e.g., Hunt and Lucas, 1991; Martz et al., 2013).

Phytosaur morphotypes occurring in higher strata exhibit a trend of depressing the parietal–squamosal bar so that the supratemporal fenestrae descend below the level of the skull roof, and a broadening of the postorbital/squamosal bar in dorsal view (Figs. 6B–D). Curiously, these striking evolutionary trends in skull morphology were not fully reflected by phytosaur alpha systematics for much of the 20th century, even among workers who recognized that these trends occurred. Indeed, there was a common tendency to lump more derived phytosaur skull morphotypes into one or two genera based on other criteria (Table 2). That trend began with Camp (1930), who assigned most

Table 2 Systematic history of all phytosaur-type specimens from western North America

North American phytosaur species type specimens with originally assigned genus (Genoholotypes indicated with an asterisk)	Camp (1930)	Gregory (1962)	Westphal (1976)	Ballew (1989)	Hunt and Lucas (1991[1], 1993[2]) and Lucas (1998[3])	Long and Murry (1995)	Irmis, (2005[1]) and Parker and Irmis (2006[2])	Stocker (2010[1], 2012[2], 2013[6]), Parker et al. (2013[3]), Hungerbühler et al. (2013[4]), Stocker and Butler (2013[5]), and Kammerer et al. (2015[7])
Belodon buceros Cope 1881 (AMNH 2318)	*Machaeroprosopus buceros*	*Rutiodon buceros*	*Rutiodon buceros*	*Pseudopalatus buceros*		*Arribasuchus buceros* (g)	*Pseudopalatus buceros*[1,2]	*Machaeroprosopus buceros*[3,4,5]
Belodon scolopax Cope 1881 (AMNH FR 2322)								Phytosauria (*nomen dubium*)[5]
Palaeoctonus dumblianus Cope 1893 (TMM 18563-1)		*Phytosaurus dumblianus*						Phytosauria (*nomen dubium*)[5]
Palaeoctonus orthodon Cope 1893 (TMM 18563-2)		*Phytosaurus orthodon*						Phytosauria (*nomen dubium*)[5]
Belodon superciliosus Cope 1893 (TMM 18562-1)		*Phytosaurus superciliosus*						Phytosauria (*nomen dubium*)[5]
Heterodontosuchus ganei Lucas 1898 (USNM V 4136)		*Phytosaurus ganei*						Phytosauria (*nomen dubium*)[5]
Paleorhinus bransoni Williston 1904 (CFMNH UC 632)*	*Paleorhinus bransoni*	*Paleorhinus bransoni*	*Paleorhinus bransoni*	*Paleorhinus*	*Paleorhinus bransoni*[1,3]	*Paleorhinus bransoni*		*Parasuchus bransoni*[7]
Angistorhinus grandis Mehl 1913 (CFMNH 631)*	*Angistorhinus grandis*	*Angistorhinus grandis*	*Angistorhinus grandis*	*Angistorhinus grandis*	*Angistorhinus grandis*[3]	*Angistorhinus grandis*		*Angistorhinus grandis*[5]
Angistorhinus gracilis Mehl 1915		*Angistorhinus gracilis*	*Angistorhinus gracilis*					Phytosauria incertae sedis[5]
Machaeroprosopus validus Mehl 1916 (UW 3807, lost)	*Machaeroprosopus validus*	?*Rutiodon validus*	?*Rutiodon validus*					Phytosauria incertae sedis[5]

Continued

Table 2 Systematic history of all phytosaur-type specimens from western North America—cont'd

North American phytosaur species type specimens with originally assigned genus (Genoholotypes indicated with an asterisk)	Camp (1930)	Gregory (1962)	Westphal (1976)	Ballew (1989)	Hunt and Lucas (1991[1], 1993[2]) and Lucas (1998[3])	Long and Murry (1995)	Irmis, (2005[1]) and Parker and Irmis (2006[2])	Stocker (2010[1], 2012[2], 2013[6]), Parker et al. (2013[3]), Hungerbühler et al. (2013[4]), Stocker and Butler (2013[5]), and Kammerer et al. (2015[7])
Phytosaurus doughtyi Case 1920 (AMNH 4919)	*Machaeroprosopus(?) doughtyi*	*Rutiodon doughtyi*	*Rutiodon doughtyi*			*Leptosuchus crosbiensis*		*Phytosauria incertae sedis*[5]
Promystriosuchus ehlersi Case 1922 (UMMP V7487)*	*Promystriosuchus ehlersi*	*Paleorhinus ehlersi*	*Paleorhinus ehlersi*		*Paleorhinus* sp.[1]	*Paleorhinus ehlersi*		*Phytosauria incertae sedis*[5,7]
Leptosuchus crosbiensis Case 1922 (UMMP 7522)*	*Leptosuchus crosbiensis*	*Rutiodon crosbiensis*	*Rutiodon crosbiensis*	*Rutiodon crosbiensis*	*Rutiodon crosbiensis*[3]	*Leptosuchus crosbiensis*	*Leptosuchus crosbiensis*[1]	*Leptosuchus crosbiensis*[1,2,5]
Leptosuchus imperfecta Case 1922 (UMMP 7523)	*Leptosuchus imperfecta*	*Rutiodon imperfecta*	*Rutiodon imperfecta*	*Rutiodon crosbiensis*		?*Leptosuchus adamanensis*		*Phytosauria (nomen dubium)*[5]
Machaeroprosopus andersoni Mehl 1922 (FMNH UC 396)	*Machaeroprosopus andersoni*	*Rutiodon andersoni*	*Rutiodon andersoni*	*Pseudopalatus pristinus*	*Pseudopalatus andersoni*[3]	*Arribasuchus buceros*		*Phytosauria incertae sedis*[5]
Paleorhinus parvus Mehl 1928 (MU 530)		*Paleorhinus parvus*	*Paleorhinus parvus*		*Paleorhinus bransoni*[1]			*"Paleorhinus" parvus*[5,7]
Angistorhinus maximus Mehl 1928 (U. Mo. 531)		*Angistorhinus maximus*	*Angistorhinus maximus*	*Angistorhinus maximus*		*"Angistorhinus" maximus*		*Angistorhinus maximus*[5]
Brachysuchus megalodon Case 1929 (UMMP 100336)*		*Phytosaurus megalodon*	*Angistorhinus megalodon*		*Brachysuchus megalodon*[3]	*Angistorhinus megalodon*		*Brachysuchus megalodon*[1,2,5]
Pseudopalatus pristinus Mehl 1928 (U. Mo. 525VP)*		*Rutiodon pristinus*	*Rutiodon pristinus*	*Pseudopalatus pristinus*	*Pseudopalatus pristinus*[3]	*Pseudopalatus pristinus*	*Pseudopalatus pristinus*[1,2]	*Machaeroprosopus pristinus*[3,4,5]

Specimen								
Machaeroprosopus lithodendrorum Camp 1930 (UCMP 26688)	***Machaeroprosopus lithodendrorum* (s)**	*Rutiodon lithodendrorum*	*R. lithodendrorum*	*Rutiodon lithodendrorum*		*Leptosuchus crosbiensis*		*Smilosuchus lithodendrorum*[1,2,5]
Machaeroprosopus adamanensis Camp 1930 (UCMP 7038/26699)	***Machaeroprosopus adamanensis* (s)**	*Rutiodon adamanensis*	*Rutiodon adamanensis*	*Rutiodon adamanensis*		*Leptosuchus adamanensis*	*Leptosuchus adamanensis*[1,2]	*Smilosuchus adamanensis*[1,2,5]
Machaeroprosopus zunii Camp 1930 (UCMP 7307/27036)	***Machaeroprosopus zunii* (s)**	*Rutiodon zunii*	*Rutiodon zunii*			?*Leptosuchus adamanensis*	*Leptosuchus zunii*[2]	*Phytosauria incertae sedis*[5]
Machaeroprosopus gregorii Camp 1930 (UCMP A272/27200)	***Machaeroprosopus gregorii* (s)**	*Phytosaurus gregorii*	*Nicrosaurus gregorii*	*Rutiodon gregorii*	*Smilosuchus gregorii*[3]	***Smilosuchus gregorii* (g)**	*Leptosuchus gregorii*[1,2]	*Smilosuchus gregorii*[1,2,5]
Machaeroprosopus tenuis Camp 1930 (UCMP 27018)	***Machaeroprosopus tenuis* (s)**	*Rutiodon tenuis*	*Rutiodon tenuis*			*A. buceros* and *P. pristinus*	*Pseudopalatus pristinus*[2]	*Machaeroprosopus tenuis*[3]
Leptosuchus studeri Case and White 1934 (UMMP 14267)		*Rutiodon studeri*	*Rutiodon studeri*	*Rutiodon crosbiensis*		*Leptosuchus crosbiensis*		*Leptosuchus studeri*[1,2,5]
Angistorhinus alticephalus Stovall and Wharton 1936 (OMNH 733)		*Angistorhinus alticephalus*	*Angistorhinus alticephalus*			?*Angistorhinus megalodon*		*Angistorhinus alticephalus*[5]
Paleorhinus scurriensis Langston 1949 (TTU-P00539)		*Paleorhinus scurriensis*	*Paleorhinus scurriensis*		*Paleorhinus bransoni*[1]			***Wannia scurriensis* (g)**[6,7]
Angistorhinus aeolamnis Eaton 1965 (KU 11659)			*Angistorhinus aeolamnis*					*Angistorhinus aeolamnis*[5]
Pseudopalatus maccauleyi Ballew 1989 (UCMP 126999)				***Pseudopalatus maccauleyi* (s)**		?*Arribasuchus maccauleyi*	*Pseudopalatus maccauleyi*[1]	*Machaeroprosopus maccauleyi*[3,4,5]
Redondasaurus bermani Hunt and Lucas 1993 (MC 69727)					***Redondasaurus bermani*[2]**	*Arribasuchus buceros*	*Redondasaurus bermani*[2]	*Redondasaurus bermani*[5]; *Machaeroprosopus bermani*[4]

Continued

Table 2 Systematic history of all phytosaur-type specimens from western North America—cont'd

North American phytosaur species type specimens with originally assigned genus (Genoholotypes indicated with an asterisk)	Camp (1930)	Gregory (1962)	Westphal (1976)	Ballew (1989)	Hunt and Lucas (1991[1], 1993[2]) and Lucas (1998[3])	Long and Murry (1995)	Irmis, (2005[1]) and Parker and Irmis (2006[2])	Stocker (2010[1], 2012[2], 2013[6]), Parker et al. (2013[3]), Hungerbühler et al. (2013[4]), Stocker and Butler (2013[5]), and Kammerer et al. (2015[7])
Redondasaurus gregorii Hunt and Lucas 1993 (YPM 3294)*					***Redondasaurus gregorii*[2]**	*Pseudopalatus pristinus*	*Redondasaurus gregorii*[2]	*Redondasaurus gregorii*[.5]; *Machaeroprosopus gregorii*[4]
Paleorhinus sawini Long and Murry 1995 (TMM 31213–16)					*Paleorhinus bransoni*[1]	***Paleorhinus sawini* (s)**		"*Paleorhinus*"[1,5,6,7]
Pseudopalatus jablonskiae Parker and Irmis 2006 (PEFO 31207)							***Pseudopalatus jablonskiae* (s)[2]**	*Machaeroprosopus jablonskiae*[3,4,5]
Pravusuchus hortus Stocker 2010 (AMNH FR 30646)								***Pravusuchus hortus*[1,2,5]**
Protome batalaria Stocker 2012 (PEFO 34034)								***Protome batalaria*[2,5]**
Machaeroprosopus lottorum Hungerbühler et al. 2013 (TTU-P10076)								*Machaeroprosopus lottorum* (s)[3,4]

Original names and type specimens for all species are given on the far left, using the originally applied binomial. The remaining columns show the taxonomic referrals of type specimens made by various major phytosaur studies. Taxon names in boldface indicate an alpha taxon named in the study heading the column; if the name in bold is followed by a (g), than the genus was new and a prior species name was retained; if the name in bold is followed by an (s), than the species was new and a prior genus name was retained; if neither a (g) or an (s) is displayed after a boldfaced name, then both binomens were new. Some columns combine more than one study, with superscripts indicating publications that used the name; new genera and species were named in the oldest publication for which there is a superscript.

derived phytosaur skull morphotypes to the genus *Machaeroprosopus* Mehl 1916. This was a particularly puzzling move given that Camp (1930) also established that:

1. *Machaeroprosopus buceros* Cope 1881 (Fig. 6C) and *M. tenuis* Camp 1930 are distinguished from other species by extremely reduced and depressed supratemporal fenestrae and extremely broad and highly sculpted postorbital-squamosal bars.

2. *M. tenuis* occurs stratigraphically higher in the Chinle Formation of Arizona than "*M.*" *lithodendrorum* Camp 1930 (Fig. 6B) and "*M.*" *adamanensis* Camp 1930, which have somewhat less depressed supratemporal fenestrae and narrower squamosals.

In the years after Camp's (1930) monograph, workers tended to unite derived taxa based on the presence or absence of a prenarial crest. Crested forms were often assigned to the German genera *Phytosaurus* Jaeger 1828 or *Nicrosaurus* O. Fraas 1866 (the former of which is a *nomen dubium*; see Hunt, 1994, and Hungerbühler, 1998, for reviews of the history of German phytosaur taxonomy) and the uncrested forms to genus *Rutiodon* Emmons (1856) from eastern North America (e.g., Gregory, 1962; Westphal, 1976; Elder, 1978; Chatterjee, 1986; Hunt, 1994). Puzzlingly, Gregory (1962) used this classification even though like Camp (1930) he recognized that the species assigned to *Phytosaurus* and *Rutiodon* could be segregated both morphologically and stratigraphically based on posterior skull morphology (Gregory, 1957, 1972; Colbert and Gregory, 1957). As a result, these systematic schemes hampered the use of distinct phytosaur taxa to characterize different biozones.

Although Huene (1926) and Camp (1930) offered stratigraphic subdivisions and correlations of Upper Triassic vertebrate faunas in western North America that largely presaged later biostratigraphic studies (Fig. 7A), Gregory (1957, 1972; Colbert and Gregory, 1957, Table 3) was the first to recognize all four of the stratigraphically sequential faunal associations that would later become the Late Triassic land vertebrate "faunachrons" (Fig. 7B). Gregory (1957, 1972; Colbert and Gregory, 1957) identified the lowest faunal association, the "*Paleorhinus* fauna" (Gregory, 1972; containing *Parasuchus*, *Promystriosuchus*, and *Angistorhinus*) in the Popo Agie Formation of Wyoming and the lower Dockum Group in western Texas. Two other faunal associations within the Dockum Group and Chinle Formation interpreted as being stratigraphically higher than the "*Paleorhinus* fauna" were both placed within the "*Rutiodon* fauna" (Gregory, 1972). These two subdivisions of the "*Rutiodon* fauna" contained more derived phytosaurs with depressed supratemporal fenestrae (Fig. 6B and C) that Gregory (1957, 1972; Colbert and Gregory, 1957) assigned to *Machaeroprosopus*, *Phytosaurus*, and/or *Rutiodon*. *Leptosuchus crosbiensis* Case 1922 and *M. tenuis* Camp 1930 occurred respectively in the lower and upper portions of the "*Rutiodon* fauna." Gregory (1957, 1972; Colbert and Gregory, 1957) also recognized a fourth, uppermost "advanced fauna" (Gregory, 1972) in the Redonda and Sloan Canyon formations of New Mexico that included a highly derived phytosaur with extremely broad squamosals and supratemporal fenestrae that are completely concealed in dorsal view (Fig. 6D).

Camp (1930)	Gregory (1957), Colbert and Gregory (1957), and Gregory (1972)	Long and Ballew (1985), and Long and Padian (1986)	Lucas (1993), and Lucas and Hunt (1993)	Lucas (1998)	Hunt (2001), and Hunt et al. (2005)	Martz (2008), Parker and Martz (2011), and Martz et al. (2013, 2014)	Current study
Quasibiozones (faunal associations tied to lithostratigraphic units)	*Quasibiozones (faunal associations tied to lithostratigraphic units)*	*Quasibiozones (faunal associations tied to lithostratigraphic units)*	*Quasibiochrons (time units without explicit taxon-based boundaries)*	*Eubiochrons (time units with explicit taxon-based boundaries)*	*Eubiochrons (time units with explicit taxon-based boundaries)*	*Eubiozones (lithologic units with explicit taxon-based boundaries)*	*Eubiozones (teilzones), estimated eubiochronozones, and estimated eubiochrons*
Upper Chinle / "Phytosaur with completely concealed superior fenestra"	"Phytosaur with completely concealed superior fenestra"	*Typothorax, Rutiodon* B(?)	Apachean land vertebrate faunachron/ = faunachron D	Apachean land vertebrate faunachron (FAD *Redondasaurus*)	Apachean land vertebrate faunachron (FAD *Redondasaurus*)	Apachean biozone (LSDk/LOk "*Redondasaurus*")	Apachean tielzone, estimated holochronozone and estimated holochron (LOk, OKR, and eFHA "*Redondasaurus*")
Machaeroprosopus tenuis zone	Upper fauna/*Machaeroprosopus tenuis* and *M. andersoni* (*Rutiodon* fauna)	*Typothorax, Rutiodon* B/ upper unit of Petrified Forest Member	Revueltian land vertebrate faunachron/ = faunachron C	Revueltian land vertebrate faunachron (FAD *Pseudopalatus*)	Revueltian lvf: Lucianoan sub-lvf (FAD *Lucianosaurus*); Barrancan sub-lvf (FAD *Pseudopalatus*); Rainbow-forestan sub-lvf (FAD *T. coccinarum*)	Revueltian biozone (LSDk/LOk *Pseudopalatus*)	Revueltian tielzone, estimated holochronozone and estimated holochron (LOk, OKR, and eFHA Pseudopalatinae)
Machaeroprosopus lithodendrorum zone (Lower Chinle)	Tecovas shale and lower Chinle fauna/*Machaeroprosopus adamenensis*, '*Phytosaurus*' *doughtyi*, and *Leptosuchus crosbiensis*	*Calyptosuchus, Desmatosuchus, Rutiodon* A/ lower unit of Petrified Forest Member	Adamanian land vertebrate faunachron/ = faunachron B	Adamanian land vertebrate faunachron (FAD *Rutiodon*)	Adamanian lvf: Lamyan sub-lvf (FAD *Typothorax antiquum*); St. Johnsian sub-faunachron (FAD *Rutiodon*)	Adamanian biozone: Lower Sonsela Member fauna; Blue Mesa Member fauna (LSDk/LOk *Leptosuchus*)	Adamanian tielzone, estimated holochronozone and estimated holochron (LOk, OKR, and eFHA Leptosuchomorpha)
Machaeroprosopus adamenensis zone							
Popo Agie (Dockum) / Basal fauna/*Paleorhinus* fauna	Basal fauna/*Paleorhinus* fauna		Otischalkian land vertebrate faunachron/ = faunachron A	Otischalkian land vertebrate faunachron (FAD *Paleorhinus*)	Otischalkian biozone (FAD *Paleorhinus*)	Otischalkian biozone (LSDk/LOk *Paleorhinus*)	Otischalkian tielzone, estimated holochronozone and estimated holochron (LOk, OKR, and eFHA Phytosauria)
(A)	(B)	(C)	(D)	(E)	(F)	(G)	(H)

Fig. 7 Changing terminology and concepts for the Late Triassic land vertebrate biostratigraphy and biochronology in western North America. (A) Camp (1930), quasibiozones tied to lithostratigraphic units; the equivalence of most of Camp's subdivisions to later biostratigraphic and biochronologic schemes is tentative. (B) Colbert and Gregory (1957) and Gregory (1972), quasibiozones in the form of faunal associations tied to lithostratigraphic units. (C) Long and Ballew (1985) and Long and Padian (1986), quasibiozones in the form of faunal associations tied to lithostratigraphic units. (D) Lucas (1993a,b) and Lucas and Hunt (1993), quasibiochrons without explicit taxon-based boundary definitions. (E) Lucas (1998), eubiochrons with explicit boundary definitions, but lacking detailed biostratigraphic documentation. (F) Hunt et al. (2005), eubiochrons with explicit taxon-based boundary definitions, but lacking detailed biostratigraphic documentation. (G) Martz (2008), Parker and Martz (2011), and Martz et al. (2013), eubiozones with explicit boundary definitions supported by detailed biostratigraphic documentation. (H) Current study, eubiozones, estimated eubiochronozones (holochronozones), and estimated eubiochrons (holochrons), with explicit boundary definitions supported by detailed biostratigraphic documentation.

Long and Ballew (1985), Colbert (1985), and Long and Padian (1986) later recognized that the subdivisions of the "*Rutiodon* fauna" were present in the lower and upper parts of the Chinle Formation of Petrified Forest National Park (Fig. 7C), as Camp (1930) and Gregory (1957) had also previously observed, although Camp (1930) had assigned all of these forms to *Machaeroprosopus*. Long and Ballew (1985) called the lower form of "Rutiodon" (including Camp's 1930 *M. lithodendrorum* and *M. adamanensis*) "*Rutiodon* Group A" and the stratigraphically higher form (including Camp's 1930 *M. tenuis*) "*Rutiodon* Group B"; the latter was referred to as *Rutiodon tenuis* by Colbert (1985). Moreover, Long and Ballew (1985) and Long and Padian (1986) demonstrated that these lower and upper forms of *Rutiodon* occurred with slightly distinct vertebrate assemblages (particularly different aetosaur taxa), which Long and Ballew (1985) referred to as the "C–D–RA assemblage" and the "T–RB assemblage" (Fig. 7C). However, none of these authors made a clear connection between these lower and upper assemblages in the Chinle Formation of Arizona and the lower and upper parts of the "*Rutiodon* fauna" recognized by Gregory (1957, 1972; Colbert and Gregory, 1957) in the Dockum Group and Chinle Formation of Texas and New Mexico.

The abandonment of phytosaur taxonomic schemes based on snout morphology that would eventually pave the way for more robust phytosaur biozonation began with Eaton (1965) and Westphal (1976) synonymizing the crested genus *Brachysuchus* with the uncrested genus *Angistorhinus*. Even more important was Ballew's (1989) landmark computer-aided phylogenetic analysis, the first applied to phytosaurs. Ballew's (1989) phylogeny showed that several crested and uncrested skull morphotypes were more closely related to each other than to other crested and uncrested forms (Table 2). Moreover, Ballew's (1989) phylogeny showed a clear phylogenetic trend reflecting posterior skull morphology: *Parasuchus* (=*Paleorhinus*) and *Angistorhinus* formed consecutive sister taxa to all other phytosaurs (Fig. 6E). More derived taxa with a partially depressed supratemporal fenestra (Long and Ballew's, 1985 "*Rutiodon* Group A," including *Leptosuchus crosbiensis*, "*Machaeroprosopus*" *lithodendrorum*, and "*M.*" *adamanensis*) were retained in the genus *Rutiodon* (Fig. 6B). However, taxa with fully depressed supratemporal fenestrae and broadened squamosals (Long and Ballew's, 1985 "*Rutiodon* Group B," including *M. tenuis*) formed a monophylum in Ballew's (1989) phylogenetic analysis that she assigned to the genus *Pseudopalatus* Mehl 1928 (Fig. 6C). This broad phylogenetic pattern has endured through increasingly detailed phylogenetic analyses up to the present day (Fig. 6E; Hungerbühler, 2002; Parker and Irmis, 2006; Stocker, 2010, 2012, 2013; Hungerbühler et al., 2013; Butler, 2013; Kammerer et al., 2015).

This systematic scheme assigning all western North American phytosaurs to five or six genera received only minor modifications during the 1990s. Hunt and Lucas (1993) provided the genus name *Redondasaurus* (Fig. 6D and Table 2) for the derived phytosaur from the Redonda and Sloan Canyon formations of New Mexico assigned by Gregory (1972) to the "advanced fauna." Long and Murry (1995) assigned the skulls from western

North America that Ballew (1989) had placed in *Rutiodon* to *Leptosuchus* and a new genus, *Smilosuchus* (Fig. 6B); these generic reassignments have become accepted in recent years, although the referral of particular species to one genus or another have varied (e.g., Parker and Irmis, 2006; Irmis, 2005; Stocker, 2010, 2012; Parker and Martz, 2011; Stocker and Butler, 2013). Long and Murry (1995) also attempted to partially resurrect the snout-based classification by assigning crested forms of *Pseudopalatus* to a new genus, *Arribasuchus*, although this taxon has never gained widespread use in pseudopalatine taxonomy (e.g., Hungerbühler, 2002; Parker and Irmis, 2006; Hungerbühler et al., 2013). However, in the past few years, this alpha taxonomic scheme for western North American phytosaurs has undergone major disruptions, which are discussed in the following.

Quasibiochrons Characterized (but not Defined) by Multiple Taxa

Although the composition and relative superpositional order of these North American Late Triassic vertebrate faunas was recognized before 1990, they were not discussed as biozones or biochrons, but rather as faunal associations (Fig. 7A–C; e.g., Camp, 1930; Colbert and Gregory, 1957; Gregory, 1972; Long and Ballew, 1985; Long and Padian, 1986). However, because these faunal associations were linked with particular lithostratigraphic units (Figs. 7 and 8A), they could arguably be considered to be quasibiostratigraphic units.

Lucas's (1990) innovative suggestion was to formalize these faunal associations as vertebrate biochrons, following the lead of the North American Land Mammal "Ages" (Figs. 7D and 8A). The Late Triassic land vertebrate "faunachrons" were originally simply designated A, B, C, and D (Lucas, 1993b; Fig. 7D). These were explicitly temporal units, representing "distinct intervals of Late Triassic time" (Lucas, 1993b, p. 31). Lucas did not provide type assemblages, although he did list those formations and members of the western United States that contained the diagnostic taxa of the "faunachrons" (Fig. 8A; Lucas, 1993b, Fig. 5). He also did not specify any taxa whose range limits defined the "faunachrons," only a list of characteristic taxa (mostly phytosaurs and aetosaurs). These included taxa that were restricted to the "faunachron," but also taxa with ranges that extended into more than one "faunachron." Because Lucas (1993b) did not present precise boundary definitions, the "faunachrons" were quasibiochronologic units.

Lucas and Hunt (1993) assigned formal names to the "faunachrons": the Otischalkian ("A"), Adamanian ("B"), Revueltian ("C"), and Apachean ("D") (Figs. 7D and 8A). These "faunachrons" were still not formally bounded by the range limits of taxa, but were defined by a "type fossil assemblage" or "type fauna" from a particular area (Lucas and Hunt, 1993, p. 327). Lucas and Hunt (1993) again identified characteristic taxa, which included forms that were found within, but not necessarily restricted to, the "faunachron"; this list was slightly expanded from Lucas (1993b). However, Lucas and Hunt (1993) also specified index taxa, which Lucas (1992) had used in the sense of Wood et al. (1941, p. 97) as being "known only from deposits of the age in question"

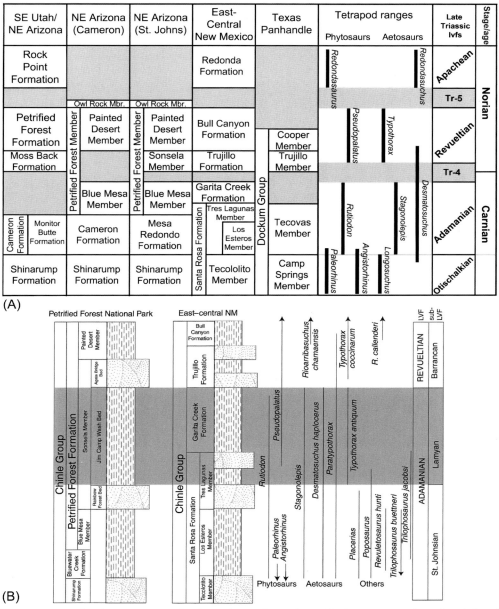

Fig. 8 Concepts for the Late Triassic land vertebrate "faunachrons" lacking support from detailed and explicit biostratigraphic documentation. (A) Late Triassic land vertebrate "faunachrons" as quasibiochronologic units tied to lithostratigraphic units, adapted from Lucas (1993a,b, Fig. 5). Note that the boundaries of "faunachrons" and taxon ranges often precisely correspond to lithostratigraphic boundaries, a situation that rarely occurs with carefully plotted biostratigraphic data. (B) Eubiochronologic units, from Hunt et al. (2005, Fig. 5); note that the ranges given are composites for the two local study areas shown.

and were therefore confined to particular "faunachrons" (Fig. 8A). This change was consistent with the prior opinion offered by Lucas (1992) that listing multiple taxa that characterize the "faunachrons" is preferable to providing a precise single taxon boundary definition. As such, it was not clear exactly how the "faunachrons" were bounded, and as a result, they continued to be quasibiochronologic units. Moreover, by using stratigraphic units to characterize the "faunachrons," Lucas and Hunt (1993) were arguably also treating them as quasibiozones or quasibiochronozones.

The way Lucas (1993b) treated the relationships between "faunachrons," individual taxa, lithostratigraphic units, and geochronologic units is worth noting. Lucas (1993b, Figs. 3 and 5) showed the beginnings and ends of ranges of individual taxa as precisely corresponding to those of other taxa, and usually to "faunachron" boundaries (Fig. 8A). Lucas (1993b, Figs. 3 and 5) also showed "faunachron" boundaries as precisely equating in many cases to those of lithostratigraphic units. This depiction of multiple taxa, "faunachrons," and/or lithostratigraphic units precisely sharing (presumably time-equivalent) boundaries would be repeated (e.g., Lucas and Hunt, 1993, Fig. 1; Heckert and Lucas, 1996, Fig. 3; Lucas and Heckert, 1996, Fig 5; Lucas, 1997, Figs. 23.5 and 23.12; Heckert and Lucas, 2015, Fig. 4), and even extended to geochronologic ages (Lucas, 1998, Fig. 14; Lucas et al., 2012, Fig. 7; Heckert and Lucas, 2015, Fig. 4). It must be concluded therefore that the Late Triassic land vertebrate "faunachrons" were temporally analogous to "paleontologically distinct lithozones" *sensu* Walsh (2001). As discussed earlier, such units result when precise and detailed biostratigraphic information is lacking.

Lucas (1993b, p. 41) argued that significant erosional hiatuses, which he named the Tr-4 and Tr-5 unconformities, separated lithostratigraphic units within the Chinle Formation and Dockum Group (Fig. 8A). Such unconformities could provide partial justification for these precisely equivalent boundaries in that they could potentially truncate the biostratigraphic ranges of taxa, causing the beginnings and/or ends of multiple-taxon ranges to precisely coincide with the boundaries of lithostratigraphic units. However, detailed stratigraphic studies of the Chinle Formation and Dockum Group conducted within the past decade demonstrated that the Tr-4 and Tr-5 unconformities probably do not exist as significant erosional hiatuses that can be traced throughout the western United States (Woody, 2006; Martz, 2008, pp. 93–94; Martz and Parker, 2010, p. 21; Parker et al., 2011). Moreover, this recent work also demonstrated that lithostratigraphic, biostratigraphic, and chronostratigraphic units do not usually have equivalent boundaries when the stratigraphic positions of fossils and samples for radioisotopic dates are plotted with high precision (e.g., Parker, 2006; Martz, 2008; Parker and Martz, 2011; Martz et al., 2013, 2014; Olsen et al., 2011; Parker et al., 2011). It should be noted that the Tr-3 unconformity of Pipiringos and O'Sullivan (1978) at the base of the Chinle Formation is indeed a major erosional hiatus (e.g., Lucas, 1993b; Dubiel, 1994) that may make the base of the Chinle Formation roughly equivalent to the base of the Norian stage

(Olsen et al., 2011; Ramazani et al., 2014), and potentially the base of the Otischalkian biochronozone (see the following).

In subsequent papers, Lucas and his colleagues characterized the "faunachrons" in more detail. Lucas et al. (1997a) provided lists of first and last appearances of taxa for the Otischalkian and Adamanian, specifying (Lucas et al., 1997a, p. 35) that they were emulating the NALMAs in doing so (e.g., Krishtalka et al., 1987; Lucas, 1992; Robinson et al., 2004). Lucas et al. (1997a, p. 37) identified index taxa as "fossils restricted to the time interval for which they provide an index and which are relatively common and readily identified." However, because it was not specified which taxa marked the boundaries, the "faunachrons" were still quasibiochrons, specifically temporal analogs to "paleontologically distinct 'fuzzy' zones" *sensu* Walsh (2001).

Interval Eubiochrons Defined Using the FHAs of Particular Phytosaur Taxa

Lucas (1998, 1999) supplanted the multiple taxon characterizations previously advocated for the Late Triassic land vertebrate "faunachrons" (Lucas, 1992, 1993b) with explicit eubiochronologic definitions based on the FADs of individual taxa that had previously only been index taxa (Fig. 7E). Lucas (1998, 1999) used the FAD of "*Paleorhinus*" (=*Parasuchus*; see the following) to define the base of the Otischalkian (Fig. 7A), the FAD of "*Rutiodon*" (=*Leptosuchus* and *Smilosuchus sensu* Long and Murry, 1995; see the following) to define the base of the Adamanian (Fig. 7B), the FAD of "*Pseudopalatus*" (Fig. 7C; =*Machaeroprosopus*; see the following) to define the base of the Revueltian, and the FAD of *Redondasaurus* to define the base of the Apachean (Fig. 7E). The use of single taxa to define "faunachron" boundaries was used in subsequent papers (e.g., Lucas et al., 2007b; Lucas, 2010), and has been applied to proposed subdivisions of the "faunachrons" (e.g., Hunt, 2001; Hunt et al., 2005). As already discussed, defining biozones and biochrons using the lowest or oldest occurrences of different taxa generally avoids the problem of potential gaps and overlaps (Figs. 3 and 4; e.g., Repenning, 1967; Woodburne, 1977).

Because Lucas apparently used the term "faunachron" in the ontological and temporal sense, it is assumed that "FAD" is equivalent to FHA *sensu* Walsh (1998): an ontological paleobiological event involving the evolution or immigration of a taxon. Lucas (1998) referred to the newly defined eubiochronologic "faunachrons" as "interval (*assemblage*) biochrons." This is a somewhat misleading label, because "assemblage" implies that multiple taxa rather than single taxa are being used to define the biochron (e.g., Walsh, 1998, 2001). As redefined by Lucas (1998, 1999), the "faunachrons" actually represent FHA-FHA interval biochrons *sensu* Walsh (1998), and are eubiochrons as defined here. However, because the given length of these time intervals was based on biostratigraphic information, it is best to think of the "faunachrons" of Lucas (1998) as epistemological estimates of FHA-FHA interval biochrons (eFHA-eFHA interval biochrons; Table 1).

The method of using the oldest occurrences of particular taxa to define "faunachron" boundaries also explicitly divorces them from lithostratigraphic units (at least in principle) by giving them distinct boundary criteria. Reiterating the point made previously by Prothero (1990, p. 240), Lucas (1998, p. 349) stated that it is a mistake to "imply that the [faunachron] refers to the duration of deposition of the formation, not just to the duration of the vertebrate fossil assemblage, which is often much shorter." At least some subsequent authors also recognized that "faunachron" boundaries need not precisely correspond to the range limits of nondefining taxa or lithologic boundaries (e.g., Heckert and Lucas, 2003, Fig. 4; Heckert et al., 2007, Fig. 4). Unfortunately, with some exceptions (e.g., Lucas and Heckert, 1996, Fig. 4; Heckert and Lucas, 1997, Figs. 3 and 4, 2003, Fig. 2), Lucas and his colleagues rarely plotted the precise positions of vertebrate localities on stratigraphic sections, instead giving generalized ranges (Fig. 8B). This made the eubiochronologic scheme nearly impossible to test.

This problem also impacts proposed subdivisions of the "faunachrons." Hunt (2001) proposed subdividing the Revueltian "faunachron" into the ?Rainbowforestan (R0), Barrancan (R1), and Lucianoan (R2) sub-"faunachrons," all bounded by vertebrate FADs (Fig. 7F), again without presenting detailed biostratigraphic range data. Moreover, Heckert and Lucas (2002) noted that these sub-"faunachrons" were based on the misidentification of a *Pseudopalatus* specimen as *Nicrosaurus*, as well as a tooth-based taxon (*Lucianosaurus*) that is not known beyond its type locality, making identification of strata deposited during the Lucianoan sub-"faunachron" rather difficult. Similarly, Hunt et al. (2005) subdivided the Adamanian into a lower St. Johnsian sub-"faunachron" and an upper Lamyan sub-"faunachron" defined by vertebrate FADs (Figs. 7F and 8B). Specifically, the base of the St. Johnsian, also being the base of the Adamanian, was defined by the FAD of *Rutiodon* (*Leptosuchus* and *Smilosuchus sensu* Long and Murry, 1995), and the base of the Lamyan was defined by the FAD of the aetosaur *Typothorax antiquum* (see discussion by Parker and Martz, 2011, pp. 235–240). Because Hunt et al. (2005) thought that the FAD of *Pseudopalatus* occurred in the Lamyan, they considered that taxon unsuitable for defining the Revueltian, which they redefined as the FAD of the aetosaur *Typothorax coccinarum*. Unfortunately, Hunt et al. (2005) presented only a composite biostratigraphic range chart combining data from two different areas (Fig. 8B) rather than individual range charts plotting the precise stratigraphic level of localities for each local study area. This makes it difficult to test whether vertebrate biostratigraphic (and consequently biochronologic) patterns are really consistent between the two regions (see Parker, 2006, pp. 57–58; Parker and Martz, 2011, p. 249; Martz et al., 2013, pp. 358–359). Moreover, Hunt et al. (2005) determination that the FAD of *Pseudopalatus* occurs below that of *Typothorax* have not survived more rigorous biostratigraphic studies, making their revised definition of the Revueltian unnecessary (Parker, 2006; Parker and Martz, 2011, pp. 245, 248, and 249; Martz et al., 2013). As noted by McKenna and Lillegraven (2006), if a system of biozonation is faltering, revising biozone definitions will not solve the problem, and may even create an illusion of biostratigraphic stability where none exists.

In summary, the historical development of the Late Triassic faunachron concept has partially emulated improvements to mammalian biochronology by modifying vaguely bounded Upper Triassic vertebrate faunal associations tied to lithostratigraphic units into a set of more clearly defined interval eubiochrons bounded by the appearances of individual taxa. However, the supporting work discussed previously has not presented the kind of detailed biostratigraphic data needed to test these biochronologic hypotheses. As a result, proposed advancement of the Late Triassic land vertebrate "faunachrons," while laudable in principle, has in practice been stymied by the absence of detailed and well-documented lithostratigraphic and biostratigraphic data (Parker and Martz, 2011; Martz et al., 2013, 2014) and has therefore failed to improve the level of biochronologic detail and precision beyond what was known in the previous century (Gregory, 1957, 1972; Long and Ballew, 1985; Long and Padian, 1986). The Late Triassic land vertebrate "faunachrons" have therefore remained effectively quasibiochronologic units in spite of their eubiochronologic reformulation.

Interval Eubiozones Defined Using the LO$_k$s of Particular Phytosaur Taxa

Over the past decade, numerous investigations have attempted to improve the methodological rigor applied to the study of Late Triassic vertebrates. Whereas some studies have provided more careful scrutiny to the systematics and identification of vertebrate fossils (Nesbitt et al., 2007; Irmis et al., 2007; Parker, 2007, 2014; Stocker, 2010, 2012, 2013; Stocker and Butler, 2013; Hungerbühler et al., 2013; Butler, 2013), others have developed detailed lithostratigraphic models on which biostratigraphic and geochronologic data have been plotted with care (e.g., Heckert and Lucas, 2002; Parker, 2006; Martz, 2008; Irmis et al., 2007, 2010, 2011; Martz and Parker, 2010; Olsen et al., 2011; Parker and Martz, 2011; Parker et al., 2011; Ramezani et al., 2011, 2014; Martz et al., 2013, 2014; Atchley et al., 2013). These studies have enabled the development of increasingly detailed, rigorous, and well-documented biostratigraphic models on which to base vertebrate biochronology. Indeed, several of these studies have treated the Late Triassic land vertebrate "faunachrons" as biozones rather than biochrons (Fig. 7G; Parker, 2006; Martz, 2008; Parker and Martz, 2011; Martz et al., 2013, 2014), recognizing that biochronology is meaningless without detailed and accurate biostratigraphic data (e.g., Tedford, 1970; Emry, 1973; Woodburne, 1977, 2004).

So far, there are three local study areas (*sensu* Parker and Martz, Chapter 1 in this volume) in western North America with fairly extensive biostratigraphic data that have been subjected to this rigorous analysis (Fig. 9): the Chinle Formation of Petrified Forest National Park in northern Arizona (Fig. 10) (Heckert and Lucas, 2002; Parker, 2006; Martz and Parker, 2010; Parker and Martz, 2011; Martz et al., 2012), the Dockum Group of southern Garza County, western Texas (Fig. 11) (Martz, 2008; Martz et al., 2013), and

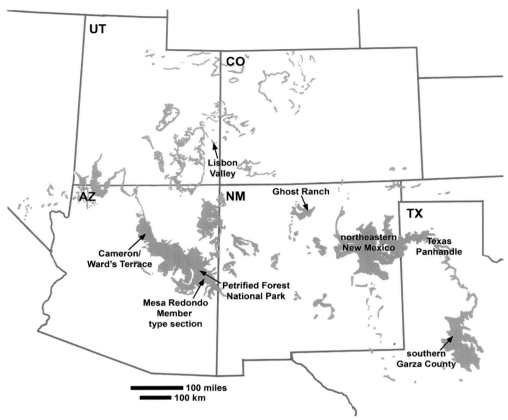

Fig. 9 Map showing local and regional study areas within the western United States for which stratigraphic data is discussed in detail (see Fig. 14). Outcrops of Upper Triassic strata are shown in light gray.

the Chinle Formation of Lisbon Valley in southeastern Utah (Fig. 12; Milner et al., 2006; Martz et al., 2014). In all of these areas, the Upper Triassic land vertebrate biozones have been treated as eubiozones (Martz, 2008, p. 289; Parker and Martz, 2011, p. 235; Martz et al., 2013, p. 357) with lower boundaries defined by the same individual phytosaur taxa whose FHAs were previously used to define the "faunachrons" (e.g., Lucas, 1998). Using the terminology of Walsh (2001), these are "single-taxon LO-LO interval fossilzones," although we refer to them as "single-taxon LO_k-LO_k interval eubiozones" according to the terminology we established earlier.

These more recent studies plot the stratigraphic positions of vertebrate specimens and geochronologic samples precisely enough to demonstrate that biostratigraphic, lithostratigraphic, and chronostratigraphic units have separate boundaries (Figs. 10–12;

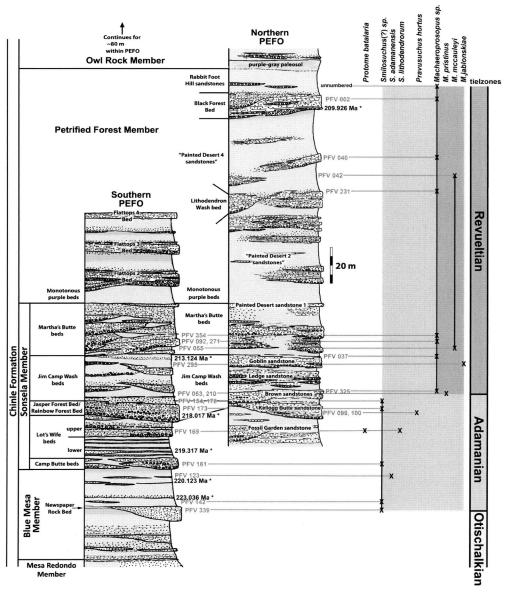

Fig. 10 Phytosaur biostratigraphy of the Chinle Formation in Petrified Forest National Park, northern Arizona. Sections for the northern and southern parts of the park are correlated lithostratigraphically using related sequences of facies in the Sonsela Member. Fossil localities are in gray, preceded by a PFV prefix, and used to plot the known occurrences and ranges of particular taxa. All stratigraphic and fossil data from Parker and Martz (2011). Radioisotopic dates from Ramezani et al. (2011, 2014) and Atchley et al. (2013) are shown in black with an asterisk.

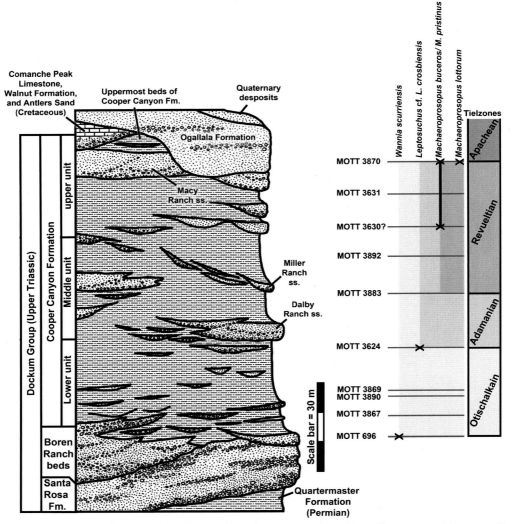

Fig. 11 Phytosaur biostratigraphy of the Dockum Group in southern Garza County, west Texas. Fossil localities are in gray, preceded by a MOTT prefix, and used to plot the known occurrences and ranges of particular taxa. All stratigraphic and fossil data from Martz (2008) and Martz et al. (2013) with nomenclatural modifications after Stocker (2013) and Parker et al. (2013).

Martz, 2008; Parker and Martz, 2011; Ramezani et al., 2011, 2014; Martz et al., 2013, 2014). Just as critical has been appreciation for the need to distinguish biostratigraphic and chronostratigraphic correlation, especially on the global scale, as well as resolving the alpha taxonomy of fossils in different parts of the world to determine whether or not global biostratigraphic correlation using these fossils is even possible

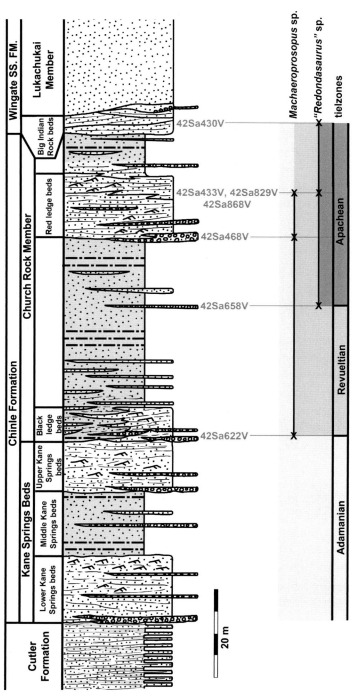

Fig. 12 Phytosaur biostratigraphy of the Chinle Formation in Lisbon Valley, southeastern Utah. Fossil localities are in gray, preceded by a BLM prefix (42Sa), and used to plot the known occurrences and ranges of particular taxa. All stratigraphic and fossil data from Martz et al. (2014). Note that the range for *Machaeroprosopus* includes specimens referable to the subgenus "*Redondasaurus.*"

(e.g., Rayfield et al., 2005, 2009; Irmis et al., 2010, 2011; Desojo and Ezcurra, 2011; Desojo et al., 2013; Butler, 2013; Kammerer et al., 2015).

Recent Complications to Phytosaur Systematics
Alpha Taxonomy and Phylogeny

As discussed previously, a relatively straightforward and convenient alpha taxonomic scheme for phytosaurs had emerged by the end of the 1990s, which endured, with only slight modifications, until very recently. In most publications, no more than five or six genera of phytosaurs were recognized in western North America (e.g., Long and Murry, 1995; Lucas, 1998, revised in 2010; Long and Murry, 1995; Irmis, 2005; Parker and Irmis, 2006; Parker and Martz, 2011), with the genera occurring in a particular strati-graphic sequence (Fig. 6) used to define the lower boundaries of the "faunachrons" (Lucas, 1998). Although complications in the form of purported overlaps between the defining phytosaur taxa and putative index fossils were alleged by some workers (e.g., Woody and Parker, 2004; Hunt et al., 2005; Lehman and Chatterjee, 2005; Parker, 2006), more detailed studies have shown that these were due to the use of erroneous lithostratigraphic correlations prior to determining biostratigraphic ranges, and that the "faunachrons" do indeed have largely distinct taxonomic compositions in which phytosaurs appear in the stratigraphic order consistent with their phylogenetic position (Martz, 2008; Parker and Martz, 2011; Martz et al., 2013, 2014).

However, recent phylogenetic analyses have made phytosaur alpha taxonomy much more complicated (Stocker, 2010, 2012, 2013; Hungerbühler et al., 2013; Kammerer et al., 2015), with important ramifications for biochronology. It should be noted that the phylogenetic taxonomy utilized here follows Parker and Irmis (2006) and Stocker (2010, 2012, 2013; Stocker and Butler, 2013); the proposed revisions of Kammerer et al. (2015) will be addressed afterwards.

The complications to biochronology caused by recent phylogenies are ironic for two reasons:

1. In spite of the increasing number of characters and taxa included in phylogenetic ana-lyses over the past 27 years, the broad picture of western North American phytosaur phylogeny has remained remarkably stable. The phylogenetic relationships of the genera *Parasuchus* (=*"Paleorhinus"*), *Angistorhinus*, *Leptosuchus*, *Machaeroprosopus* (=*"Pseudopalatus"*), and *"Redondasaurus"* are broadly the same in the most recent phy-logenies (e.g., Parker and Irmis, 2006; Hungerbühler et al., 2013; Stocker and Butler, 2013; Kammerer et al., 2015) as in the original phylogeny of Ballew (1989), and the stratigraphic sequence in which these forms appear has become increasingly well cor-roborated as biostratigraphic resolution has improved (Fig. 6E; e.g., Parker and Martz, 2011; Martz et al., 2013).

2. The quality of the phytosaur fossil record in western North America is excellent, and has only improved. Numerous skulls have been discovered representing

morphological and phylogenetic intermediates between previously known forms (Fig. 6E); for example, between *Leptosuchus* and *Machaeroprosopus* (=*Pseudopalatus*) (e.g., *Pravusuchus hortus*; Stocker, 2010), and between species of *Machaeroprosopus* assigned to *Pseudopalatus* and *Redondasaurus* (e.g., *Machaeroprosopus lottorum*; Hungerbühler et al., 2013; Martz et al., 2014). Again, these taxa generally appear in the stratigraphic order that would be predicted based on their phylogenetic positions (Martz, 2008; Parker and Martz, 2011; Martz et al., 2013, 2014).

The problems come from the sheer diversity of phytosaur morphotypes, bound by the requirement that genera are monophyletic (e.g., de Querioz and Gauthier, 1992; Padian et al., 1994; Angielczyk and Kurkin, 2003), and possibly by a degree of endemism within western North America. Specific difficulties are as follows:

1. Phytosaurs that are basal to Phytosauridae (*sensu* Doyle and Sues, 1995; Mystriosuchinae *sensu* Kammerer et al., 2015) consist of forms commonly assigned to *Parasuchus* Lydekker 1885 or *Paleorhinus* Williston 1904 (Fig. 6E and Table 2; Hunt and Lucas, 1991; Long and Murry, 1995; Lucas et al., 2007a; Stocker, 2013; Kammerer et al., 2015). Stocker (2013) identified "*Paleorhinus*" *scurriensis* Long and Murry, 1995 from the Camp Springs Formation of Scurry County, Texas, as the most basal of these forms. Based on the presence of at least five autapomorphies in this species, and the fact that uniting "*Paleorhinus*" *scurriensis* with other species traditionally assigned to *Paleorhinus* (*P. bransoni* Williston 1904 and *P. sawini* Long and Murry 1995) would make the genus paraphyletic, Stocker (2013) erected a new genus, *Wannia*. Subsequently, Kammerer et al. (2015) found that *Parasuchus hislopi* Lydekker 1885 formed a monophylum with *P. bransoni* and *Francosuchus angustifrons* Kuhn 1936, allowing these three species to be united within a monophyletic *Parasuchus* (see also Butler, 2013). However, *Wannia scurriensis* remained an outgroup to this clade, and *Ebrachosuchus neukami* Kuhn 1936 and "*Paleorhinus*" *sawini*, two species often considered co-generic with *P. bransoni* (Table 2; Hunt and Lucas, 1991; Long and Murry, 1995; Lucas et al., 2007a), were found to be sister taxa to more derived clades (Stocker, 2013; Kammerer et al., 2015). As a result, *Parasuchus/Paleorhinus* as used inclusively (e.g., Long and Murry, 1995; Lucas et al., 2007a) is strongly paraphyletic (Fig. 6E), supporting its separation into distinct genera. This raises the question of which North American alpha taxon should be used to define the base of the Otischalkian in western North America: *Wannia*, *Parasuchus*, or "*Paleorhinus*" *sawini*?

2. Phytosaurs recovered within Leptosuchomorpha (Stocker, 2010) but outside Pseudopalatinae (*sensu* Parker and Irmis, 2006; Mystriosuchini *sensu* Kammerer et al., 2015) consist mostly of species traditionally assigned to *Leptosuchus* Case 1922, *Smilosuchus* Long and Murry 1995, and/or *Rutiodon* Emmons 1856 (e.g., Long and Murry, 1995; Lucas, 1998; Irmis, 2005; Parker and Irmis, 2006; Parker and Martz, 2011; Stocker and Butler, 2013; Kammerer et al., 2015). These nonpseudopalatine leptosuchomorphs have also been found to be paraphyletic (Stocker, 2010, 2013; Stocker and Butler, 2013;

Kammerer et al., 2015), forming consecutive outgroups to Pseudopalatinae within Leptosuchomorpha. Stocker (2010) therefore restricted the genera *Leptosuchus* and *Smilosuchus* to separate monophyla (the latter being sister taxon to *Pravusuchus* and Pseudopalatinae; see Fig. 6E). Interestingly, there is a hint of geographical segregation in this phylogeny. The more basal species ("*Phytosaurus*" *doughty* Case 1920, *Leptosuchus crosbiensis* Case 1922, and *L. studeri* Case and White 1934) are mostly known from the Dockum Group of western Texas, whereas more derived species (*Smilosuchus adamanensis* Camp 1930, *S. lithodendrorum* Camp 1930, *S. gregorii* Camp 1930, and *Pravusuchus hortus* Stocker 2010) are all known from the Chinle Formation of northern Arizona. Therefore, specifying any particular alpha taxon as defining the base of the Adamanian would only allow it to be identified in one state.

3. Pseudopalatinae *sensu* Parker and Irmis (2006) (Mystriosuchini *sensu* Kammerer et al., 2015), the most derived phytosaur clade (Fig. 6C–E), has an exceptionally rich North American fossil record (e.g., Long and Murry, 1995; Hungerbühler, 2002; Zeigler et al., 2003; Parker and Irmis, 2006; Hungerbühler et al., 2013; Spielmann and Lucas, 2012; Martz et al., 2014). Pseudopalatinae presents a slightly different problem from the taxa just discussed in that the clade is monophyletic (e.g., Hungerbühler, 2002; Parker and Irmis, 2006; Hungerbühler et al., 2013; Kammerer et al., 2015), with all North American species (including all traditionally assigned to "*Pseudopalatus*" Mehl 1928 and "*Redondasaurus*" Hunt and Lucas 1993) having been recently placed in the genus *Machaeroprosopus* Camp 1930 (Parker et al., 2013; Hungerbühler et al., 2013). The type species of *Machaeroprosopus* is *M. buceros* Cope 1881, formerly considered to be the type species of "*Pseudopalatus*" (Ballew, 1989). Unfortunately, "*Pseudopalatus*" in the traditional sense (e.g., Ballew, 1989; Long and Murry, 1995; Parker and Irmis, 2006) is paraphyletic with respect to "*Redondasaurus*" (Hungerbühler et al., 2013). Moreover, the two named species of "*Redondasaurus*" ("*R.*" *gregorii* Hunt and Lucas 1993 and "*R.*" *bermani* Hunt and Lucas 1993) may actually be polyphyletic (Hungerbühler et al., 2013; i.e., members of the genus do not share a most recent common ancestor that is ancestral only to species assigned to the same genus). This means that maintaining genus monophyly requires either assigning most species of *Machaeroprosopus* to distinct genera, or sinking "*Redondasaurus*" into *Machaeroprosopus*. Hungerbühler et al. (2013) opted for the latter solution, which permits the FHA of *Machaeroprosopus* to define the base of the Revueltian, but prevents "*Redondasaurus*" from defining the base of the Apachean.

One solution might be to only use the original type species for "*Redondasaurus*," "*R.*" *gregorii* Hunt and Lucas 1993 to define the base of the Apachean. However, other derived pseudopalatines in western Texas and southern Utah possessing the characters traditionally used to diagnose "*Redondasaurus*" may belong to distinct species from both "*R.*" *gregorii* and "*R.*" *bermani* (Hungerbühler et al., 2013; Martz et al., 2014; but see Spielmann and Lucas, 2012). This means that using only "*R.*" *gregorii* to define the base of the Apachean means that the "faunachron" may not be identifiable outside of the Redonda Formation of New Mexico. Martz et al. (2014) proposed treating "*Redondasaurus*" as an informal subgenus, and continued to use it to define the base of the Apachean. However, this is a decidedly arbitrary taxonomic solution that does not solve the problem of possible polyphyly within the subgenus.

In summary, improved understanding of phytosaur diversity makes it highly impractical, if not impossible, to present a simple definition for any "faunachron" boundaries relying on a single species or monophyletic genus. Next, we therefore propose an alternative solution for the definition of the faunachrons utilizing monophyletic groups.

Comments on the Clade Name Revisions of Kammerer et al. (2015)

We have chosen to not adopt the taxonomic revisions for phytosaur clades proposed by Kammerer et al. (2015), who have attempted to adapt rules for the application of family rank mandated by the ICZN (1999) to phylogenetic taxonomy. Kammerer et al.'s (2015) revisions have supplanted two well-established clade names, Phytosauridae Meyer 1861 (*sensu* Doyle and Sues, 1995) and Pseudopalatinae Long and Murry, 1995 (*sensu* Parker and Irmis, 2006), following the guidelines of the ICZN (Article 37.2) that states that family-level names must be based on valid taxa. *Phytosaurus* Jaeger 1828 and *Pseudopalatus* Mehl 1928 are respectively a *nomen dubium* and a junior synonym (e.g., Parker et al., 2013; Stocker and Butler, 2013). Kammerer et al. (2015) therefore suggested replacing the clade names Phytosauridae and Pseudopalatine with Mystriosuchinae von Huene 1915 and Mystriosuchini von Huene 1915, respectively; both names are based on the valid German taxon *Mystriosuchus* Meyer 1863.

However, one of the biggest advantages to abandoning the Linnaean taxonomic system is the abandonment of supra-generic ranks, which are inherently arbitrary and tend to proliferate (e.g., de Queiroz and Gauthier, 1992). Regardless of their original formulation, Phytosauridae (*sensu* Doyle and Sues, 1995) and Pseudopalatinae (*sensu* Parker and Irmis, 2006) are clades, not Linnaean families, rendering their original family and subfamily ranks moot. Moreover, phylogenetic definitions can cause clades with suprafamilial suffixes to fall within those with family or subfamily suffixes, rendering the distinction between names based on different Linnaean ranks even more meaningless; for example, in Kammerer et al.'s (2015) phylogeny, Leptosuchomorpha Stocker, 2010 actually falls inside of Mystriosuchinae *sensu* Kammerer et al. (2015)! There is therefore no compelling reason to treat clade names that have a family or subfamily suffix (-idae or -inae) differently from those with any other type of Linnaean-rank suffix that have no particular requirement to be based on valid alpha taxa under the ICZN. Although we applaud the exploration of how ICZN and phylogenetic nomenclatural codes might be reconciled, we feel that it is counterproductive to disrupt a well-established nomenclature utilizing the phylogenetic definitions of Phytosauridae and Pseudopalatinae (e.g., Doyle and Sues, 1995; Hungerbühler and Hunt, 2000; Hungerbühler, 2002; Parker and Irmis, 2006;

Stocker, 2010, 2012, 2013; Hungerbühler et al., 2013; Parker et al., 2013; Stocker and Butler, 2013) to bring one of the most problematic aspects of Linnaean taxonomy (that of arbitrary ranks) back into phylogenetic taxonomy.

PROPOSED REFORMULATION OF THE LATE TRIASSIC LAND VERTEBRATE "FAUNACHRONS"

Proposals for Biozone, Biochronozone, and Biochron Definitions and Applications

Redefining "Faunachrons" Boundaries Using Multiple Taxa

Two possible solutions to the problem of traditional "faunachron"-defining phytosaur alpha taxa being nonmonophyletic will be considered in turn:

1. Breaking each "faunachron" into several geographically localized "faunachrons," each defined by a single, locally occurring alpha taxon. For example, the Adamanian might be broken into two localized "faunachrons" in Texas and Arizona, with the base of the Texas "faunachron" being defined by the FHA of *Leptosuchus*, and the base of the Arizona "faunachron" being defined by the FHA of *Smilosuchus*.

2. Using *multiple* alpha taxa to define the "faunachron" boundaries, with the boundary for each "faunachron" being defined by different alpha taxa in different local study areas; i.e., changing the "faunachrons" from single-taxon interval biozones or biochrons into multiple-taxon interval biozones or biochrons *sensu* Walsh (2001) (Fig. 3B and Table 2).

The first solution is here rejected as premature, because local variation in alpha taxonomy among Upper Triassic vertebrates in western North America is poorly understood. Many Late Triassic taxa known from one or a very few localities are known from only a few specimens. Examples include the phytosaurs *Pravusuchus hortus* Stocker 2010 and *Protome batalaria* Stocker 2012, and the aetosaurs *Desmatosuchus smalli* Parker 2005 and *Sierritasuchus macalpini* Parker et al. 2008. Moreover, even local studies where known biostratigraphic ranges have been precisely documented have stratigraphic intervals that are poorly sampled (Figs. 10–13; e.g., Martz, 2008; Parker and Martz, 2011; Martz et al., 2013, 2014). For example, within North America nonphytosaurid phytosaurs are currently unknown from outside the Popo Agie Formation of Wyoming and the Dockum Group of Texas (e.g., Long and Murry, 1995; Lucas et al., 2007a; Stocker, 2013; Kammerer et al., 2015), but the equivalent stratigraphic interval where they might occur in the Chinle Formation of the Colorado Plateau, in the Shinarump and Mesa Redondo members, has produced few vertebrate specimens that can be assigned to alpha taxa (e.g., Parrish, 1999; Martz et al., 2015, p. 32). It is therefore difficult to assess whether taxa that are currently known from geographically restricted samples are truly endemic or merely appear so due to incomplete sampling. Eventually, with more extensive collecting, regional patterns of variation may become clear enough to allow local modifications of "faunachron" definitions.

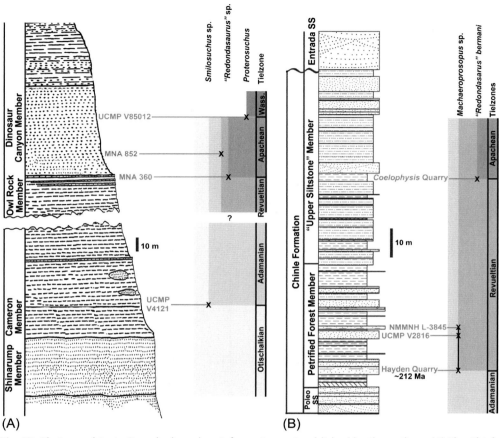

Fig. 13 Phytosaur biostratigraphy based on information not published by the authors. (A) The Chinle Formation in the Cameron/Ward's Terrace area of northern Arizona, section modified from Colbert and Mook (1951, Fig. 2), stratigraphic information from Colbert and Mook (1951), Colbert (1952), Clark and Fastovsky (1986), Kirby (1991), and Parker et al. (2011). Note that the base of the Revueltian teilzone cannot be delineated as the LO$_k$ of *Machaeroprosopus* is higher than the LO$_k$ of "*Redondasaurus*." (B) The Chinle Formation in the Ghost Ranch area of northern New Mexico, section modified from Irmis et al. (2007, Fig. 1), stratigraphic information from Colbert (1989) and Irmis et al. (2007).

This leaves multiple-taxon definitions as the best alternative. Using the phylogenetic framework for phytosaurs provided within the past decade (Fig. 6; Hungerbühler, 2002; Parker and Irmis, 2006; Stocker, 2010, 2012, 2013; Hungerbühler et al., 2013; Butler, 2013; Kammerer et al., 2015), biozones, estimated biochronozones, and estimated biochrons can be defined using the lowest stratigraphic occurrences of nested phytosaur clades, specifically Phytosauria (*sensu* Doyle and Sues, 1995), Leptosuchomorpha (Stocker, 2010), Pseudopalatinae (*sensu* Hungerbühler, 2002; in North America, essentially equivalent to the lowest occurrence of *Machaeroprosopus sensu* Parker et al., 2013 and Hungerbühler et al., 2013), and "*Redondasaurus*" if the latter is considered a potentially

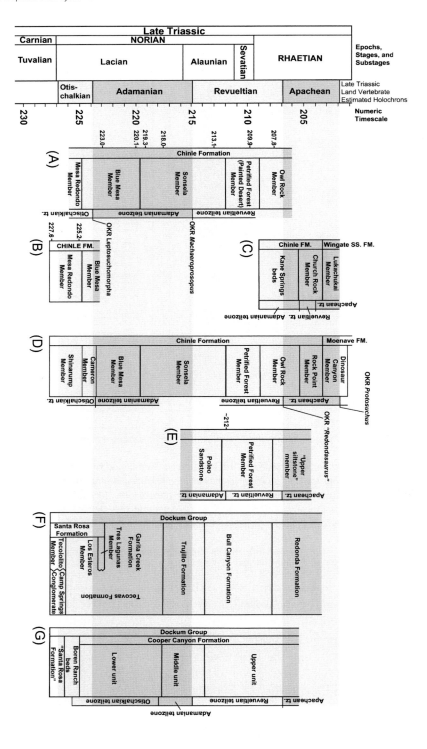

monophyletic subgenus within *Machaeroprosopus* (Martz et al., 2014, pp. 432–433). The alpha taxa used to define the base of each "faunachron" in any particular local study area represent the most basal members of each clade present in that study area. This allows for the stratigraphic appearance of the clade to be locally constrained as precisely as possible. As already discussed, this strategy is viable because the stratigraphic appearance of phytosaur taxa within the Chinle Formation and Dockum Group corresponds fairly well to their phylogenetic position (Figs. 6, and 10–14). Moreover, this strategy avoids the problems imposed by endemism by using multiple alpha taxa, including those endemic to different areas rather than relying on any particular one.

Multiple-taxon definitions are also advantageous given the weak Bremer support for phytosaur clades. Most nodes in recent phylogenetic analyses (Stocker, 2010, 2012, 2013; Hungerbühler et al., 2013; Kammerer et al., 2015) only have a Bremer support of one or two, meaning that the precise phylogenetic relationships of taxa could easily shift slightly within particular phylogenies. If only the present earliest branching species for each clade (e.g., *Wannia scurriensis*) was used to bound the base of each biozone, estimated biochronozone, and estimated biochron, any revisions to the phylogenetic position of that taxon could cause its lowest stratigraphic appearances to represent the appearance of a more or less inclusive clade. Using multiple taxa occurring in the same general region of the phylogenetic tree makes it less likely that revisions to a phylogeny will shift local teilzone boundary placement, and by extension the lower boundaries of estimated biochronozones and estimated biochrons. This strategy is therefore advantageous for providing boundary definitions with a degree of stability.

The Use of Provisional Biozone Reference Sections Instead of Biozone Type Sections

Because quasibiozones are often characterized in terms of the lithostratigraphic units that contain them, they have often been associated with a type section, or principal reference section (e.g., Wood et al., 1941; Lucas and Hunt, 1993). However, reformulating these quasibiozones as interval estimated eubiochrons and estimated eubiozones (e.g., Lucas, 1998; Parker and Martz, 2011) creates the potential problem of two different definitions:

Fig. 14 Regional chronostratigraphic correlations between the Chinle Formation and Dockum Group in parts of Arizona, New Mexico, and Texas. Sections show the placement of local teilzone boundaries. The gray banding on the measured sections represent estimated holochronozone ranges. (A) The Chinle Formation of Petrified Forest National Park (see Fig. 10). Note that the LO$_k$s of Pseudopalatinae and Leptosuchomorpha in PEFO are also the OKRs for the entire western United States. (B) The lower Chinle Formation at the Mesa Redondo–type section, northern Arizona; radioisotopic dates from Ramazani et al. (2014). (C) The Chinle Formation of Lisbon Valley, southeastern Utah (see Fig. 12). (D) The Chinle Formation of the Navajo Nation near Cameron and Ward's Terrace (differs from Fig. 13A in showing the full Chinle Formation section), lithostratigraphy modified from Stewart et al. (1972) and Lucas (1993a,b). (E) The Chinle Formation of the Ghost Ranch area of northern New Mexico (Fig. 13B). (F) The Dockum Group of northeastern New Mexico and the Texas Panhandle; lithostratigraphy after Lucas et al. (1994) and Lehman (1994), approximate holochronozone correlations based on data from Hunt (1994), Lucas (1998), and Spielmann and Lucas (2012). (G) The Dockum Group of southern Garza County, western Texas (Fig. 11).

one based on the lowest occurrences of particular taxa, and one based on a fossil assemblage of multiple taxa occurring in a particular lithostratigraphic unit at a particular place (Fig. 4). These definitions can easily come into conflict (Walsh, 1998, pp. 169–170). With the precise plotting of known stratigraphic ranges, and new fossil discoveries extending the range of the taxon forming the upper boundary (Fig. 4B), it is possible that a type assemblage for a eubiozone can actually be found to fall partially or entirely outside the eubiozone (i.e., above the lowest occurrence of the taxon forming the upper boundary)! This would also impact purported index fossils (fossils considered restricted to, and therefore diagnostic of, a biozone), because they could be found to occur partially, or even entirely outside of the biozone they are thought to diagnose.

Indeed, the NACSN (2005, Article 54e) noted that biozones do not have true stratotypes, although "it is desirable to designate a reference section in which the biostratigraphic unit is characteristically developed." Reference sections are ideally constrained by detailed biostratigraphic data demonstrating that the section falls within the eubiozone, and should also contain enough fossil localities producing taxonomically identifiable specimens to allow the fossil assemblage of the biozone to be characterized. If the reference section is later found to fall outside the eubiozones (Fig. 4B), it might be abandoned, or simply considered the reference section for another biozone. Because this paper deals with the definition of biozones, the characterization of biozones through the use of reference sections, local faunas, and index fossils (e.g., Tedford, 1970; Woodburne, 2004) will be discussed elsewhere.

Combining Local and Regional Biostratigraphic and Geochronologic Information: Teilzones, Estimated Holochronozones, and Estimated Holochrons

One frustrating aspect of the Late Triassic land vertebrate "faunachrons" is the difficulty of discussing multiple stratigraphic and chronologic concepts under a single term. Just as the use of the term "faunachron" (Lucas, 1993a,b) makes it possible to discuss ontological intervals of time but neglects the biostratigraphic foundation of biochronology, so reformulating these units as biozones (Parker and Martz, 2011; Martz et al., 2013, 2014) allows the discussion of epistemological stratigraphic intervals, but not the intervals of time that they represent. We therefore propose dividing each of the Late Triassic land vertebrate "faunachrons" into several related but distinct stratigraphic and chronologic units (Figs. 7F and 14). This simple system outlined in the following provides a practical "starter kit" for building a system of rigorous biochronology when biostratigraphic and geochronologic data is scarce. Although we only demonstrate the utility of this system for the Late Triassic of western North America, we believe that it may prove applicable to other nascent vertebrate biochronologic models.

From the bewildering variety of terms available (e.g., Walsh, 1998, 2001), we select three: one biostratigraphic unit (a teilzone), one chronostratigraphic unit (an estimated holochronozone), and one geochronologic unit (an estimated holochron). All are explicitly epistemological, being based exclusively on known stratigraphic and geochronologic data:

1. A *teilzone* (Arkell, 1933) is a biozone identified in a particular local study area (Figs. 1, 2, and 10–14). We employ a multiple-taxon LO_k-LO_k interval assemblage eubiozone concept (Figs. 3B and 4; Walsh, 2001). Several taxa may be used to define each biozone boundary, and in any particular local study area the lowest known stratigraphic occurrence from among these taxa provides the lower boundary of each teilzone. We use the broader term *biozone* to encompass not only individual teilzones, but also the total body of sedimentary strata represented by all teilzones where boundary-defining taxa can be identified. It should be emphasized that teilzones are strictly epistemological units based strictly on known fossil occurrences from particular localities, and therefore subject to the bias of the fossil record; the LO_k of a taxon should never be assumed to represent the first individual to live in an area, or even the first to be preserved as a fossil (Walsh, 1998), although it may be considered to approximate those individuals if the stratigraphic interval has been sufficiently well sampled (i.e., if confidence intervals for stratigraphic range limits are short; Marshall, 1997; Solow, 2003). As a result, both the lower and upper boundaries of teilzones may be expected to descend stratigraphically as fossil sample sizes increase (Fig. 4). Particularly when sample sizes are poor, the stratigraphic position of these boundaries may differ radically from the boundaries of the estimated biochronozone (estimated holochronozone; see item 2).

   In some local study areas, fossils demarcating only one boundary of a teilzone may be known, in which case we refer to the teilzone as being "open-ended." An open-ended teilzone where the LO_k of the upper boundary defining taxon is absent is referred to as a "topless teilzone," whereas an open-ended teilzone where the LO_k of the lower boundary defining taxon is absent is referred to as a "bottomless teilzone."

2. An *estimated holochronozone* is an estimated biochronozone, specifically the total body of strata estimated to have been deposited between two particular paleobiological events of regional or global significance. In other words, it is the best estimate of the "real" stratigraphic interval demarcated by the stratigraphic appearance of phytosaur clades. We employ a multiple-taxon OKR-OKR interval assemblage eubiochronozone concept, in which the oldest known OKRs from among the same taxa defining the teilzone boundaries are used to define the lower boundaries of the estimated holochronozone (Fig. 14). The estimated holochronozones therefore represent the best estimate for the total body of strata deposited between the evolution or immigration of the taxa defining the upper and lower boundaries of the estimated holochronozone. Estimated holochronozone boundaries are correlated between local study areas by estimating the stratigraphic levels that are approximately isochronous with the OKRs (Figs. 2 and 14), using radioisotopic, lithostratigraphic, and/or biostratigraphic correlations. Given the current limitations of local biostratigraphic and geochronologic data, there is little point in attempting to recognize estimated

teilchronozones (and teilchrons) for individual local study areas for reasons discussed previously.

3. An *estimated holochron* is an estimated biochron, specifically an estimate of the interval of time represented by an estimated holochronozone. We employ a multiple-taxon $eFHA_o$-$eFHA_o$ interval eubiochron concept, in which estimated oldest first historical appearances (the temporal equivalents of the OKRs) mark the beginning and end of the holochron. Estimated holochrons are discussed in reference to absolute ages and/or geochronologic systems, series, and stages.

These units must be established in the proper sequence. Empirically determined local teilzones are determined first and combined into regional eubiozones, which are used to estimate regional (or even global) estimated holochronozones. Combining estimated holochronozones with available absolute age dates provide estimated holochrons.

New Definitions for the Upper/Late Triassic Land Vertebrate Teilzones, Estimated Holochronozones, and Estimated Holochrons of Western North America

The Otischalkian

The Otischalkian teilzone (biozone) is defined here for any particular local study area as the stratigraphic interval between the lowest stratigraphic appearances of Phytosauria and Leptosuchomorpha (Fig. 6E). It is identified as the stratigraphic interval between the LO_k of any species of the nonphytosaurid phytosaur genera *Wannia* Stocker 2013 or *Parasuchus* (*sensu* Kammerer et al., 2015; Fig. 6A), and the lowest LO_k of any species of the nonpseudopalatine leptosuchomorph genera *Smilosuchus* (Fig. 6B) or *Leptosuchus*, or of "*Phytosaurus*" *doughtyi*. *Promystriosuchus ehlersi* Case 1922 is currently excluded from being used for the lower boundary identifier of the Otischalkian; although *Promystriosuchus* is likely a basal member of Phytosauria because of the retention of plesiomorphic character states that also occur in *Wannia* and *Parasuchus* (e.g., Hunt and Lucas, 1991; Long and Murry, 1995); it is badly in need of redescription and has not been incorporated into any recent phylogenetic analyses (e.g., Kammerer et al., 2015; M. Stocker, personal communication, 2016). "*Paleorhinus*" *sawini* Long and Murry 1995 is also excluded as a lower boundary identifier, because the taxon may fall within Phytosauridae (Kammerer et al., 2015), and known stratigraphic occurrences of the taxon are in the Colorado City Member/"pre-Tecovas horizon" (Lucas, 1998; Long and Murry, 1995), well above the LO_ks of *Wannia* and *Parasuchus* (see the following). The nonpseudopalatine leptosuchomorph *Pravusuchus hortus* is excluded as an upper boundary identifier because it is not only more derived than other nonpseudopalatine leptosuchomorphs (Fig. 6E; Stocker, 2010), but the only known occurrences in northern Arizona are stratigraphically well above the lowest occurrences of *Smilosuchus* (Fig. 10; Parker and Martz, 2011).

The Otischalkian estimated holochronozone is defined here as the stratigraphic interval between the oldest OKRs of any of the taxa defining the bases of the Otischalkian and Adamanian teilzones.

The Otischalkian estimated holochron is defined here as the time interval equivalent to this estimated holochronozone, bounded by the oldest eFHA (temporal equivalent of the OKR) of any of the taxa defining the bases of the Otischalkian and Adamanian holochronozones, and calibrated using available radioisotopic dates.

The Adamanian

The Adamanian teilzone (biozone) is defined here for any particular local study area as the stratigraphic interval between the lowest stratigraphic appearances of Leptosuchomorpha and *Machaeroprosopus* (Fig. 6E). It is identified as the stratigraphic interval between the lowest LO_k of any species of the nonpseudopalatine leptosuchomorph genera *Smilosuchus* (Fig. 6B) or *Leptosuchus*, or of *"Phytosaurus" doughtyi*, and the lowest LO_k of the pseudopalatine phytosaur species *M. buceros* (Fig. 6C), *M. validus*, *M. andersoni*, *M. pristinus*, *M. tenuis*, *M. maccauleyi*, or *M. jablonskiae* (not all of these species may be valid; see Long and Murry, 1995; Hungerbühler et al., 2013; Parker et al., 2013). *Machaeroprosopus lottorum* Hungerbühler et al. 2013, *M. gregorii* Hunt and Lucas 1993, *M. bermani* Hunt and Lucas 1993, and other specimens assigned by some authors to the genus or subgenus *"Redondasaurus"* (Spielmann and Lucas, 2012; Martz et al., 2014) are excluded from this definition; they are known to occur phylogenetically well above the base of Pseudopalatinae, and stratigraphically much higher than the LO_ks of other species of *Machaeroprosopus* (e.g., Martz, 2008; Martz et al., 2014).

The Adamanian estimated holochronozone is defined here as the stratigraphic interval between the oldest OKRs of any of the taxa defining the bases of the Adamanian and Revueltian teilzones.

The Adamanian estimated holochron is defined here as the time interval equivalent to this estimated holochronozone, bounded by the oldest eFHA (temporal equivalent of the OKR) of any of the taxa defining the bases of the Adamanian and Revueltian holochronozones, and calibrated using available radioisotopic dates.

The Revueltian

The Revueltian teilzone (biozone) is defined here for any particular local study area as the stratigraphic interval between the lowest stratigraphic appearance of *Machaeroprosopus* (Fig. 6E) and *"Redondasaurus"* (*sensu* Spielmann and Lucas, 2012; Martz et al., 2014). It is identified as the stratigraphic interval between the lowest LO_k of any species of the pseudopalatine phytosaur species *M. buceros* (Fig. 6C), *M. validus*, *M. andersoni*, *M. pristinus*, *M. tenuis*, *M. maccauleyi*, or *M. jablonskiae* (e.g., Long and Murry, 1995; Hungerbühler et al., 2013; Parker et al., 2013) and the lowest LO_k of any species of *"Redondasaurus"* (*sensu* Spielmann and Lucas, 2012; Martz et al., 2014), specifically *"R." gregorii* (Fig. 6D), *"R." bermani*, (Hunt and Lucas, 1993; Spielmann and Lucas, 2012); *"R." lottorum* (Hungerbühler et al., 2013), and unnamed species from Utah (e.g., UMNH VP 21898, UMNH VP 24236; Martz et al., 2014). As has been discussed, the taxonomic status and diagnosis of "*Redondasaurus*" is subjective and in need of revision (Hungerbühler et al., 2013; Martz et al., 2014).

The Revueltian estimated holochronozone is defined here as the stratigraphic interval between the oldest OKRs of any of the taxa defining the bases of the Revueltian and Apachean teilzones.

The Revueltian estimated holochron is defined here as the time interval equivalent to this estimated holochronozone, bounded by the oldest eFHA (temporal equivalent of the OKR) of any of the taxa defining the bases of the Revueltian and Apachean holochronozones, and calibrated using available radioisotopic dates.

The Apachean

The Apachean teilzone (biozone) is defined here for any particular study area as the stratigraphic interval between the lowest stratigraphic appearance of all species assigned to the subgenus "*Redondasaurus*" (*sensu* Spielmann and Lucas, 2012; Martz et al., 2014; Fig. 6D) and the basal crocodyliform *Protosuchus*. It is identified using the lowest LO_k of all species of "*Redondasaurus*" and the LO_k of *Protosuchus richardsoni* (Sues et al., 1996), which designates the base of the Wassonian teilzone (formerly the Wassonian "faunachron"; Lucas, 1998; Lucas and Tanner, 2007). Using the definitions given here, the top of the Apachean biozone is the only Upper Triassic land vertebrate biozone in western North America to have a boundary defined by the LO_k of a single taxon, or a taxon that is not a phytosaur.

The Apachean estimated biochronozone is defined here as the stratigraphic interval between the OKRs of any of the taxa defining the bases of the Apachean and Wassonian biozones.

The Apachean estimated holochron is defined here as the time interval equivalent to this estimated holochronozone, bounded by the oldest eFHA (temporal equivalent of the OKR) of any of the taxa defining the bases of the Apachean and Wassonian holochronozones, and calibrated using available radioisotopic dates.

APPLICATION OF TEILZONE, ESTIMATED HOLOCHRONOZONE, AND ESTIMATED HOLOCHRON DEFINITIONS

Local Teilzones

Here we will present the Upper Triassic land vertebrate teilzone boundaries for all local study areas in western North America for which the precise biostratigraphic positions of localities have been established (Fig. 9). Three local study areas have been documented by the authors of the current study: Petrified Forest National Park in northern Arizona (Parker and Martz, 2011; Fig. 10), southern Garza County in western Texas (Martz, 2008; Martz et al., 2013; Fig. 11), and Lisbon Valley in southeastern Utah (Martz et al., 2014; Fig. 12). Additionally, there are two other local study areas in western North America where boundaries for Upper Triassic land vertebrate teilzones can be precisely placed based on published biostratigraphic data: the Cameron/Ward Terrace area of northern Arizona (Colbert, 1947, 1952; Colbert and Mook, 1951; Parker et al., 2011; W.G. Parker, unpublished data; Fig. 13A), and the Ghost Ranch area of northern New Mexico (Colbert, 1989; Irmis et al., 2007; Fig. 13B). The type section of the Mesa Redondo Member in northern Arizona (Cooley, 1958) and the Dockum Group of north-eastern New Mexico (e.g., Lehman, 1994; Lucas et al., 1994) also are correlated to the estimated biochronozones using tentative lithostratigraphic correlations (Fig. 14), even though explicit biostratigraphic data is mostly unavailable for these local study areas. Specimen and locality data for biostratigraphically significant voucher specimens are given in Table 3.

At Petrified Forest National Park, northern Arizona (Fig. 10), the boundaries of the Upper Triassic land vertebrate teilzones within the Chinle Formation are placed as follows:

1. The base of the Otischalkian teilzone cannot be identified anywhere in Arizona, because nonphytosaurid phytosaurs have not been identified anywhere in the state. The Otischalkian teilzone in Petrified Forest National Park is therefore open ended (bottomless).

2. The base of the Adamanian teilzone at Petrified Forest National Park is the stratigraphic horizon in the top of the Newspaper Rock Bed of the Blue Mesa Member that produced a specimen (TMM 43685-261; "*Leptosuchus*" *sensu* Parker and Martz, 2011, p. 241) that is assigned to *Smilosuchus* because the squamosal is deep and plate-like and has a ventral margin that slopes gently anteroventrally from the posterior edge to the opisthotic process (Stocker, 2010). This specimen therefore represents the LO_k of Leptosuchomorpha.

3. The base of the Revueltian teilzone at Petrified Forest National Park is the stratigraphic horizon in the lowermost Jim Camp Wash beds of the Sonsela Member, a few meters above the Jasper Forest bed, that produced two specimens of *Machaeroprosopus pristinus* (Fig. 10; PEFO 34042 and AMNH FR 7222; Parker and Martz, 2011, pp. 242–243), representing the LO_k of *Machaeroprosopus*.

4. The base of the Apachean teilzone at Petrified Forest National Park cannot be placed, because "*Redondasaurus*" has not been identified there. The Revueltian biozone in PEFO is therefore open ended (topless), and the Apachean teilzone cannot be identified.

In southern Garza County, western Texas (Fig. 11), the boundaries of the Upper Triassic land vertebrate teilzones within the Dockum Group are placed as follows:

1. The base of the Otischalkian teilzone in southern Garza County is placed within the basal conglomeratic sandstone of the Dockum Group referred to as the Santa Rosa Sandstone or Camp Springs Conglomerate (see Lehman, 1994; Lucas et al., 1994 for nomenclatural discussion). Although no phytosaurs have been recovered from this horizon in southern Garza County, the type locality of *Wannia scurriensis* (MOTT 696; Langston, 1949) lies at this stratigraphic level just across the county line in Scurry County. TTU P-00539, the holotype of *Wannia scurriensis* (Langston, 1949; Stocker, 2013), is therefore considered to represent the LO_k of Phytosauria in southern Garza County.

Table 3 Voucher specimens for all LO$_k$s used to delineate teilzones

Taxon	Clade/taxon	Specimen #	Locality #	Stratigraphic level	Local study area	References
Machaeroprosopus	Pseudopalatinae	**PEFO 34042**	PFV 210	Sonsela Member of the Chinle Formation, lowermost Jim Camp Wash beds, just above the "persistent red silcrete"	Petrified Forest National Park, AZ	Parker and Martz (2011)
Smilosuchus	Leptosuchomorpha	**TMM 43685-261**	PFV 339	Blue Mesa Member of the Chinle Formation, top of the Newspaper Rock Bed	Petrified Forest National Park, AZ	Parker and Martz (2011)
Machaeroprosopus	Pseudopalatinae	UMNH VP 21874	42Sa662V	Church Rock Member of the Chinle Formation, 2 m above the top of the Kane Springs beds	Lisbon Valley, UT	Martz et al. (2014)
"*Redondasaurus*"	"*Redondasaurus*"	UMNH VP 21898	42Sa658V	Church Rock Member of the Chinle Formation, 15.8 m below the red ledge beds	Lisbon Valley, UT	Martz et al. (2014)
"*Redondasaurus*"	"*Redondasaurus*"	**MNA V3495**	MNA 360	Owl Rock Member of the Chinle Formation, third of the four major carbonate ledges	Cameron/Ward's Terrace, AZ	Kirby (1991) and Parker et al. (2011)
Protosuchus richardsoni	*Protosuchus*	**UCMP 97638**	UCMP V85012	Dinosaur Canyon Member of the Moenave Formation, 9–16 m below the top of the formation	Cameron/Ward's Terrace, AZ	Clark and Fastovsky (1986)
Smilosuchus gregorii	Leptosuchomorpha	AMNH 3060	Ward Bone Bed	Cameron or Blue Mesa Member of the Chinle Formation, about 30 m above the Shinarump Member	Cameron/Ward's Terrace, AZ	Colbert (1952)

Taxon	Specimen number	Clade	MOTT VPL	Stratigraphic/locality	Location	Reference
Machaeroprosopus sp.	GR 233	Pseudopalatinae	Hayden Quarry	Petrified Forest Member of the Chinle Formation, a few meters above the Poleo Sandstone	Ghost Ranch, NM	Irmis et al. (2007)
"*Redondasaurus*" *bermani*	CM 69727	"*Redondasaurus*"	*Coelophysis* Quarry	"Upper siltstone member" of the Chinle Formation, about 37 m below the base of the Entrada Sandstone	Ghost Ranch NM	Colbert (1989) and Hunt and Lucas (1993)
Leptosuchus cf. *L. crosbiensis*	TTU P-9234	Leptosuchomorpha	MOTT VPL 3624	Cooper Canyon Formation of the Dockum Group, lower part of the formation about 8 m below the Cooper Creek beds	Southern Garza County, TX	Martz et al. (2013)
Machaeroprosopus sp.	TTU P-11880	Pseudopalatinae	MOTT VPL 3892	Cooper Canyon Formation of the Dockum Group, middle part of the formation a few meters below the Miller Ranch beds	Southern Garza County, TX	Martz (2008)
"*Redondasaurus*" *lottorum*	TTU P-10077	"*Redondasaurus*"	MOTT VPL 3870	Cooper Canyon Formation of the Dockum Group, Macy Ranch sandstone in the uppermost part of the formation	Southern Garza County, TX	Martz (2008) and Hungerbühler et al. (2013)
Wannia scurriensis	TTU P-00539	Phytosauria	MOTT VPL 696	Camp Springs Conglomerate of the Dockum Group. More detailed information is unavailable	Scurry County, TX	Langston (1949) and Stocker (2013)

Specimen numbers in bold are also the OKR for an estimated holochronozone.

2. The base of the Adamanian teilzone in southern Garza County is the Post Quarry, a stratigraphic horizon about 8 m below the base of the Cooper Creek beds (middle Cooper Canyon Formation) that produced a skull and associated mandible referred to *Leptosuchus* cf. *L. crosbiensis* (TTU P-9234) and representing the LO_k of Leptosuchomorpha (Martz, 2008, p. 290; Martz et al., 2013, pp. 242, 346).

3. The base of the Revueltian teilzone in southern Garza County is the stratigraphic horizon in the middle unit of the Cooper Canyon Formation, a few meters below the Miller Ranch beds (in the middle Cooper Canyon Formation). This horizon produced an isolated broad and highly sculptured posterior process of a squamosal (TTU P-11880; Martz, 2008, pp. 247, 290; Martz et al., 2013, Fig. 2) that would have extended far posterior to the opisthotic process and is therefore referable to *Machaeroprosopus* (Hungerbühler, 2002, character 31; Stocker, 2010, character 24). This specimen represents the LO_k of *Machaeroprosopus*.

4. The base of the Apachean teilzone in southern Garza County is the stratigraphic horizon about 5 m above the base of the Macy Ranch sandstone at the Patricia Site (Frelier, 1987; Lehman and Chatterjee, 2005; Martz, 2008, pp. 250, 290, 343) that produced a referred specimen of *Machaeroprosopus lottorum* (TTU P-10077; Hungerbühler et al., 2013). This taxon meets the tentative criterion given by Martz et al. (2014, p. 432) distinguishing the subgenus "*Redondasaurus*" from other derived *Machaeroprosopus* specimens in that the squamosals are extremely broad with posterolaterally curving medial margins (Hungerbühler et al., 2013). Moreover, within the *Machaeroprosopus* clade (Hungerbühler et al., 2013), *M. lottorum* forms a monophylum with *Machaeroprosopus* ("*Redondasaurus*") *gregorii* (Hunt and Lucas, 1993).

5. The basal crocodyliform *Protosuchus*, the LO_k of which defines the top of the Apachean biozone, is not known from southern Garza County. The Apachean teilzone in southern Garza County is therefore open ended (topless).

In Lisbon Valley in southeastern Utah (Fig. 12), the boundaries of the Upper Triassic land vertebrate teilzones within the Chinle Formation are placed as follows:

1. The base of the Otischalkian and Adamanian teilzones cannot be identified because nonphytosaurid phytosaurs and nonpseudopalatine leptosuchomorphs are unknown from southern Utah. There is therefore no Otischalkian teilzone recognized in Lisbon Valley, and the Adamanian teilzone is open ended (bottomless).

2. The base of the Revueltian teilzone in Lisbon Valley is the stratigraphic horizon in the lowermost Church Rock Member, about two meters above the Kane Springs beds, that produced an isolated squamosal (UMNH VP 21874) referable to *Machaeroprosopus* sp., due to being extremely broad relative to nonpseudopalatine phytosaurs and projecting posteriorly beyond the opisthotic process. This specimen represents the LO_k of *Machaeroprosopus* but does not meet the criteria for the subgenus "*Redondasaurus*" suggested by Martz et al. (2014, p. 432) in that the squamosals are

not as broad as typically seen in "*Redondasaurus*" and the medial margins are straight rather than curving posterolaterally.

3. The base of the Apachean teilzone in Lisbon Valley is the stratigraphic horizon about 15.8 m below the red ledge beds in the lowermost Church Rock Member that produced an isolated squamosal (UMNH VP 21898) referable to "*Redondasaurus*" (*Machaeroprosopus*) sp., in being extremely broad and having a posterolaterally curving medial margin (Martz et al., 2014, p. 432). This specimen represents the LO_k of the subgenus.

Published biostratigraphic information is available for the Chinle Formation and Moenave Formation at Ward Terrace and other localities near Cameron in northern Arizona (Fig. 13A), including for specimens of *Smilosuchus*, "*Redondasaurus*," and *Protosuchus* that are stratigraphically significant (Colbert, 1947, 1952; Colbert and Mook, 1951; Parker et al., 2011; W.G. Parker, unpublished data). The boundaries of the teilzones in the Cameron area are placed as follows:

1. The base of the Otischalkian teilzone cannot be identified, because nonphytosaurid phytosaurs are unknown from Arizona. The Otischalkian teilzone near Cameron is therefore open ended (bottomless).

2. The base of the Adamanian teilzone occurs at the stratigraphic level of Ward's Bonebed (UCMP V4121; Fig. 13A), which occurs about 30 m above the Shinarump Member (Colbert, 1952, pp. 565–566), placing it either high in the Cameron Member or in the overlying Blue Mesa Member (see Lucas, 1993a,b, p. 42). The Ward's Bonebed locality produced an excellent specimen of *Smilosuchus gregorii* (AMNH FR 3060) identified and described by Colbert (1947, 1952, p. 566; see also Long and Murry, 1995; Stocker, 2010, p. 1018), which is the LO_k of Leptosuchomorpha in the Cameron area. The Blue Mesa Member at Cameron is probably only correlative to the upper part of the Blue Mesa Member in Petrified Forest National Park (Irmis et al., 2011, supplemental data), indicating that the Ward's Bonebed specimen (AMNH FR 3060) may occur close to the level of the better constrained LO_k of *Smilosuchus* in Petrified Forest National Park.

3. The base of the Revueltian teilzone in the Cameron area is problematic; although there is abundant phytosaur cranial material known from the area, most of it has not been assigned to an alpha taxon. The holotype of *Machaeroprosopus validus* (Camp, 1930; UW 3807) was recovered stratigraphically above the Ward Bonebed (Camp, 1930, p. 19), but the precise stratigraphic level is unclear, and in any case the specimen is lost (Westphal, 1976). A specimen of *Machaeroprosopus* not referable to "*Redondasaurus*" (MNA V1595; Kirby, 1991; W.G. Parker, unpublished data) was recovered from locality MNA 852, about 15 m above the "third ledge" in the upper part of the Owl Rock Member (Kirby, 1991); technically, this represents the LO_k for *Machaeroprosopus* in the Cameron area. However, this specimen is stratigraphically higher than the LO_k of "*Redondasaurus*" (see item 4), which would place the base

of the Revueltian teilzone higher than the base of the Apachean teilzone. Because the Revueltian teilzone is defined as having the LO_k of *Machaeroprosopus* at the base and the LO_k of "*Redondasaurus*" at the top, the base of the teilzone cannot be currently be delineated. The Revueltian teilzone is therefore bottomless.

4. The base of the Apachean teilzone in the Cameron area is the stratigraphic horizon in the Owl Rock Member that produced a mostly complete skull (MNA V3495) referable to "*Redondasaurus*" in having extremely broad squamosals that appear to have posterolaterally curving medial margins and posterior tips that are not knoblike (Kirby, 1991; Parker et al., 2011; Hungerbühler et al., 2013; Martz et al., 2014, p. 432). The specimen represents the LO_k for "*Redondasaurus*." This locality (MNA 360) was indicated by Kirby (1991) to occur at the level of the third of four prominent carbonate ledges in the upper part of the Owl Rock Member at Ward Terrace, about 80 m above the base of the member.

5. The top of the Apachean teilzone in the Cameron area is represented by several specimens of *Protosuchus richardsoni* (UCMP 97638, UCMP 125358) collected 2–4 m below the type locality for that taxon (locality UCMP V85012) (Colbert and Mook, 1951; Colbert, 1952; Crompton and Smith, 1980; Clark and Fastovsky, 1986). This stratigraphic level lies 9–16 m below the top of the Dinosaur Canyon Member of the Moenave Formation (Clark and Fastovsky, 1986, p. 291; *contra* Colbert and Mook, 1951).

Published data are also available for Ghost Ranch and the surrounding area in northern New Mexico (Fig. 13B), which contains the famous Hayden Quarry and *Coelophysis* Quarry, which have played a critical role in the study of North American Triassic dinosauromorphs (e.g., Colbert, 1989; Nesbitt et al., 2007; Irmis et al., 2007). These localities, as well as the nearby Canjilon Quarry and Snyder Quarry (localities UCMP V2816 and NMMNH L-3845, respectively; Long and Murry, 1995; Martz, 2002; Zeigler et al., 2003; Nesbitt and Stocker, 2008) all contain the remains of pseudopalatine (=mystriosuchine *sensu* Kammerer et al., 2015) phytosaurs; indeed, the *Coelophysis* Quarry is the type locality for "*Redondasaurus*" *bermani* (Hunt and Lucas, 1993). The measured section for the Chinle Formation in the Ghost Ranch area and stratigraphic information provided by Irmis et al. (2007) allows boundaries of the teilzones to be placed as follows:

1. The Otischalkian teilzone and the base of the Adamanian teilzone cannot be identified, because nonphytosaurid phytosaurs and nonpseudopalatine leptosuchomorphs are unknown from northern New Mexico. The Adamanian teilzone at Ghost Ranch is therefore open ended (bottomless).

2. The base of the Revueltian teilzone is at the Hayden Quarry, which contains pseudopalatine phytosaurs referable to *Machaeroprosopus* (="*Pseudopalatus*"; e.g., partial posterior skull GR 233; Irmis et al., 2007, supplemental data) because the squamosals are extremely broad and extend posterior to the opisthotic process. The Hayden Quarry lies near the base of the Petrified Forest Member (Petrified Forest Formation *sensu* Lucas et al., 2003, 2005;

Mesa Montosa Member *sensu* Zeigler et al., 2008; but see Cather et al., 2013), only a few meters above the Poleo Sandstone (Irmis et al., 2007, Fig. 1).

Lucas et al. (2003) provided the name "Mesa Montosa Member" for the lower-most 20–30 m of the Petrified Forest Formation, and *Machaeroprosopus pristinus* (NMMH P-11076 from NMMNH L-911) has been reported from the very top of the Mesa Montosa Member (Lucas and Hunt, 1992; Zeigler et al., 2005), about 20–30 m above the top of the Poleo Sandstone (Zeigler et al., 2008). However, Zeigler et al. (2008) gave a total thickness of 170–180 m for the Petrified Forest Formation in the Chama Basin, which is nearly five times greater than the thickness given by Irmis et al. (2007). We are therefore uncertain where to place locality NMMNH L-911 stratigraphically relative to the Hayden Quarry.

3. The base of the Apachean teilzone at Ghost Ranch is at the *Coelophysis* Quarry, which contains the holotype of "*Redondasaurus*" *bermani* (CM 69727; Hunt and Lucas, 1993; Spielmann and Lucas, 2012). The *Coelophysis* Quarry lies in the "upper siltstone" member (Stewart et al., 1972) of the Chinle Formation, about 65 m above the Hayden Quarry (Irmis et al., 2007), and 37 m below the base of the Entrada Sandstone (Colbert, 1989, p. 13).

4. The top of the Apachean teilzone at Ghost Ranch cannot be placed, because *Protosuchus* is unknown from the Ghost Ranch area. The Apachean teilzone is therefore open ended (topless).

Estimated Holochronozones

The best current estimate and correlation of the *Otischalkian estimated holochronozone* (Fig. 14) is obtained as follows:

1. Out of all the nonphytosaurid phytosaur taxa in western North America that may define the base of the Otischalkian biozone, two specimens occur in the Camp Springs Formation (=Santa Rosa Sandstone *sensu* Lehman, 1994; Lehman and Chatterjee, 2005) at the base of the Dockum Group. These are the holotype of *Wannia scurriensis* (TTU P-00539; Langston, 1949; Stocker, 2013), and a skull assigned to *P. bransoni* (PPM 2217; Hunt and Lucas, 1991). At current levels of lithostratigraphic and chronostratigraphic resolution, these specimens can be considered to be about the same age, and to be stratigraphically lower (and therefore older) than other specimens of nonphytosaurid phytosaurs in the Dockum Group of Texas (Hunt and Lucas, 1991; Long and Murry, 1995; Martz, 2008; Martz et al., 2013). Therefore, the OKR approximating the base of the Otischalkian holochronozone occurs in the Camp Springs Conglomerate/Santa Rosa Formation. Note that specimens of *Parasuchus* from the Popo Agie Formation of Wyoming (Hunt and Lucas, 1991; Long and Murry, 1995) are excluded from consideration, because their stratigraphic positions and ages relative to the Camp Springs Conglomerate are unclear.

It should be noted that *Parasuchus* (*sensu* Kammerer et al., 2015), unlike the other phytosaur genera identified below, currently is considered to occur outside of western North America (Lucas et al., 2007a; Kammerer et al., 2015). If the Otischalkian estimated holochronozone is considered to be global in extent, the oldest known occurrence of the genus worldwide would technically define the base of the unit. However, it is unclear at this time exactly how the ages of the Indian-type species *Parasuchus hislopi* Lydekker 1885, and the German and Morrocan referred species *P. angustifrons* Kuhn 1936 and *P. magnoculus* Dutuit 1977 compare with that of the American species, *P. bransoni*. Correlations between these units have been made almost solely on vertebrate biostratigraphy, especially the presence of these basal phytosaurs (e.g., Hunt and Lucas, 1991; Lucas, 1998; Lucas et al., 2007a), making any statements about their relative ages based on the occurrence of *Parasuchus* is inherently circular (Irmis et al., 2010). However, Walsh (1998) indicated that biochronozones need not be global in scope, and could be specified to be restricted to a particular region. For the time being, the OKR of *Parasuchus* given here to define the base of the Otischalkian holochronozone is considered to be representative only of western North American occurrences.

2. Out of all the nonpseudopalatine leptosuchomorph taxa that may define the base of the Adamanian biozone, the oldest OKR is more difficult to determine. Numerous specimens of *Leptosuchus* have been recovered from the Tecovas Formation (Dockum Group) in the Canadian River Valley and southern Crosby County, Texas (Case, 1922; Gregory, 1957, 1972; Long and Murry, 1995; Heckert, 2004; Lehman and Chatterjee, 2005; Parker et al., 2008), but their exact stratigraphic positions within the formation are unclear because precise locality information is unavailable (Long and Murry, 1995, pp. 13–14). Only one specimen of *Leptosuchus* (TTU-P09234; Martz et al., 2013) is known from the lower Cooper Canyon Formation (Dockum Group) of Garza County, Texas (Fig. 11), occurring above a very poorly sampled interval (Martz, 2008; Martz et al., 2013). The two best candidates for the oldest known nonpseudopalatine leptosuchomorph both occur in northern Arizona: the specimen of *Smilosuchus* (TMM 43685-261, formerly assigned to "*Leptosuchus*" by Parker and Martz, 2011) from the Newspaper Rock Bed (Blue Mesa Member; Fig. 10), and the specimen of *Smilosuchus gregorii* from the Ward Bonebed (AMNH FR 3060; Colbert, 1947; Long and Murry, 1995; Fig. 13A) from the upper Cameron Member or Blue Mesa Member. The Cameron Member is probably correlative with the lower part of the Blue Mesa Member–type section at Petrified Forest National Park (Irmis et al., 2011, supplemental data), indicating that these specimens may be from nearly the same stratigraphic level. However, radioisotopic dates for this stratigraphic interval only are known at Petrified Forest National Park (Ramezani et al., 2011; Atchley et al., 2013), and in the vicinity of St. Johns, Arizona (Ramazani et al., 2014). TMM 43685-261 from Petrified Forest National Park is therefore somewhat arbitrarily considered the oldest OKR for any of the taxa defining

the base of the Adamanian biozone. It should be noted that the stratigraphic interval below both TMM 43685-261 and AMNH FR 3060 is poorly sampled, making placement of the base of the Adamanian teilzone in both areas sensitive to stratigraphically lower phytosaur discoveries. As a result, placement of the base of the Adamanian estimated biochronozone should be considered highly tentative.

3. Therefore, the best current estimate for the Otischalkian estimated holochronozone is the stratigraphic interval between the Camp Springs Conglomerate/Santa Rosa Formation and all age equivalent strata, and the upper part of the Newspaper Rock Bed of the Blue Mesa Member and all age-equivalent strata (Fig. 14). The Camp Springs Formation and Santa Rosa Formation of the Dockum Group in New Mexico and Texas, and the Shinarump Member of the Chinle Formation in northern Arizona, were probably deposited more or less contemporaneously by the same river system (e.g., Riggs et al., 1996), allowing tentative correlation of the holochronozone across Texas and into the Colorado Plateau. In western Texas, the top of the Otischalkian holochronozone may occur in the lower part of the Tecovas Formation (Heckert, 2004, p. 37) and the lower part of the lower unit of the Cooper Canyon Formation (Martz et al., 2013, Fig. 2), but biostratigraphically significant fossils from these stratigraphic intervals are poorly documented in the former case (e.g., the specimens of *Leptosuchus* described by Case, 1922, and Case and White, 1934), and poorly sampled in the latter (Martz, 2008; Martz et al., 2013). In northern Arizona, where nonphytosaurid phytosaurs are currently unknown, the Otischalkian holochronozone may extend roughly from the Shinarump Member to the middle of the Blue Mesa Member–type section and the upper parts of the Cameron and Bluewater Creek members (see Irmis et al., 2011, supplemental information).

The best current estimate and correlation of the *Adamanian estimated holochronozone* (Fig. 14) is obtained as follows:

1. As discussed previously, the best candidate for the oldest OKR among all taxa defining the base of the Adamanian is TMM 43685-261 from the upper Newspaper Rock Bed of the Blue Mesa Member (Chinle Formation) at Petrified Forest National Park (Fig. 10; Parker and Martz, 2011).

2. Compared to other phytosaur taxa, local lowest stratigraphic occurrences of *Machaeroprosopus* are relatively well documented and widespread, being known from the Trujillo and middle Cooper Canyon formations of western Texas (Fig. 11; Hunt, 2001; Martz et al., 2013), the Bull Canyon Formation of eastern New Mexico (Hunt, 2001), the upper Sonsela Member of northern Arizona (Parker and Martz, 2011), and the lower Petrified Forest Member or Mesa Montosa Member of northern New Mexico (Irmis et al., 2007; Parker and Martz, 2011; Zeigler et al., 2005). The best candidate for an OKR occurs in Petrified Forest National Park (Fig. 10), where the stratigraphically lowest known specimens of *Machaeroprosopus* are specimens of *M. pristinus* (PEFO 34042 and AMNH FR 7222) collected from the lower

Jim Camp Wash beds of the Sonsela Member (Chinle Formation) (Parker and Martz, 2011). This stratigraphic interval is extremely well sampled, with multiple localities at several stratigraphic levels within a few meters of these specimens demonstrating that *Machaeroprosopus* is probably absent lower in the section (Parker and Martz, 2011).

3. Therefore, the best current estimate for the Adamanian holochronozone is the stratigraphic interval between the Newspaper Rock Bed of the Blue Mesa Member and all age-equivalent strata, and the lowermost Jim Camp Wash beds of the Sonsela Member and all age-equivalent strata (Fig. 14). As the OKRs used to bound the biochronozone also happen to be the LO_ks for Leptosuchomorpha and *Machaeroprosopus* in Petrified Forest National Park, this makes the park the only local study area where the Adamanian teilzone and the Adamanian estimated holochronozone are precisely equivalent. In the Dockum Group of eastern New Mexico and western Texas, the stratigraphically equivalent interval can be estimated using both lithostratigraphy and biostratigraphy to extend from somewhere in the lower Tecovas, Garita Creek, and Cooper Canyon formations to the Trujillo and middle Cooper Canyon Formations (Lucas, 1993a,b; 1998; Dubiel, 1994; Heckert, 2004; Martz et al., 2013). The Trujillo Formation, middle Cooper Canyon, and Sonsela Member all represent sandstone-dominated units occurring in the middle of the Dockum Group and the Chinle Formation (e.g., Lucas, 1993a,b; Dubiel, 1994) that may represent regional tectonic uplift (e.g., Lucas, 1993a,b; Lehman and Chatterjee, 2005, p. 329; Martz, 2008, pp. 107–108; Howell and Blakey, 2013), and all contain the local lowest known occurrences of *Machaeroprosopus* (Hunt, 2001; Parker and Martz, 2011; Martz et al., 2013).

The best current estimate and correlation of the *Revueltian estimated holochronozone* (Fig. 14) is obtained as follows:

1. As discussed previously, the best candidate for the oldest OKR among all taxa defining the base of the Revueltian are specimens of *Machaeroprosopus pristinus* (PEFO 34042 and AMNH FR 7222) occurring in the lower Jim Camp Wash beds (Sonsela Member) of the Chinle Formation at Petrified Forest National Park (Fig. 10; Parker and Martz, 2011).

2. Identifying the OKR for "*Redondasaurus*" is considerably more problematic, partially due to the extreme subjectivity in determining which derived specimens of *Machaeroprosopus* should be assigned to the subgenus (Hungerbühler et al., 2013; Martz et al., 2014), and also because many characters considered diagnostic of the taxon (e.g., Spielmann and Lucas, 2012; Martz et al., 2014) have not been subjected to phylogenetic analysis. The stratigraphic positions of "*Redondasaurus*" specimens within the Redonda Formation in northeastern New Mexico, including the holotype of "*R.*" *gregorii* (YPM 3294), are placed fairly precisely within the uppermost Quay and Duke Ranch members (Spielmann and Lucas, 2012). The *Machaeroprosopus* specimens from the Patricia Site (TTU VPL 3870) in the Cooper Canyon Formation in western Texas, including the holotype and paratype of *Machaeroprosopus lottorum*

(TTU-P10076 and TTU-P10077, respectively; Hungerbühler et al., 2013; considered by the current authors to be referable to "*Redondasaurus*" based on the criteria recommended by Martz et al., 2014), are also precisely placed stratigraphically in the Macy Ranch sandstone near the top of the formation (Fig. 10). The holotype of "*Redondasaurus*" *bermani* (CM 69727) occurs within the "upper siltstone" member at Ghost Ranch (Fig. 13B). A specimen of *Machaeroprosopus* from Ward's Terrace in Arizona (MNA V3495; Fig. 12A) occurring about 80 m above the base of the Owl Rock Member was also assigned by Parker et al. (2011) to "*Redondasaurus*."

Unfortunately, there is extreme uncertainty about the precise lithostratigraphic correlations between these units containing the local LO_ks of "*Redondasaurus*." The uppermost Cooper Canyon Formation in Texas usually is considered to underlie the Redonda Formation of New Mexico (e.g., Lehman, 1994; Lucas et al., 1994), but parts of these units may be laterally equivalent (Martz, 2008), and the precise lithostratigraphic relationships between the upper units of the Chinle Formation and Dockum Group are not at all clear. The only case where relative superpositional relationships between these local study areas containing "*Redondasaurus*" may be known with some confidence is between Lisbon Valley and Ward's Terrace (Figs. 9, 12, and 13). The Lisbon Valley LO_k of "*Redondasaurus*" occurs above the Kane Springs beds (Fig. 12), which have been correlated lithostratigraphically to the Owl Rock Member (e.g., Stewart et al., 1972; Blakey and Gubitosa, 1983; Lucas et al., 1997b; Martz et al., 2014), placing the LO_k of "*Redondasaurus*" in Lisbon Valley higher than the LO_k of the taxon at Ward's Terrace. These uncertainties in stratigraphic correlation make selection of an OKR highly subjective; however, given that the lower ledge–forming carbonates of the Owl Rock Member have been radioisotopically dated to 207.8 Ma (Ramezani et al., 2011), the "*Redondasaurus*" specimen from slightly higher within that unit (MNA V3495; Parker et al., 2011) seems to be the most convenient choice for the OKR (Fig. 13A).

Other authors (e.g., Chavez, 2010; Spielmann and Lucas, 2012) assigned derived pseudopalatine specimens to "*Redondasaurus*" that occur at lower stratigraphic levels than those discussed earlier, but we do not consider those referable to the subgenus. TTU P-9425 (Martz, 2008, p. 191; Chavez, 2010) from the upper Cooper Canyon Formation several meters below the Macy Ranch sandstone (Martz, 2008; Chavez, 2010) is a derived pseudopalatine that shows a combination of characters found in *M. buceros* and "*Redondasaurus*" *gregorii* (Axel Hungerbühler, personal communication, 2015), but does not have the extremely widened squamosals with diverging medial margins used by Martz et al. (2014) to diagnose "*Redondasaurus*." UCMP V78034/119436 from the Petrified Forest Member of New Mexico has very broad squamosals, but the posterior ends are knob-like as in *M. buceros* and *M. pristinus*; moreover, the skull is in a poor state of preservation and is largely reconstructed (Axel Hungerbühler, personal communication, 2015; J.W.M. and W.G.P., personal observation).

3. Therefore, the best current estimate for the Revueltian holochronozone is the stratigraphic interval between the lower Jim Camp Wash beds of the Sonsela Member and all age-equivalent strata, and the lower ledge-forming carbonates of the Owl Rock Member and all age-equivalent strata. The lower Kane Springs beds are considered to fall in the upper part of the holochronozone on the basis of lithostratigraphic correlations with the Owl Rock Member (Blakey and Gubitosa, 1983, 1984; Martz et al., 2014), but lithostratigraphic correlations between the Owl Rock Member, the Chinle and Redonda formations of northern New Mexico, and the upper Cooper Canyon Formation of western Texas are not possible due to the lack of continuous outcrop between these areas. Tentative correlation of the top of the Revueltian estimated holochronozone must therefore depend on correlating the top of the Revueltian teilzones (i.e., the local lowest known occurrences of "*Redondasaurus*").

The best current estimate and correlation of the *Apachean estimated holochronozone* (Fig. 13) is obtained as follows:

1. As discussed, the best candidate for the OKR of "*Redondasaurus*" (MNA V3495) occurs in the lower Owl Rock Member of the Chinle Formation in northern Arizona (Parker et al., 2011).

2. The top of the Apachean holochronozone is the OKR of *Protosuchus richardsoni*, which occurs 9–16 m below the top of the Dinosaur Canyon Member of the Moenave Formation near Cameron in northern Arizona (Figs. 13A and 14; Clark and Fastovsky, 1986).

3. Therefore, the best current estimate for the Apachean holochronozone is the stratigraphic interval between the lower Owl Rock Member in northern Arizona and all age-equivalent strata, and the uppermost Dinosaur Canyon Formation in northern Arizona and all age-equivalent strata. Across the Four Corners region, the top of the Apachean biochronozone probably occurs in the lower Wingate Sandstone Formation, which is correlative with the Moenave Formation (e.g., Blakey, 1994). The "upper siltstone member" of the Chinle Formation in northern New Mexico is truncated and unconformably capped by the Middle Jurassic Entrada Sandstone (e.g., Colbert, 1989; Irmis et al., 2007), and the Dockum Group of eastern New Mexico and western Texas is also truncated unconformably and overlain by Cretaceous and Cenozoic deposits (e.g., Barnes et al., 1993, 1994). The top of the Apachean estimated holochronozone in these areas is therefore the disconformity at the top of these truncated Upper Triassic units.

Estimated Holochrons

The best estimate for the *Otischalkian estimated holochron* (Fig. 14) is as follows. As explained previously, the Camp Springs Formation/Santa Rosa Formation of western Texas, which contains the OKR of *Wannia* and *Parasuchus*, probably correlates with the Shinarump Member of Arizona, which is also correlative with the Mesa Redondo Member (Irmis et al., 2011, supplemental data; Riggs et al., 2016). A radioisotopic date of about 227.6 Ma is available from the lower Mesa Redondo Formation (Atchley et al., 2013). In Petrified Forest National Park (Fig. 10), the Newspaper Rock Bed of the Blue Mesa Member, which contains the OKR of *Smilosuchus*, has produced a radioisotopic age of about 224 Ma (Ramezani et al., 2011, 2014). This gives an estimated duration for the Otischalkian holochron of about 227–224 Ma (about 3 Ma), which is early Lacian (earliest early Norian) (Muttoni et al., 2004; Furin et al., 2006; Hüsing et al., 2011).

The best estimate for the *Adamanian estimated holochron* (Fig. 14) is as follows. As explained previously, the OKR of *Smilosuchus* is about 224 Ma. A date of about 215 Ma for the lower Jim Camp Wash beds in Petrified Forest National Park (Dunlavey et al., 2009) is consistent with other published dates for the Sonsela Member from the park of about 220–213 Ma (Ramezani et al., 2011, 2014; Atchley et al., 2013; Nordt et al., 2015; Fig. 9). As the OKR of *Machaeroprosopus* occurs in the lower Jim Camp Wash beds (Parker and Martz, 2011), this gives an estimated duration for the Adamanian holochron of about 224–215 Ma (about 9 Ma), which extends from the late Lacian (late early Norian) to possibly the earliest Sevatian (early middle Norian) (Muttoni et al., 2004; Hüsing et al., 2011). The possible unnamed upper subdivision of the Adamanian holochron proposed by Martz et al. (2013, p. 359) may have had a duration of about 220–215 Ma (about 5 Ma), although this is poorly constrained because there is not yet a precise lower boundary definition for this putative estimated subholochron.

The best estimate for the *Revueltian estimated holochron* (Fig. 14) is as follows. As explained earlier, the lower Jim Camp Wash beds, which contain the OKR of *Machaeroprosopus*, have been dated to 215 Ma (Dunlavey et al., 2009). The OKR of "*Redondasaurus*" in the lower part of the Owl Rock Member at Ward Terrace (Parker et al., 2011) came from resistant ledge-forming carbonate beds in the upper part of the member (Kirby, 1991). In Petrified Forest National Park, a radioisotopic date of about 207.8 Ma (Ramezani et al., 2011; W.G. Parker, unpublished data) was produced from immediately below the lowermost ledge-forming carbonate beds, suggesting that the LO_k of "*Redondasaurus*" at Ward's Terrace is younger. This gives a duration for the Revueltian biochron of about 215–207 Ma (about 8 Ma), which extends from the earliest Sevatian (early middle Norian) to the Rhaetian (Muttoni et al., 2004; Hüsing et al., 2011).

The best estimate for the *Apachean estimated holochron* (Fig. 14) is as follows. As has been discussed, the OKR of "*Redondasaurus*" may be about 207 Ma. No radioisotopic dates are available from the lower Dinosaur Canyon Member of northern Arizona, which contains the OKR *Protosuchus richardsoni* in western North America (e.g., Sues et al., 1996). The Triassic-Jurassic boundary, which is well calibrated at 201.3 Ma (Schoene et al., 2010), may lie between the Dinosaur Canyon Member and the overlying Whitmore Point Member (e.g., Kirkland et al., 2014) or even higher in the Whitmore Point

Member (Donohoo-Hurley et al., 2010). Moreover, *Protosuchus micmac* from the McCoy Brook Formation of Nova Scotia (Sues et al., 1996) may be latest Triassic (Rhaetian) based on magnetostratigraphy (Whiteside et al., 2007, 2011). The OKR of *Protosuchus* may therefore slightly pre-date the Triassic-Jurassic boundary, giving an estimated duration for the Apachean holochron of about 207–202 Ma (about 5 Ma).

DISCUSSION

Interpreting the Discrepancy Between Teilzones and Holochronozones

The vertebrate fossil record is, at best, a fragmentary trace of the biological reality of the past preserved through inconstant and highly selective taphonomic biases, and biostratigraphy is as distorted by taphonomic processes as any other branch of paleontology (e.g., Behrensmeyer et al., 2000). The stratigraphic distribution of vertebrate fossils in fluvial deposits is sporadic, varies between parts of the channel system (Behrensmeyer, 1987, 1988), and is influenced by multiple other variables that include tectonic uplift, climate, rate of basin subsidence, stage in the history of basin filling, local geochemical conditions, environmental catastrophes such as drought, flooding, and wildfires, and the size, anatomy, biochemistry, and behavior of organisms (e.g., Behrensmeyer and Kidwell, 1985; Aslan and Behrensmeyer, 1996; Behrensmeyer et al., 2000; Carpenter, 2005; Rogers and Kidwell, 2007; Brown et al., 2013). Moreover, fossil collecting exerts its own particular biases that further skew our already fragmentary perception of the past (e.g., McKenna and Lillegraven, 2006; Dunhill et al., 2012; Hart, 2012).

Our incomplete perception of the fossil record must always be remembered when considering the apparent biostratigraphic ranges of organisms, and therefore the demarcation of lowest and/or highest occurrences that bound eubiostratigraphic units (McKenna and Lillegraven, 2005, 2006), and by extension, estimates of eubiochronozones and eubiochrons derived from them. We can never be certain that the known lowest stratigraphic occurrence of a taxon in a particular local study area approximates, or even approaches, the stratigraphic level at which the taxon first appeared in the local area, and the region. To put it another way, there will almost certainly always be a discrepancy between a local teilzone and the estimated holochronozone, and between the estimated (epistemological) holochronozone and the actual (ontological) holochronozone.

However, one goal of biostratigraphy is to construct empirically determined stratigraphic ranges for fossil taxa that are densely sampled enough to approximate the true ranges (e.g., teilzones that approximate holochronozones). This goal is aided by the detailed plotting of biostratigraphic data, which reveals stratigraphic intervals in which fossils are poorly sampled, encouraging concentrated collecting efforts within those intervals. In other words, the discrepancy between teilzones and holochronozones identifies the stratigraphic intervals in local study areas where improved collecting may be predicted to extend the biostratigraphic ranges of particular alpha taxa by identifying new LO_ks that occur lower in the section than currently known specimens.

This current study provides several examples of where identifying such intervals is possible. For example, the base of the Otischalkian holochronozone in northern Arizona is here estimated to lie well below known diagnostic phytosaur occurrences in the state, and indeed the taxa delimiting the base of the Otischalkian teilzone are completely unknown in Arizona. This indicates that nonphytosaurid phytosaurs may be discovered in the Shinarump and Mesa Redondo members, if the proposed correlations of these units to each other (Irmis et al., 2011, supplemental data; Riggs et al., 2016), and to the Camp Springs Conglomerate/Santa Rosa Formation (e.g., Lucas, 1993a,b; Dubiel, 1994; Riggs et al., 1996), are correct.

Likewise, the base of both the Revueltian and Apachean holochronozones in Lisbon Valley (Martz et al., 2014) are estimated here to occur well below the LO_ks of *Machaeroprosopus* and "*Redondasaurus*" (i.e., well below the Revueltian and Apachean teilzones in Lisbon Valley), indicating that fossils of these taxa may be discovered in the lower Church Rock Member and Kane Springs beds with improved sampling. The incompleteness of the Lisbon Valley stratigraphic record is highlighted by the fact that the current data places the Kane Springs beds within the Adamanian teilzone, in spite of the fact that, based on lithostratigraphic correlations of the Kane Springs beds to the Owl Rock Member, the Adamanian holochronozone is probably completely absent there!

This current study also demonstrates that the discrepancy between teilzones and estimated holochronozones may be exacerbated when the ranges of boundary-defining taxa overlap because this creates the possibility of the LO_ks for boundary-defining taxa to occur out of sequence. For example, at Ward Terrace near Cameron, the lowest known occurrence of "*Redondasaurus*" is below that of *Machaeroprosopus* within the Owl Rock Member. Until additional specimens of *Machaeroprosopus* lacking the diagnostic characters of "*Redondasaurus*" are found lower in the section, ignoring the LO_k of *Machaeroprosopus* that cannot be assigned to "*Redondasaurus*" and treating the Revueltian as bottomless (as has been done here) is the only way to avoid this problem. Stratigraphic overlap between "*Redondasaurus*" and specimens of *Machaeroprosopus* not referable to "*Redondasaurus*" has also been documented in the Church Rock Member at Lisbon Valley (Martz et al., 2014), and the upper Cooper Canyon Formation in western Texas (Martz, 2008), demonstrating that the possibility of lowest occurrences for occurring out of the expected stratigraphic sequence when biostratigraphic data is incomplete exists in other areas.

To add further uncertainty, estimated holochronozones are cobbled together using these heavily biased local teilzones, and are therefore themselves almost certainly shorter than the true ontological holochronozones. This is particularly evident when considering the Adamanian teilzone in both Petrified Forest National Park, the only local study area where the Adamanian teilzone and Adamanian estimated holochronozone are precisely equivalent (i.e., the LO_ks of both *Smilosuchus* and *Machaeroprosopus* within PEFO are also the OKRs). In spite of having the most completely known biostratigraphic range data for *Smilosuchus*, no phytosaur specimens assignable to alpha taxa are known from below the LO_k of *Smilosuchus* in northern Arizona. It is therefore not at all clear how closely the

OKR for that taxon may approximate the stratigraphic level at which it truly appeared. The problem is nearly as severe in southern Garza County, where only a single specimen of *Leptosuchus* is known, and there is an enormous stratigraphic gap separating this specimen from the nonphytosaurid phytosaurs delineating the base of the Otischalkian teilzone lower in the section (Martz, 2008; Martz et al., 2013). In both cases, placement of the base of the Adamanian holochronozone may be considerably lowered by improved sampling.

Improving the Analysis of Biostratigraphic Data

In recent decades, several methods have been developed for evaluating the quality of biostratigraphic data through the calculation of confidence intervals (Strauss and Sadler, 1989; Marshall, 1990; Solow, 1996), including variants that consider possible nonrandom stratigraphic distributions of fossil taxa (Marshall, 1994, 1997; Holland, 2003). These methods have also been applied toward evaluating possible relationships between the extinction of taxa and environmental changes. Several authors (Springer, 1990; Marshall, 1995; Solow and Smith, 2000; Wang et al., 2012; Alroy, 2014) have evaluated putative mass extinctions, calculating the likelihood that the true ranges of multiple taxa have a common termination. Of particular note is the work of Barry et al. (2002), which quantitatively explored the limitations of known biostratigraphic data on our understanding of the relationships between faunal change and environmental change. Unfortunately, such quantitative methods are rarely, if ever, applied to Mesozoic terrestrial vertebrates.

An extinction event involving multiple taxa has been alleged to occur in the Colorado Plateau region at the transition between the Adamanian and Revueltian holochrons (Parker and Martz, 2011), and a far more devastating global catastrophe is thought to have occurred at or near the Triassic-Jurassic boundary (e.g., Olsen et al., 2002; Whiteside et al., 2007). These events, if they are indeed real, have important ramifications for understanding the early development of Mesozoic ecosystems. Unfortunately, the ability of a capricious vertebrate fossil record to create the illusion of biostratigraphic order has been well established (McKenna and Lillegraven, 2005, 2006), and fossil sampling density across both the Adamanian-Revueltian turnover and the Triassic-Jurassic event leave much to be desired. Quantitative methods coupled with thorough sampling regimens will help ascertain whether or not these transitions truly represent abrupt faunal turnovers, or are merely artifacts of the incomplete fossil record.

Expansion of Systematically Informative Characters in Phytosaurs

For the past 25 years, the alpha taxonomy of phytosaurs has been based almost exclusively on cranial material, specifically for the posterior region of the skull (e.g., Ballew, 1989; Hunt and Lucas, 1991; Long and Murry, 1995; Irmis, 2005; Parker and Martz, 2011; Stocker and Butler, 2013). Even increasingly detailed phylogenetic analyses of phytosaurs have not greatly improved this situation, at best identifying more cranial characters that may be diagnostic for alpha taxa (e.g., Hungerbühler, 2002; Stocker, 2010, 2012, 2013; Hungerbühler et al., 2013; Butler, 2013; Kammerer et al., 2015). Only a handful of authors have attempted to identify mandibular and/or postcranial variation between phytosaur alpha taxa (e.g., Camp, 1930; Hunt, 1994; Lucas et al., 2002), and potential character variations have never been rigorously explored using detailed comparative descriptions or phylogenetic analyses restricted to Phytosauria. Interestingly, similar biases in which particular elements are used disproportionately to diagnose taxa exist for other Upper Triassic vertebrates; aetosaur alpha taxa are identified almost exclusively using osteoderms (e.g., Long and Ballew, 1985; Heckert and Lucas, 2000; Parker, 2007, 2015), and large metoposaurid alpha taxa are identified almost exclusively using the position of the lacrimal relative to the orbit (Hunt, 1993; Long and Murry, 1995), although a few studies suggest that other cranial and postcranial characters in metoposaurids may prove systematically informative (e.g., Colbert and Imbrie, 1956; Long and Murry, 1995; Sulej, 2002, 2007; Brusatte et al., 2015).

This is a regrettable situation given that phytosaur specimens that include the posterior region of the skull represent only a fraction of all phytosaur specimens collected. Numerous phytosaur postcrania exist, but presently they cannot be assigned to an alpha taxon unless they co-occur with diagnostic cranial material. Complicating matters further is that at face value phytosaur postcranial and mandibular characters appear to be conserved throughout the entire clade (M. Stocker, personal communication, 2016); However, this has not been tested phylogenetically beyond the analysis of Nesbitt (2011), which only incorporated three phytosaur alpha taxa as it was primarily concerned with phylogenetic relationships across Archosauriformes rather than resolving phytosaur topology. The ability to assign phytosaur mandibles and postcrania to alpha taxa would be enormously beneficial to Upper Triassic vertebrate biostratigraphy, vastly extending biostratigraphic sample sizes and potentially biostratigraphic ranges for phytosaur alpha taxa. For example, in Lisbon Valley, phytosaur postcranial elements are known from below the stratigraphically lowest specimens that can be assigned to the alpha taxa *Machaeroprosopus* and "*Redondasaurus*" (Martz et al., 2014). The assignment of these postcranial elements to alpha taxa could conceivably lower the base of the Revueltian and/or Apachean teilzones, bringing the boundaries of the teilzones closer to those of the holochronozones. Detailed descriptions of phytosaur mandibular and postcranial elements, and their incorporation into phylogenetic analyses, should be considered priorities for phytosaur research.

Improving Biozone, Biochronozone, and Biochron Definition and Characterization

Biozone boundary concepts can change with improved sampling (e.g., McKenna and Lillegraven, 2006). As discussed, previously proposed biostratigraphic subdivisions of the Revueltian and Adamanian land vertebrate biozones (Hunt, 2001; Hunt et al.,

2005) have been based on erroneous taxonomic identifications and poorly resolved biostratigraphic data (see discussions in Parker, 2006; Parker and Martz, 2011). Nonetheless, the recent detailed biostratigraphic studies discussed here have revealed that stratigraphic variations in vertebrate taxa within biozones may occur (e.g., Martz et al., 2013, 2014; Parker, 2014), making the identification of subbiozones, subbiochronozones, and subbiochrons a possibility.

For example, Martz et al. (2013) noted that it may be possible to subdivide the Adamanian into upper and lower subbiozones characterized by slightly different vertebrate faunal assemblages, with several taxa, including the nonpseudopalatine leptosuchomorphs *S. lithodendrorum* and *Pravusuchus hortus* (Stocker, 2010), and the aetosaurs *Paratypothorax* sp., *Typothorax coccinarum*, and *Desmatosuchus smalli* (Parker, 2005) being confined to the upper subbiozone (Parker and Martz, 2011). *Scutarx delatylus*, a new aetosaur taxon closely related to *Calyptosuchus wellesi*, may also be restricted to the upper part of the Adamanian (Parker, 2014, 2016). Moreover, it is becoming clear that morphological variation within the genus *Machaeroprosopus* is far greater than was once realized, with a wide range of forms existing that are intermediate in morphology and stratigraphic position between the type species *M. buceros* and *Machaeroprosopus* ("*Redondasaurus*") *gregorii* (Hungerbühler et al., 2013; Martz et al., 2014). It is possible that, with better taxonomic resolution, it may be possible to identify subbiozones, subbiochronozones, and subbiochrons for the Revueltian and/or Apachean based on different species or clades within *Machaeroprosopus*.

Phytosaurs are far from being the only Upper Triassic vertebrates with biostratigraphic utility. Aetosaurs also have a long history as biostratigraphically useful taxa (e.g., Long and Ballew, 1985; Long and Padian, 1986; Heckert and Lucas, 2000; Desojo et al., 2013). Aetosaurs are extremely abundant in Upper Triassic deposits, and indeed locally they are even more common than phytosaur material, as at the Placerias Quarry in northern Arizona (e.g., Long and Murry, 1995; Parker, 2007) and the Post Quarry in western Texas (Small, 1989; Martz et al., 2013). Aetosaur osteoderms can often be assigned to particular alpha taxa (e.g., Heckert and Lucas, 2000; Parker, 2007), provided that they are sufficiently complete to assess character states other than ornamentation (Martz and Small, 2006). Lucas and Heckert (1996) and Lucas and Huber (2003, p. 147) even suggested for these reasons that aetosaurs are preferable Upper Triassic index fossils compared to phytosaurs and metoposaurids. However, it should be noted that osteoderms have much greater taxonomic utility for desmatosuchine and typothoracisine aetosaurs (*sensu* Parker, 2007; Desojo et al., 2013) than for aetosaur taxa falling outside of these clades because these taxa are united only by a radial ornamentation pattern that is plesiomorphic for the clade (e.g., Desojo and Ezcurra, 2011; Desojo et al., 2013; Small and Martz, 2013; Parker, 2014, 2015). Although aetosaur-based biozone definitions can be applied no more readily than phytosaur-based biozone definitions without detailed regional biostratigraphic range charts, the Late Triassic land vertebrate biochrons all contain characteristic aetosaur taxa (e.g., Heckert and Lucas, 2000; Heckert et al., 2007; Parker and Martz, 2011) that could hypothetically be used for biostratigraphy and biochronology, although it requires a evaluating biostratigraphic data on the LO_ks of aetosaur taxa with the same degree of care applied to here to the LO_ks of phytosaur taxa.

The issue of poor sample sizes also has important ramifications for developing a more detailed and meaningful biozonation. As discussed previously, possible endemic alpha taxa present difficulty in correlating biozones and biochronozones bounded by single-taxa, as many taxa are known from geographically restricted areas. With improved sampling, it is possible that these apparently endemic taxa may be discovered across a wider range. This would not only allow biozones and biochronozones to be subdivided, but potentially permit existing multiple-taxon biozone, biochronozone, and biochron definitions to be modified into units bounded by single-taxa. Alternately, more detailed treatment of known specimens may demonstrate that endemism is even more pronounced than currently appreciated (as occurred in the case of specimens traditionally lumped into "*Rutiodon*" and "*Leptosuchus*"; Stocker, 2010). This may serve to entrench the multiple alpha taxon-based definitions advocated here, or alternately lead to the establishment of distinct biozonations and biochronology in different parts of the western United States. In carefully articulating exactly how and why biozone and biochron definitions are defined and applied, we hope to make it easier to adapt the system to future discoveries and insights.

CONCLUSIONS

The Late Triassic land vertebrate "faunachrons" (the Otischalkian, Adamanian, Revueltian, and Apachean) are reformulated as follows, with the term "faunachron" being abandoned:

1. The "faunachrons" are defined by the stratigraphically lowest and phylogenetically basal-most representatives of three major phytosaurian clades (Phytosauria, Leptosuchomorpha, and Pseudopalatinae), one dubious phytosaur "subgenus" ("*Redondasaurus*"), and one crocodyliform genus (*Protosuchus*). Given the potentially endemic nature of some phytosaur alpha taxa, and the instability of particular nodes within phytosaur phylogenies, these multiple-taxon definitions are essential for allowing correlation of biozones and biochronozones across the western United States.

2. The "faunachrons" are broken into three distinct units bounded by these taxa: a teil-zone (a biostratigraphic unit), an estimated holochronozone (a chronostratigraphic unit), and an estimated holochron (a biochronologic unit). These units allow us to discuss respectively the known biostratigraphic data for local study areas, the best estimate for the true interval of strata between the appearance of phytosaur clades, and the best estimate for the timing of the appearance of these clades in western North America.

The best current estimates for the Late Triassic land vertebrate estimated holochrons and Upper Triassic land vertebrate holochronozones (the sedimentary strata deposited over these intervals) are as follows (Fig. 14):

1. The Otischalkian estimated holochron extends from about 227 to 224 Ma (an interval of 3 million years) during the earliest Lacian (earliest early Norian), beginning with the earliest appearance of Phytosauria in the western United States. The Otischalkian estimated holochronozone encompasses the Camp Springs Conglomerate, Santa Rosa Formation, Colorado City Member, the lowermost Cooper Canyon, the lowermost Tecovas Formation, and the lowermost Garita Creek Formation of the Dockum Group, and the Shinarump Member, Mesa Redondo Member, lowermost Cameron Member, and the lowermost Blue Mesa Member type section of the Chinle Formation.

2. The Adamanian estimated holochron extends from about 224 to 215 Ma (an interval of about 9 million years) during the early Lacian and earliest Alaunian (early Norian to earliest middle Norian), beginning with the earliest appearance of Leptosuchomorpha in the western United States. The Adamanian estimated holochronozone encompasses most of the lower and part of the middle Cooper Canyon Formation, most of the Tecovas Formation, most of the Garita Creek Formation, and the lower Trujillo Formation of the Dockum Group, and most of the Cameron Member, most of the Blue Mesa Member–type section, and the lower Sonsela Member of the Chinle Formation.

3. The Revueltian estimated holochron extends from about 215 to 207 Ma (an interval of about 8 million years) during the Alaunian and Sevatian (middle and late Norian), beginning with the appearance of Pseudopalatinae and the genus *Machaeroprosopus* in the western United States. The Revueltian estimated holochronozone encompasses the upper part of the middle Cooper Canyon Formation, the upper Trujillo Formation, the upper Cooper Canyon Formation, the Bull Canyon Formation, and the Redonda Formation of the Dockum Group, and the upper Sonsela Member, the Petrified Forest (Painted Desert) Member, and possibly the lower parts of the Owl Rock Member and Kane Springs beds of the Chinle Formation.

4. The Apachean estimated holochron extends from about 207 to 202 Ma (an interval of about 5 million years) during the Rhaetian, beginning with the appearance of the subgenus "*Redondasaurus*," and ending with the appearance of the crocodyliform *Protosuchus*. The Apachean holochronozone encompasses most of the Redonda Formation of the Dockum Group, most of the Owl Rock Member, Rock Point Member, and Church Rock Member and at least the upper part of the "upper siltstone member" of the Chinle Formation, and the lower Moenave Formation of the Glen Canyon Group.

The estimation of these units from raw stratigraphic data highlights several problems with the current nature of Upper Triassic vertebrate fossil sample sizes, vertebrate systematics, and the number of available radioisotopic age dates. We must continue to improve our biostratigraphic data sets and begin to evaluate the quality of these data quantitatively. Otherwise, like McKenna and Lillegraven's (2006) fictitious Minister of Biostratigraphy, we risk discovering that our biochronologic models are houses of cards distorting our understanding of faunal succession.

ACKNOWLEDGMENTS

The tremendous database of fossil material that made this paper possible is the result of years of labor by innumerable fossil collectors. We thank in particular those individuals, too numerous to name, who collected the material that was the focus of our case studies in Utah, Arizona, and Texas. We also thank the reviewers Michelle Stocker, Barry Albright, and Michael Woodburne for their extremely helpful reviews, which allowed us to greatly improve the manuscript. This is Petrified Forest Paleontological Contribution #41.

REFERENCES

Alroy, J., 2014. Accurate and precise estimates of origination and extinction rates. Paleobiology 40 (3), 374–397.

Angielczyk, K.D., Kurkin, A.A., 2003. Has the utility of *Dicynodon* for Late Permian terrestrial biostratigraphy been overstated? Geology 31 (4), 363–366.

Archibald, D.J., Gingerich, P.D., Lindsay, E.H., Clemens, W.A., Krause, D.W., Rose, K.D., 1987. First North American land mammal ages of the Cenozoic era. In: Woodburne, M.O. (Ed.), Cenozoic Mammals of North America: Geochronology and Biostratigraphy. University of California Press, Berkeley, CA, pp. 18–23.

Arkell, W.J., 1933. The Jurassic System in Great Britain. Clarendon Press, Oxford.

Aslan, A., Behrensmeyer, A.K., 1996. Taphonomy and time resolution of bone assemblages in a contemporary fluvial system: the East Fork River, Wyoming. Palaios 11, 411–421.

Atchley, S.C., Nordt, L.C., Dworkin, S.I., Ramezani, J., Parker, W.G., Ash, S.R., Bowring, S.A., 2013. A linkage among Pangean tectonism, cyclic alluviation, climate change, and biologic turnover in the Late Triassic: the record from the Chinle Formation, southwestern United States. J. Sediment. Res. 83, 1147–1161.

Ballew, K.L., 1989. A phylogenetic analysis of Phytosauria from the late Triassic of the western United States. In: Lucas, S.G., Hunt, A.P. (Eds.), Dawn of the Age of Dinosaurs in the American Southwest. New Mexico Museum of Natural History, Albuquerque, NM, pp. 309–339.

Barnes, V.E., Eifler Jr., G.K., Frye, J.C., Leonard, A.B., Hentz, T.F., 1993. Geologic Map of Texas, Lubbock Sheet (Revised). 1:250,000. Texas Bureau of Economic Geology, The University of Texas, Austin, TX.

Barnes, V.E., Eifler Jr., G.K., Frye, J.C., Leonard, A.B., Hentz, T.F., 1994. Geologic Map of Texas, Big Springs Sheet (Revised). 1:250,000. Texas Bureau of Economic Geology, The University of Texas, Austin, TX.

Barry, J.C., Morgan, M.E., Flynn, L.J., Pilbeam, D., Behrensmeyer, A.K., Raza, S.M., Khan, I.A., Badgley, C., Hicks, J., Kelley, J., 2002. Faunal and environmental change in the Miocene Siwaliks of northern Pakistan. Paleobiology 28 (2), 1–71.

Behrensmeyer, A.K., 1982. Time resolution in fluvial vertebrate assemblages. Paleobiology 8 (3), 211–227.

Behrensmeyer, A.K., 1987. Taphonomy and the fossil record. Am. Sci. 72, 558–566.

Behrensmeyer, A.K., 1988. Vertebrate preservation in fluvial channels. Palaeogeogr. Palaeoclimatol. Palaeoecol. 63, 183–199.

Behrensmeyer, A.K., 1991. Terrestrial vertebrate accumulations. In: Allison, A., Briggs, D.E.G. (Eds.), Taphonomy: Releasing Data Locked in the Fossil Record. Plenum Press, New York, NY, pp. 291–335.

Behrensmeyer, A.K., Barry, J.C., 2005. Biostratigraphic surveys in the Siwaliks of Pakistan: a method for standardized surface sampling of the vertebrate fossil record. Palaeontol. Electron. 8 (1). 24 p.

Behrensmeyer, A.K., Kidwell, S.M., 1985. Taphonomy's contributions to paleobiology. Paleobiology 11 (1), 105–119.

Behrensmeyer, A.K., Kidwell, S.M., Gastaldo, R.A., 2000. Taphonomy and paleobiology. In: Erwin, D., Wing, S. (Eds.), Deep Time, Paleobiology's Perspective. Paleobiology, vol. 26 (Suppl. 4), University of Chicago Press, Chicago, pp. 103–147.

Berggren, W.A., Van Couvering, J.A., 1978. Biochronology. In: Cohee, G.V., Glaessner, M.F., Hedberg, H.D. (Eds.), Contributions to the Geologic Time Scale. American Association of Petroleum Geologists Studies in Geology, vol. 6, American Association of Petroleum Geologists, Tulsa, OK, pp. 39–55.

Berry, W.B.N., 1966. Zones and zones-with exemplification from the Ordovician. Bull. Am. Assoc. Pet. Geol. 50 (7), 1487–1500.

Blakey, R.C., 1994. Paleogeographic and tectonic controls on some lower and middle Jurassic erg deposits, Colorado Plateau. In: Caputo, M.V., Peterson, J.A., Franczyk, K.J. (Eds.), Mesozoic Systems of the Rocky Mountain Region, USA. Rocky Mountain Section, SEPM, Denver, CO, pp. 273–298.

Blakey, R.C., Gubitosa, R., 1983. Late Triassic paleogeography and depositional history of the Chinle Formation, southern Utah and northern Arizona. In: Reynolds, M.W., Dolly, E.D. (Eds.), Mesozoic Paleogeography of West-Central United States. Rocky Mountain Section, Society of Economic Paleontologists and Mineralogists, Denver, CO, pp. 57–76.

Blakey, R.C., Gubitosa, R., 1984. Controls of sandstone body geometry and architecture in the Chinle Formation (Upper Triassic), Colorado Plateau. Sediment. Geol. 38, 51–86.

Brown, C.M., Evans, D.C., Campione, N.E., O'Brien, L.J., Eberth, D.A., 2013. Evidence for taphonomic size bias in the Dinosaur Park Formation (Campanian, Alberta), a model Mesozoic terrestrial alluvial-paralic system. Palaeogeogr. Palaeoclimatol. Palaeoecol. 372, 108–122.

Brusatte, S.L., Butler, R.J., Mateus, O., Steyer, J.S., 2015. A new species Metoposaurus from the Late Triassic of Portugal and comments on the systematics and biogeography of metoposaurid temnospondyls. J. Vertebr. Paleontol. 35 (3). e912988-1–e912988-23.

Butler, R.J., 2013. "Francosuchus" trauthi is not Paleorhinus: implications for Late Triassic vertebrate biostratigraphy. J. Vertebr. Paleontol. 33 (4), 858–864.

Camp, C.L., 1930. A study of the phytosaurs with description of new material from western North America. Memoirs of the University of California, vol. 10, University of California Press, Berkeley, pp. 1–174.

Carpenter, K., 2005. Experimental investigation of the role of bacteria in bone fossilization. Neues Jb. Geol. Paläontol. Monat. 2005 (2), 83–94.

Case, E.C., 1920. Preliminary description of a new suborder of phytosaurian reptiles with a description of a new species of Phytosaurus. J. Geol. 28, 524–535.

Case, E.C., 1922. New reptiles and stegocephalians from the Upper Triassic of western Texas. Carnegie Institution of Washington Publication, vol. 321, Carnegie Institution of Washington, Washington, DC, pp. 1–84.

Case, E.C., 1929. Description of the skull of a new form of phytosaur, with notes on the characters of described North American phytosaurs. Memoirs of the University of Michigan Museums, Museum of Paleontology, vol. 2, University of Michigan, Ann Arbor, pp. 1–56.

Case, E.C., White, T.E., 1934. Two new specimens of phytosaurs from the Upper Triassic of western Texas. Contributions from the Museum of Paleontology, University of Michigan, vol. 4, no. 9, University of Michigan Press, Ann Arbor, pp. 133–142.

Cather, S.M., Zeigler, K.E., Mack, G.H., Kelley, S.A., 2013. Toward standardization of Phanerozoic stratigraphic nomenclature in New Mexico. Rocky Mount. Geol. 48 (2), 101–124.

Chatterjee, S., 1978. A primitive parasuchid (phytosaur) reptile from the Upper Triassic Maleri Formation of India. Palaeontology 21 (1), 83–127.

Chatterjee, S., 1986. The Late Triassic Dockum vertebrates: their stratigraphic and paleobiogeographic significance. In: Padian, K. (Ed.), The Beginning of the Age of Dinosaurs: Faunal Change Across the Triassic-Jurassic Boundary. Cambridge University Press, Cambridge, pp. 161–169.

Chavez, C., 2010. A New Phytosaur (Archosauria: Pseudosuchia) From the Late Triassic Dockum Group of Texas (M.S. thesis). Texas Tech University, Lubbock, TX. 77 p.

Cifelli, R.L., Eberle, J.J., Lofgren, D.L., Lillegraven, J.A., Clemens, W.A., 2004. In: Woodburne, M.O. (Ed.), Late Cretaceous and Cenozoic Mammals of North America. Columbia University Press, New York, NY, pp. 21–42.

Clark, J.M., Fastovsky, D.E., 1986. Vertebrate biostratigraphy of the Glen Canyon Group in northern Arizona. In: Padian, K. (Ed.), The Beginning of the Age of Dinosaurs: Faunal Change Across the Triassic-Jurassic Boundary. Cambridge University Press, Cambridge, pp. 285–301.

Colbert, E.H., 1947. Studies of the phytosaurs *Machaeroprosopus* and *Rutiodon*. Bull. Am. Mus. Nat. Hist. 88 (2), 1–96.

Colbert, E.H., 1952. A pseudosuchian reptile from Arizona. Bull. Am. Mus. Nat. Hist. 99 (10), 561–592.

Colbert, E.H., 1985. The Petrified Forest National Park and its vertebrate fauna in Triassic Pangaea. Mus. North. Ariz. Bull. 54, 33–43.

Colbert, E.H., 1989. The Triassic dinosaur *Coelophysis*. Mus. North. Ariz. Bull. 57. 160 p.

Colbert, E.H., Gregory, J.T., 1957. Correlation of continental Triassic sediments by vertebrate fossils. In: Reeside Jr., J.B., Applin, P.L., Colbert, E.H., Gregory, J.T., Hadley, H.D., Kummel, B., Lewis, P.J., Love, J.D., Maldonado-Koerdell, M., McKee, E.D., McLaughlin, D.B., Muller, S.W., Reinemund, J.A., Rodgers, J., Sanders, J., Silbering, N.J., Waagé, K. (Eds.), Correlation of the Triassic Formations of North America Exclusive of Canada. In: Bulletin of the Geological Society of America, vol. 68, United States Geological Survey, Washington, DC, pp. 1451–1514.

Colbert, E.H., Imbrie, J., 1956. Triassic metoposaurid amphibians. Bull. Am. Mus. Nat. Hist. 110 (6), 399–452.

Colbert, E.H., Mook, C.C., 1951. The ancestral crocodilian *Protosuchus*. Bull. Am. Mus. Nat. Hist. 97 (3), 149–182.

Cooley, M.E., 1958. The Mesa Redondo member of the Chinle formation, Apache and Navajo Counties Arizona. Plateau 31 (1), 7–15.

Cope, E.D., 1881. *Belodon* in New Mexico. Am. Nat. 15, 922–923.

Cope, E.D., 1884. The Vertebrata of the Tertiary formations of the west. In: Report of the United States Geological Survey of the Territories. Report no. 3, 1009 p.

Cope, E.D., 1893. A preliminary report on the vertebrate paleontology of the Llano Estacado. Fourth Annual Report of the Geological Survey of Texas, 1892. Part II. Paleontology and Natural History. Geological Survey of Texas B.C. Jones & Company, Austin, pp. 1–137.

Crompton, A.W., Smith, K.K., 1980. A new genus and species of crocodilian from the Kayenta Formation (Late Triassic?) of Northern Arizona. In: Jacobs, L.L. (Ed.), Aspects of Vertebrate History: Essays in Honor of Edwin Harris Colbert. Museum of Northern Arizona Press, Flagstaff, AZ, pp. 193–217.

De Queiroz, K., Gauthier, J.A., 1992. Phylogenetic taxonomy. Annu. Rev. Ecol. Syst. 23, 449–480.

Desojo, J.B., Ezcurra, M.D., 2011. A reappraisal of the taxonomic status of *Aetosauroides* (Archosauria, Aetosauria) specimens from the Late Triassic of South America and their proposed synonymy with *Stagonolepis*. J. Vertebr. Paleontol. 31 (3), 596–609.

Desojo, J.B., Heckert, A.B., Martz, J.W., Parker, W.G., Schoch, R.R., Small, B.J., Sulej, T., 2013. Aetosauria: a clade of armoured pseudosuchians from the Upper Triassic continental beds. In: Nesbitt, S.J., Desojo, J.B., Irmis, R.B. (Eds.), Anatomy, Phylogeny, and Paleobiology of Early Archosaurs and Their Kin. In: Special Publications of the Geological Society of London, vol. 379, Geological Society of London, London, pp. 203–239.

Donohoo-Hurley, L.L., Geissman, J.W., Lucas, S.G., 2010. Magnetostratigraphy of the uppermost Triassic and lowermost Jurassic Moenave Formation, western United States: correlation with strata in the United Kingdom, Morocco, Turkey, Italy, and eastern United States. Geol. Soc. Am. Bull. 122 (11–12), 2005–2019.

Doyle, K.D., Sues, H.-D., 1995. Phytosaurs (Reptilia: Archosauria) from the Upper Triassic New Oxford Formation of York County, Pennsylvania. J. Vertebr. Paleontol. 15, 545–553.

Dubiel, R.F., 1994. Triassic deposystems, paleogeography, and paleoclimate of the Western Interior. In: Caputo, M.V., Peterson, J.A., Franczyk, K.J. (Eds.), Mesozoic Systems of the Rocky Mountain Region, USA. Society of Economic Paleontologists and Mineralogists Rocky Mountain Section, Denver, CO, pp. 133–168.

Dunhill, A.M., Benton, M.J., Twitchett, R.J., Newell, A.J., 2012. Completeness of the fossil record and the validity of sampling proxies at outcrop level. Palaeontology 55 (6), 1155–1175.

Dunlavey, M.G., Whiteside, J.H., Irmis, R.B., 2009. Ecosystem instability during the rise of the dinosaurs: evidence from the Late Triassic in New Mexico and Arizona. Geological Society of America Abstracts With Programs, vol. 41, Geological Society of America, Boulder, p. 477.

Dutuit, J.-M., 1977. *Paleorhinus magnoculus*, phytosaure de Trias supérieur de l'Atlas Marocain. Géol. Mediterr. 4, 255–268.

Eaton, T.H., 1965. A new Wyoming phytosaur. University of Kansas, Paleontological Contributions, vol. 2, Paleontological Institute, Lawrence, KS, pp. 1–6.

Eberth, D.A., Evans, D.C., Brinkman, D.B., Therrien, F., Tanke, D.H., Russell, L.S., 2013. Dinosaur biostratigraphy of the Edmonton Group (Upper Cretaceous), Alberta, Canada: evidence of climate influence. Can. J. Earth Sci. 50, 701–726.

Elder, R.L., 1978. Paleontology and Paleoecology of the Dockum Group, Upper Triassic, Howard County, Texas (M.S. thesis). University of Texas, Austin, TX. 206 p.

Emmons, E., 1856. Geological report of the midland counties of North Carolina. North Carolina Geological Survey, 1852–1863. Putnam, New York, NY. 352 p.

Emry, R.J., 1973. Stratigraphy and preliminary biostratigraphy of the Flagstaff Rim area, Natrona County, Wyoming. Smithson. Contrib. Paleobiol. 18, 1–43.

Emry, R.J., Russell, L.S., Bjork, P.R., 1986. The Chadronian, Orellan, and Whitneyan North American land mammal ages. In: Woodburne, M.O. (Ed.), Cenozoic mammals of North America: Geochronology and Biostratigraphy. University of California Press, Berkeley, CA, pp. 18–23.

Ezcurra, M.D., 2010. Biogeography of Triassic tetrapods: evidence for provincialism and driven sympatric cladogenesis in the early evolution of modern tetrapod lineages. Proc. R. Soc. B 277, 2547–2552.

Flynn, J.J., Macfadden, B.J., McKenna, M.C., 1984. Land-Mammal Ages, faunal heterochrony, and temporal resolution in Cenozoic terrestrial sequences. J. Geol. 92, 687–705.

Fraas, O., 1866. Vor der Slindfluth! Eine Geschichte der Urwel. Hoffinann, Stuttgart (in German).

Frelier, A.P., 1987. Sedimentology, fluvial paleohydrology, and paleogeomorphology of the Dockum Formation (Triassic), West Texas (M.S. thesis). Texas Tech University, Lubbock, TX. 198 p.

Furin, S., Preto, N., Rigo, M., Roghi, G., Gianolla, P., Crowley, J.L., Bowring, S.A., 2006. High-precision U–Pb zircon age from the Triassic of Italy: implications for the Triassic time scale and the Carnian origin of calcareous nannoplankton and dinosaurs. Geology 34, 1009–1012.

Gregory, J.T., 1957. Significance of fossil vertebrates for correlation of Late Triassic continental deposits of North America. In: XX Congreso Geologico Internacional, Sección II - El Mesozoico del Hemisferio Occidental y sus Correlaciones Mundiales. International Geological Congress, Mexico City, pp. 7–25. 2.

Gregory, J.T., 1962. The genera of phytosaurs. Am. J. Sci. 260, 652–690.

Gregory, J.T., 1972. Vertebrate faunas of the Dockum Group, eastern New Mexico and West Texas. N. M. Geol. Soc. Guideb. 23, 120–130.

Gregory, J.T., Westphal, F., 1969. Remarks on the phytosaur genera of the European Trias. J. Paleontol. 43 (5), 1296–1298.

Hancock, J.M., 1977. The historic development of concepts of biostratigraphic correlation. In: Kauffman, E.G., Hazel, J.E. (Eds.), Concepts and Methods of Biostratigraphy. Dowden, Hutchinson, and Ross, Stroudsberg, PA, pp. 3–22.

Hart, M.B., 2012. Geodiversity, paleodiversity, or biodiversity: where is the place of paleobiology in an understanding of taphonomy? Proc. Geol. Assoc. 123, 551–555.

Hayden, F.V., 1869. On the geology of the Tertiary formations of Dakota and Nebraska. Philadelphia Acad. Nat. Sci. J. 7, 9–21 (2nd series).

Heckert, A.B., 2004. Late Triassic microvertebrates from the lower Chinle Group (Otischalkian-Adamanian: Carnian), southwestern U.S.A. N. M. Mus. Nat. Hist. Sci. Bull. 27, 1–170.

Heckert, A.B., Lucas, S.G., 1996. Late Triassic aetosaur biochronology. Albertiana 17, 57–64.

Heckert, A.B., Lucas, S.G., 1997. Lower Chinle Group (Adamanian: latest Carnian) tetrapod biostratigraphy and biochronology, eastern Arizona and West-Central New Mexico. In: Southwest Paleontological Symposium—Proceedings 5. Mesa Southwest Museum, Mesa, AZ, pp. 11–23.

Heckert, A.B., Lucas, S. G., 2000. Taxonomy, phylogeny, biostratigraphy, biochronology, paleobiogeography, and evolution of the Late Triassic Aetosauria (Archosauria: Crurotarsi). Zentralblatt für Geologie und Paläontologie Teil I 1998 Heft 11–12, pp. 1539–1587.

Heckert, A.B., Lucas, S.G., 2002. Revised Upper Triassic stratigraphy of the Petrified Forest National Park. N. M. Mus. Nat. Hist. Sci. Bull. 21, 1–36.

Heckert, A.B., Lucas, S.G., 2003. Stratigraphy and paleontology of the lower Chinle Group (Adamanian: latest Carnian) in the vicinity of St. Johns, Arizona. N. M. Geol. Soc. Guideb. 54, 281–288.

Heckert, A.B., Lucas, S.G., 2015. Triassic vertebrate paleontology in New Mexico. N. M. Mus. Nat. Hist. Sci. Bull. 68, 77–96.

Heckert, A.B., Spielmann, J.A., Lucas, S.G., Hunt, A.P., 2007. Biostratigraphic utility of the Upper Triassic aetosaur *Tecovasuchus* (Archosauria: Stagonolepididae), an index taxon of St. Johnsian (Adamanian: Late Carnian) time. N. M. Mus. Nat. Hist. Sci. Bull. 41, 229–240.

Hedberg, H.D., 1976. International Stratigraphic Guide: A Guide to Stratigraphic Classification, Terminology, and Procedure, first ed. John Wiley and Sons, New York, NY.

Holland, S.M., 2003. Confidence limits on fossil ranges that account for facies changes. Paleobiology 29, 468–479.

Howell, E.R., Blakey, R.C., 2013. Sedimentological constraints on the evolution of the Cordilleran arc: new insights from the Sonsela Member, Upper Triassic Chinle Formation, Petrified Forest National Park (Arizona, USA). Geol. Soc. Am. Bull. 125, 1349–1368.

Huene, F.v., 1915. On reptiles of the New Mexican Trias in the Cope collection. Bull. Am. Mus. Nat. Hist. 34, 485–507.

Huene, F.v., 1926. Notes on the age of the continental Triassic beds in North America with remarks on some fossil vertebrates. In: Proceedings of the U.S. National Museum, vol. 69, pp. 1–10.

Hungerbühler, A., 1998. Cranial Anatomy and Diversity of the Norian Phytosaurs of Southwestern Germany (Ph.D. dissertation). University of Bristol, Bristol. 453 p.

Hungerbühler, A., 2002. The Late Triassic phytosaur *Mystriosuchus westphali*, with a revision of the genus. Palaeontology 45 (2), 377–418.

Hungerbühler, A., Hunt, A.P., 2000. Two new phytosaur species (Archosauria, Crurotarsi) from the Upper Triassic of Southwest Germany. Neues Jb. Geol. Paläontol. Monat. 2000 (8), 467–484.

Hungerbühler, A., Mueller, B., Chatterjee, S., Cunningham, D.P., 2013. Cranial anatomy of the Late Triassic phytosaur *Machaeroprosopus*, with the description of a new species from West Texas. Earth Environ. Sci. Trans. R. Soc. Edinb. 103 (3–4), 269–312.

Hunt, A.P., 1993. Revision of the Metoposauridae (Amphibia: Temnospondyli) and description of a new genus from western North America. Mus. North. Ariz. Bull. 59, 67–97.

Hunt, A.P., 1994. Vertebrate paleontology and biostratigraphy of the Bull Canyon Formation (Chinle Group, Upper Triassic), East-Central New Mexico with revisions of the families Metoposauridae (Amphibia: Temnospondyli) and Parasuchidae (Reptilia: Archosauria) (Ph.D. dissertation). University of New Mexico, Albuquerque, NM. 404 p.

Hunt, A.P., 2001. The vertebrate fauna, biostratigraphy, and biochronology of the type Revueltian land-vertebrate faunachron, Bull Canyon Formation (Upper Triassic), East-central New Mexico. N. M. Geol. Soc. Guideb. 52, 123–152.

Hunt, A.P., Lucas, S.G., 1991. The *Paleorhinus* biochron and the correlation of the non-marine Upper Triassic of Pangaea. Palaeontology 34 (2), 487–501.

Hunt, A.P., Lucas, S.G., 1993. A new phytosaur (Reptilia: Archosauria) genus from the uppermost Triassic of the western United States and its biochronological significance. N. M. Mus. Nat. Hist. Sci. Bull. 3, 193–196.

Hunt, A.P., Lucas, S.G., Heckert, A.B., 2005. Definition and correlation of the Lamyan: a new biochronological unit for the nonmarine Late Carnian (Late Triassic). N. M. Geol. Soc. Guideb. 56, 357–366.

Hüsing, S.K., Deenan, M.H.L., Koopmans, J.G., Krijgsman, W., 2011. Magnetostratigraphic dating of the proposed Rhaetian GSSP at Steinbergkogel (Upper Triassic, Austria): implications for the Late Triassic time scale. Earth Planet. Sci. Lett. 302, 203–216.

ICZN, 1999. International Code of Zoological Nomenclature, fourth ed. International Trust for Zoological Nomenclature, London 306 p.

Irmis, R.B., 2005. The vertebrate fauna of the Upper Triassic Chinle Formation in northern Arizona. Mesa Southwest Mus. Bull. 9, 63–88.

Irmis, R.B., 2011. Evaluating hypotheses for the early diversification of dinosaurs. Earth Environ. Sci. Trans. R. Soc. Edinb. 101, 397–426.

Irmis, R.B., Nesbitt, S.J., Padian, K., Smith, N.D., Turner, A.H., Woody, D., Downs, A., 2007. A Late Triassic dinosauromorph assemblage from New Mexico and the rise of dinosaurs. Science 317, 358–361.

Irmis, R.B., Martz, J.W., Parker, W.G., Nesbitt, S.J., 2010. Re-evaluating the correlation between Late Triassic terrestrial vertebrate biostratigraphy and the GSSP-defined marine stages. Albertiana 38, 40–52.

Irmis, R.B., Mundil, R., Martz, J.W., Parker, W.G., 2011. High-resolution U-Pb ages from the Upper Triassic Chinle Formation (New Mexico, USA) support a diachronous rise of dinosaurs. Earth Planet. Sci. Lett. 309 (3-4), 258–267.

Jaeger, G.F., 1828. Über die fossilen Reptilien, welche in Würtemburg aufgefunden worden sind. Metzler, Stuttgart (in German).

Johnson, J.G., 1979. Intent and reality in biostratigraphic zonation. J. Paleontol. 53 (4), 931–942.

Kammerer, C.F., Butler, R.J., Bandyopadhyay, S., Stocker, M.R., 2015. Relationships of the Indian phytosaur *Parasuchus hislopi* Lydekker, 1885. Pap. Palaeontol. http://dx.doi.org/10.1002/spp2.1022.

Kirby, R.E., 1991. A vertebrate fauna from the Upper Triassic Owl Rock Member of the Chinle Formation of Northern Arizona (M.S. thesis). Northern Arizona University, Flagstaff. 476 p.

Kirkland, J.I., Martz, J.W., Deblieux, D.D., Madsen, S.K., Santucci, V.L., Inkenbradt, P., 2014. Paleontological resources inventory and Monitoring Chinle and Cedar Mountain Formations, Capitol Reef National Park. Utah Geological Survey unpublished report, 123 p.

Krishtalka, L., Stucky, R.K., West, R.M., McKenna, M.C., Black, C.C., Bown, T.M., Dawson, M.R., Golz, D.J., Flynn, J.J., Lillegraven, J.A., Turnbull, W.D., 1987. Eocene (Wasatchian through Duchesnean) biochronology of North America. In: Woodburne, M.O. (Ed.), Cenozoic Mammals of North America: Geochronology and Biostratigraphy. University of California Press, Berkeley, CA, pp. 77–117.

Kuhn, O., 1936. Weitere Parasuchier und Labyrinthodonten aus dem Blasensandstein des mittleren Keuper von Ebrach. Palaeontogr. Abt. A 83, 61–98.

Langston Jr., W.L., 1949. A new species of *Paleorhinus* from the Triassic of Texas. Am. J. Sci. 247, 324–341.

Lehman, T.M., 1994. The saga of the Dockum Group and the case of the Texas/New Mexico boundary fault. Bull. New Mex. Bur. Min. Mineral Resour. 150, 37–51.

Lehman, T.M., Chatterjee, S., 2005. The depositional setting and vertebrate biostratigraphy of the Triassic Dockum Group of Texas. J. Earth Syst. Sci. 114 (3), 325–351.

Lindsay, E.H., Tedford, R.H., 1990. Development and application of land mammal ages in North America and Europe, a comparison. In: Lindsay, E.H., Falbusch, V., Mein, P. (Eds.), European Neogene Mammal Chronology. In: NATO Advanced Science Institute Series, vol. 180. Plenum Press, New York, NY, pp. 601–624.

Lofgren, D.L., Lillegraven, J.A., Clemens, W.A., Gingerich, P.D., Williamson, T.E., 2004. Paleocene biochronology: The Puercan through Clarkfordian Land Mammal Ages. In: Woodburne, M.O. (Ed.), Late Cretaceous and Cenozoic Mammals of North America. Columbia University Press, New York, NY, pp. 43–105.

Long, R.A., Ballew, K.L., 1985. Aetosaur dermal armour from the Late Triassic of southwestern North America, with special reference to the Chinle Formation of Petrified Forest National Park. Mus. North. Ariz. Bull. 54, 45–68.

Long, R.A., Murry, P.A., 1995. Late Triassic (Carnian and Norian) tetrapods from the southwestern United States. N. M. Mus. Nat. Hist. Sci. Bull. 4, 1–254.

Long, R.A., Padian, K., 1986. Vertebrate biostratigraphy of the Late Triassic Chinle Formation, Petrified Forest National Park, Arizona: preliminary results. In: Padian, K. (Ed.), The Beginning of the Age of Dinosaurs: Faunal Change Across the Triassic-Jurassic Boundary. Cambridge University Press, Cambridge, pp. 161–169.

Lucas, F.A., 1898. A new crocodile from the Trias of southern Utah. Am. J. Sci. Ser. 4 6, 399.

Lucas, S.G., 1990. Toward a vertebrate biochronology of the Triassic. Albertiana 8, 36–41.

Lucas, S.G., 1992. Redefinition of the Duchesnean Land Mammal "Age", Late Eocene of western North America. In: Prothero, D.R., Bergrenn, W.A. (Eds.), Eocene-Oligocene Climatic and Biotic Evolution. Princeton University Press, Princeton, NJ, pp. 88–105.

Lucas, S.G., 1993a. Vertebrate biochronology of the Triassic of China. N. M. Mus. Nat. Hist. Sci. Bull. 3, 301–306.

Lucas, S.G., 1993b. The Chinle Group: revised stratigraphy and biochronology of Upper Triassic nonmarine strata in the western United States. Mus. North. Ariz. Bull. 59, 27–50.

Lucas, S.G., 1997. Upper Triassic Chinle Group, western United States: a nonmarine standard for late Triassic time. In: Dickins, J.M., Yin, H., Lucas, S.G., Acharyya, S.K. (Eds.), Late Paleozoic and Early Mesozoic Circum-Pacific Events and Their Global Correlation. Cambridge University Press, Cambridge, pp. 209–228.

Lucas, S.G., 1998. Global Triassic tetrapod biostratigraphy and biochronology. Palaeogeogr. Palaeoclimatol. Palaeoecol. 143, 347–384.

Lucas, S.G., 1999. A tetrapod-based Triassic timescale. Albertiana 22, 31–40.

Lucas, S.G., 2010. The Triassic timescale based on nonmarine tetrapod biostratigraphy and biochronology. In: Lucas, S.G. (Ed.), The Triassic Timescale. Geological Society of London Special Publications, vol. 334, Geological Society of London, London, pp. 447–500.

Lucas, S.G., Heckert, A.B., 1996. Vertebrate biochronology of the Late Triassic of Arizona. In: Boaz, D., Dierking, P., Dornan, M., McGeorge, R., Tegowski, B.J. (Eds.), Proceedings of the Fossils of Arizona Symposium, vol. 4, pp. 63–81. Mesa Southwest Museum Bulletin and the City of Mesa, Mesa.

Lucas, S.G., Huber, P., 2003. Vertebrate biostratigraphy and biochronology of the nonmarine Late Triassic. In: Letourneau, P.M., Olsen, P.E. (Eds.), The Great Rift Valleys of Pangea in Eastern North America. In: Sedimentology, Stratigraphy, and Paleontology, vol. 2. Columbia University Press, New York, NY, pp. 143–191.

Lucas, S.G., Hunt, A.P., 1992. Triassic stratigraphy and paleontology, Chama Basin and adjacent areas, north-central New Mexico. N. M. Geol. Soc. Guideb. 43, 151–167.

Lucas, S.G., Hunt, A.P., 1993. Tetrapod biochronology of the Chinle Group (Upper Triassic), western United States. N. M. Mus. Nat. Hist. Sci. Bull. 3, 327–329.

Lucas, S.G., Tanner, L.H., 2007. The nonmarine Triassic-Jurassic boundary in the Newark Supergroup of eastern North America. Earth Sci. Rev. 84, 1–20.

Lucas, S.G., Anderson, O.J., Hunt, A.P., 1994. Triassic stratigraphy and correlations, southern High Plains of New Mexico-Texas. Bull. New Mex. Bur. Min. Mineral Resour. 150, 105–126.

Lucas, S.G., Heckert, A.B., Hunt, A.P., 1997a. Stratigraphy and biochronological significance of the Late Triassic *Placerias* quarry, eastern Arizona (U.S.A.). Neues Jb. Geol. Paläontol. Abh. 203 (1), 23–46.

Lucas, S.G., Heckert, A.B., Estep, J.W., Anderson, O.J., 1997b. Stratigraphy of the Upper Triassic Chinle Group, Four Corners Region. In: Anderson, O.J., Kues, B.S., Lucas, S.G. (Eds.), Mesozoic Geology and Paleontology of the Four Corners Region. New Mexico Geological Society Guidebook, vol. 48, New Mexico Geological Society, Albuquerque, NM, pp. 81–107.

Lucas, S.G., Heckert, A.B., Kahle, R., 2002. Postcranial anatomy of *Angistorhinus*, a Late Triassic phytosaur from West Texas. In: Heckert, A.B., Lucas, S.G. (Eds.), Upper Triassic Stratigraphy and Paleontology. In: New Mexico Museum of Natural History and Science Bulletin, vol. 21, New Mexico Museum of Natural History and Science, Albuquerque, pp. 157–164.

Lucas, S.G., Zeigler, K.E., Heckert, A.B., Hunt, A.P., 2003. Upper Triassic stratigraphy and biostratigraphy, Chama Basin, north-central New Mexico. N. M. Mus. Nat. Hist. Sci. Bull. 24, 15–39.

Lucas, S.G., Zeigler, K.E., Heckert, A.B., Hunt, A.P., 2005. Review of Upper Triassic stratigraphy and biostratigraphy in the Chama Basin, Northern New Mexico. N. M. Geol. Soc. Guideb. 56, 170–181.

Lucas, S.G., Heckert, A.B., Rinehart, L., 2007a. A giant skull, ontogenetic variation and taxonomic validity of the Late Triassic phytosaur *Parasuchus*. N. M. Mus. Nat. Hist. Sci. Bull. 41, 222–228.

Lucas, S.G., Hunt, A.P., Heckert, A.B., Spielmann, J.A., 2007b. Global Triassic tetrapod biostratigraphy and biochronology: 2007 status. N. M. Mus. Nat. Hist. Sci. Bull. 41, 229–240.

Lucas, S.G., Tanner, L.H., Kozur, H.W., Weems, R.E., Heckert, A.B., 2012. The Late Triassic timescale: age and correlation of the Carnian-Norian boundary. Earth Sci. Rev. 114, 1–18.

Ludvigson, R., Westrop, S.R., Pratt, B.R., Tuffnell, P.A., Young, G.A., 1986. Dual biostratigraphy: zones and biofacies. Geosci. Can. 13, 139–154.

Lydekker, R., 1885. Maleri and Denwa reptilia and amphibia. Palaeontol. Indica 1, 1–38.

Marshall, C.R., 1990. Confidence intervals on stratigraphic ranges. Paleobiology 16, 1–10.

Marshall, C.R., 1994. Confidence intervals on stratigraphic ranges: partial relaxation of the assumption of a random distribution of fossil horizons. Paleobiology 20, 459–469.

Marshall, C.R., 1995. Distinguishing between sudden and gradual extinctions in the fossil record: predicting the position of the Cretaceous-Tertiary iridium anomaly using the ammonite fossil record on Seymour Island, Antarctica. Geology 23, 731–734.

Marshall, C.R., 1997. Confidence intervals on stratigraphic ranges with non-random distributions of fossil horizons. Paleobiology 23, 165–173.

Marshall, C.R., 1998. Determining stratigraphic ranges. In: Donovan, S.K., Paul, C.R.C. (Eds.), The Adequacy of the Fossil Record. John Wiley & Sons, Ltd, Hoboken, NJ, pp. 23–53.

Martz, J.W., 2002. The morphology and ontogeny of *Typothorax coccinarum* (Archosauria, Stagonolepididae) from the Upper Triassic of the American southwest (M.S. thesis). Texas Tech University, Lubbock, TX. 279 p.

Martz, J.W., 2008. Lithostratigraphy, chemostratigraphy, and vertebrate biostratigraphy of the Dockum Group (Upper Triassic), of southern Garza County, West Texas (Ph.D. dissertation). Texas Tech University, Lubbock, TX. 504 p.

Martz, J.W., Parker, W.G., 2010. Revised lithostratigraphy of the Sonsela Member (Chinle Formation, Upper Triassic) in the southern part of Petrified Forest National Park, Arizona. PLoS ONE 5(2), e9329. http://dx.doi.org/10.1371/journal.pone.0009329

Martz, J.W., Small, B.J., 2006. *Tecovasuchus chatterjeei*, a new aetosaur (Archosauria: Aetosauria) from the Tecovas Formation (Upper Triassic, Carnian) of Texas. J. Vertebr. Paleontol. 26, 308–320.

Martz, J.W., Parker, W.G., Skinner, L., Raucci, J.J., Umhoefer, P., Blakey, R.C., 2012. Geologic map of Petrified Forest National Park. 1:50,000. Arizona Geological Society, Phoenix, AZ.

Martz, J.W., Mueller, B., Nesbitt, S.J., Stocker, M.R., Parker, W.G., Atanassov, M., Fraser, N., Weinbaum, J., Lehane, J., 2013. A taxonomic and biostratigraphic re-evaluation of the Post Quarry vertebrate assemblage from the Cooper Canyon Formation (Dockum Group, Upper Triassic) of southern Garza County, western Texas. Earth Environ. Sci. Trans. R. Soc. Edinb. 103 (3–4), 339–364.

Martz, J.W., Irmis, R.B., Milner, A.R.C., 2014. Lithostratigraphy and biostratigraphy of the Chinle Formation (Upper Triassic) in southern Lisbon Valley, southeastern Utah. In: Maclean, J.S., Biek, R.F., Huntoon, J.E. (Eds.), Geology of Utah's Far South. Utah Geological Association Publication, vol. 43, Utah Geological Association, Salt Lake City, pp. 399–446.

Martz, J.W., Kirkland, J.I., Deblieux, D.D., Suarez, C.A., Santucci, V.L., 2015. Stratigraphy of the Chinle Formation (Upper Triassic) and Moenave Formation (Upper Triassic-Lower Jurassic) in the southwestern part of Zion National Park, Washington County, Utah. Utah Geological Survey, Salt Lake City, UT. Unpublished open file report.

Matthew, W.D., 1924. Correlation of the tertiary formations of the Great Plains. Geol. Soc. Am. Bull. 12, 19–75.

McGowran, B., 2005. Biostratigraphy: Microfossils and Geologic Time. Cambridge University Press, Cambridge. 480 p.

McKenna, M.C., Lillegraven, J.A., 2005. Problems with Paleocene palynozones in the Rockies: Hell's Half Acre revisited. J. Mamm. Evol. 12 (1/2), 23–51.

McKenna, M.C., Lillegraven, J.A., 2006. Biostratigraphic deception by the Devil, salting fossil kollinbrains into the Poobahcene section of central Myroaming. Palaeontogr. Abt. A 277, 1–17.

Mehl, M.G., 1913. *Angistorhinus*, a new genus of Phytosauria from the Trias of Wyoming. J. Geol. 21, 186–191.

Mehl, M.G., 1915. The Phytosauria of the Trias. J. Geol. 23, 129–165.

Mehl, M.G., 1916. New or little known phytosaurs from Arizona. In: Mehl, M.G., Toepelmann, W.C., Schwartz, G.M. (Eds.), New or Little Known Reptiles from the Trias of Arizona and New Mexico, with Notes from the Fossil Bearing Horizons Near Wingate, New Mexico. In: Bulletin of the University of Oklahoma, New Series 103, University Studies Series, vol. 5. University of Oklahoma, Norman, pp. 5–28.

Mehl, M.G., 1922. A new phytosaur from the Trias of Arizona. J. Geol. 30, 144–157.

Mehl, M.G., 1928. *Pseudopalatus pristinus*, a new genus and species of phytosaurs from Arizona. University of Missouri Quarterly 3, University of Missouri, Columbia, pp. 1–22.

Meyer, H. v., 1861. Reptilien aus dem Stubensandstein des oberen Keupers. Palaeontographica A 7 (1859–1861), 253–346.

Meyer, H.v., 1863. Der Schädel des *Belodon* aus dem Stubendsandstein des oberen Keupers (Dritte Folge). Palaeontographica 10, 227–246.

Milner, A.R.C., Mickelson, D.L., Kirkland, J.I., Harris, J.D., 2006. Reinvestigation of Late Triassic fish sites in the Chinle Group, San Juan County, Utah: new discoveries. In: Parker, W.G., Ash, S.R., Irmis, R.B. (Eds.), A Century of Research at Petrified Forest National Park: Geology and Paleontology. Museum of Northern Arizona Bulletin, vol. 62, Museum of Northern Arizona, Flagstaff, pp. 163–165.

Muttoni, G., Kent, D.V., Olsen, P.E., Distephano, P., Lowrie, W., Bernasconi, S.M., Hernández, F.M., 2004. Tethyan magnetostratigraphy from Pizza Mondelo (Sicily) and correlation to the Late Triassic Newark astrochronological polarity time scale. Geol. Soc. Am. Bull. 116, 1043–1058.

Nesbitt, S.J., 2011. The early evolution of archosaurs: relationships and origins of major clades. Bull. Am. Mus. Nat. Hist. 359, 1–292.

Nesbitt, S.J., Stocker, M.R., 2008. The vertebrate assemblage of the Late Triassic Canjilon Quarry (northern New Mexico, U.S.A.), and the importance of apomorphy-based assemblage comparisons. J. Vertebr. Paleontol. 28 (4), 1063–1072.

Nesbitt, S.J., Irmis, R.B., Parker, W.G., 2007. A critical re-evaluation of the Late Triassic dinosaur taxa of North America. J. Syst. Palaeontol. 5 (2), 209–243.

Nordt, L., Atchley, S., Dworkin, S., 2015. Collapse of the Late Triassic megamonsoon in western equatorial Pangea, present-day American southwest. Geol. Soc. Am. Bull. http://dx.doi.org/10.1130/B31186.1.

North American Commission on Stratigraphic Nomenclature (NACSN), 1983. North American Stratigraphic Code. Am. Assoc. Pet. Geol. Bull. 67, 841–875.

North American Commission On Stratigraphic Nomenclature (NACSN), 2005. North American Stratigraphic Code. Am. Assoc. Pet. Geol. Bull. 89 (11), 1547–1591.

Olsen, P.E., Kent, D.V., Sues, H.-D., Koeberl, C., Huber, H., Montanari, A., Rainforth, E.C., Fowell, S.J., Szajna, M.J., Hartline, B.W., 2002. Ascent of dinosaurs linked to an Iridium anomaly at the Triassic—Jurassic boundary. Science 296, 1305–1307.

Olsen, P.E., Kent, D.V., Whiteside, J.H., 2011. Implications of the Newark Supergroup-based astrochronology and geomagnetic polarity time scale (Newark-APTS) for the tempo and mode of the early diversification of the Dinosauria. Earth Environ. Sci. Trans. R. Soc. Edinb. 101, 201–229.

Opdyke, N.D., Lindsey, E.H., Johnson, N.M., Downs, T., 1977. The paleomagnetism and magnetic polarity stratigraphy of the mammal-bearing section of Anza-Borrego State Park, California. Quat. Res. 7, 316–329.

Padian, K., Lindberg, D.R., Polly, P.D., 1994. Cladistics and the fossil record: the uses and history. Annu. Rev. Earth Planet. Sci. 22, 63–89.

Parker, W.G., 2005. A new species of the Late Triassic aetosaur *Desmatosuchus* (Archosauria: Pseudosuchia). C. R. Palevol 4, 327–340.

Parker, W.G., 2006. The stratigraphic distribution of major fossil localities in Petrified Forest National Park, Arizona. Museum of Northern Arizona Bulletin, vol. 62, Museum of Northern Arizona, Flagstaff, pp. 46–62.

Parker, W.G., 2007. Reassessment of the aetosaur "*Desmatosuchus*" *chamaensis* with a revision of the phylogeny of the Stagonolepididae (Archosauria: Pseudosuchia). J. Syst. Palaeontol. 5 (1), 41–68.

Parker, W.G., 2014. Taxonomy and phylogeny of the Aetosauria (Archosauria: Pseudosuchia) including a new species from the Upper Triassic of Arizona (Ph.D. dissertation). The University of Texas, Austin, TX. 437 p.

Parker, W.G., 2015. Improved phylogenetic resolution tracks aetosaurian (Archosauria: Pseudosuchia) diversity through Late Triassic extinction events. J. Vertebr. Paleontol. 2015, 191–192. Programs and Abstracts.

Parker, W.G., 2016. Revised phylogenetic analysis of the Aetoauria (Archosauria: Pseudosuchia); assessing the effects of incongruent morphological sets. PeerJ 4 (e1583). 70 p.

Parker, W.G., Irmis, R.B., 2006. A new species of the Late Triassic phytosaur *Pseudopalatus* (Archosauria: Pseudosuchia) from Petrified Forest National Park, Arizona. Museum of Northern Arizona, vol. 62, Museum of Northern Arizona, Flagstaff, pp. 126–143.

Parker, W.G., Martz, J.W., 2011. The Late Triassic (Norian) Adamanian-Revueltian tetrapod transition in the Chinle Formation of Petrified Forest National Park, Arizona. Earth Environ. Sci. Trans. R. Soc. Edinb. 101, 231–260.

Parker, W., Martz, J.W., Dubiel, R., 2011. A newly recognized specimen of the phytosaur *Redondasaurus* from the Upper Triassic Owl Rock Member (Chinle Formation) and its biostratigraphic implications. In: Society of Vertebrate Paleontology 71st Annual Meeting Programs and Abstracts. 171 p.

Parker, W.G., Stocker, M.R., Irmis, R.B., 2008. A new desmatosuchine aetosaur (Archosauria: Suchia) from the Upper Triassic Tecovas Formation (Dockum Group) of Texas. J. Vertebr. Paleontol. 28, 692–701.

Parker, W.G., Hungerbühler, A., Martz, J.W., 2013. The taxonomic status of the phytosaurs (Archosauriformes) *Machaeroprosopus* and *Pseudopalatus* from the Late Triassic of the western United States. Earth Environ. Sci. Trans. R. Soc. Edinb. 103 (3–4), 265–268.

Parrish, J.M., 1999. Small fossil vertebrates from the Chinle Formation (Upper Triassic) of southern Utah. In: Gillette, D.D. (Ed.), Vertebrate Paleontology in Utah. In: Utah Geological Survey Miscellaneous Publication, vol. 99, no. 1, Utah Geological Survey, Salt Lake City, pp. 45–50.

Patzkowsky, M.E., Holland, S.M., 2012. Stratigraphic Paleobiology: Understanding the Distribution of Fossil Taxa in Time and Space. The University of Chicago Press, Chicago, IL. 259 p.

Pipiringos, G.N., O'Sullivan, R.B., 1978. Principal unconformities in Triassic and Jurassic rocks, Western Interior United States – a preliminary survey. Geological Survey Professional Paper no. 1035-A, 29 pp.

Prothero, D.R., 1990. Interpreting the Stratigraphic Record. W.H. Freeman & Co., New York, NY. 410 p.

Prothero, D.R., Emry, R.J., 1996. Summary. In: Prothero, D.R., Emry, R.J. (Eds.), The Terrestrial Eocene-Oligocene Transition in North America. Cambridge University Press, Cambridge, pp. 664–683.

Ramazani, J., Fastovsky, D.E., Bowring, S.A., 2014. Revised chronostratigraphy of the lower Chinle Formation strata in Arizona and New Mexico (USA): high-precision U-Pb geochronological constraints on the Late Triassic evolution of dinosaurs. Am. J. Sci. 314, 981–1008.

Ramezani, J., Hoke, G.D., Fastovsky, D.E., Bowring, S.A., Therrien, F., Dworkin, S.I., Atchley, S.C., Nordt, L.C., 2011. High-precision U-Pb zircon geochronology of the Late Triassic Chinle Formation, Petrified Forest National Park (Arizona, USA): temporal constraints on the early evolution of dinosaurs, vol. 123. Geological Society of America, Boulder, CO, pp. 2142–2159.

Rayfield, E.J., Barrett, P.M., McDonnell, R.A., Willis, K.J., 2005. A geographical information system (GIS) study of Triassic vertebrate biochronology. Geol. Mag. 142 (4), 327–354.

Rayfield, E.J., Barrett, P.M., Milner, A.R., 2009. Utility and validity of Middle and Late Triassic land vertebrate faunachrons. J. Vertebr. Paleontol. 29 (1), 80–87.

Repenning, C.A., 1967. Palaearctic-Nearctic mammalian dispersal in the late Cenozoic. In: Hopkins, D.M. (Ed.), The Bering Land Bridge. Stanford University Press, Standford, pp. 288–314.

Riggs, N.R., Lehman, T.M., Gehrels, G.E., Dickinson, W.R., 1996. Detrital zircon link between headwaters and terminus of the Upper Triassic Chinle-Dockum paleoriver system. Science 273, 97–100.

Riggs, N.R., Oberling, Z.A., Howell, E.R., Parker, W.G., Barth, A.P., Cecil, M.R., Martz, J.W., 2016. Paleotopography and evolution of the Early Mesozoic Cordilleran margin as reflected in the detrital zircon signature of the basal Upper Triassic Chinle Formation, Colorado Plateau. Geosphere 12, 439–463.

Robinson, P., Gunnell, G.F., Walsh, S.L., Clyde, W.C., Storer, J.E., Stucky, R.K., Froehlich, D.J., Ferrusquia-Villafranca, I., McKenna, M.C., 2004. In: Woodburne, M.O. (Ed.), Late Cretaceous and Cenozoic Mammals of North America. Columbia University Press, New York, NY, pp. 106–155.

Rogers, R.R., Kidwell, S.M., 2007. A conceptual framework for the genesis and analysis of vertebrate skeletal concentrations. In: Rogers, R.R., Eberth, D.A., Fiorillo, A.R. (Eds.), Bonebeds: Genesis, Analysis, and Paleobiological Significance. The University of Chicago Press, Chicago, IL, pp. 1–63.

Rowe, T.B., Sues, H.-D., Reisz, R., 2010. Dispersal and diversity in the earliest North American sauropodomorph dinosaurs, with a description of a new taxon. Proc. R. Soc. B 278 (1708), 1044–1053.

Salvador, A. (Ed.), 1994. International Stratigraphic Guide, second ed. International Union of Geological Sciences and the Geological Society of America, Boulder, CO. 214 p.

Savage, D.E., 1977. Aspects of vertebrate paleontological stratigraphy and geochronology. In: Kauffman, E.G., Hazel, J.E. (Eds.), Concepts and Methods of Biostratigraphy. Dowden, Hutchinson, & Ross, Stroudsburg, PA, pp. 427–442.

Schoch, R.M., 1989. Stratigraphy: Principles and Methods. Van Nostrand Reinhold, New York, NY. 375 p.

Schoene, B., Guex, J., Bartolini, A., Schaltegger, U., Blackburn, T.J., 2010. Correlating the end-Triassic mass extinction and flood basalt volcanism at the 100 ka level. Geology 38, 387–390.

Schultz, C.L., 2005. Biostratigraphy of the non-marine Triassic: is a global correlation based on tetrapod faunas possible? In: Koutsoukos, E.A.M. (Ed.), Applied Stratigraphy. Springer, Netherlands, pp. 123–145.

Small, B.J., 1989. Post quarry. In: Lucas, S.G., Hunt, A.G. (Eds.), Dawn of the Age of Dinosaurs in the American Southwest. University of Albuquerque Press, Albuquerque, NM, pp. 145–148.

Small, B.J., Martz, J.W., 2013. A new basal aetosaur from the Upper Triassic Chinle Formation of the Eagle Basin, Colorado, USA. In: Nesbitt, S.J., Desojo, J.B., Irmis, R.B. (Eds.), Anatomy, Phylogeny and Palaeobiology of Early Archosaurs and Their Kin. In: Geological Society, London, Special Publications, vol. 379. The Geological Society Publishing House, Bath, pp. 393–412.

Solow, A.R., 1996. Tests and confidence intervals for a common upper endpoint in fossil taxa. Paleobiology 22, 406–410.

Solow, A.R., 2003. Estimation of stratigraphic ranges when fossil finds are not randomly distributed. Paleobiology 29, 181–185.

Solow, A.R., Smith, W.K., 2000. Testing for a mass extinction without selecting taxa. Paleobiology 26, 647–650.

Spielmann, J.A., Lucas, S.G., 2012. Tetrapod fauna of the Upper Triassic Redonda Formation, East-Central New Mexico: the characteristic faunal assemblage of the Apachean land-vertebrate faunachron. N. M. Mus. Nat. Hist. Sci. Bull. 55, 1–119.

Springer, M.S., 1990. The effect of random range truncations on patterns of evolution in the fossil record. Paleobiology 16, 512–520.

Stewart, J.H., Poole, F.G., Wilson, R.F., 1972. Stratigraphy and origin of the Chinle Formation and related Upper Triassic strata in the Colorado Plateau region. US Geol. Surv. Prof. Pap. 690, 1–336.

Stocker, M.R., 2010. A new taxon of phytosaur (Archosauria: Pseudosuchia) from the Late Triassic (Norian) Sonsela Member (Chinle Formation) in Arizona, and a critical reevaluation of *Leptosuchus* Case, 1922. Palaeontology 53 (5), 997–1022.

Stocker, M.R., 2012. A new phytosaur (Archosauriformes, Phytosauria) from the Lot's Wife beds (Sonsela Member) within the Chinle Formation (Upper Triassic) of Petrified Forest National Park, Arizona. J. Vertebr. Paleontol. 32 (3), 573–586.

Stocker, M.R., 2013. A new taxonomic arrangement for *Paleorhinus scurriensis*. Earth Environ. Sci. Trans. R. Soc. Edinb. 103 (3–4), 251–263.

Stocker, M.R., Butler, R.J., 2013. Phytosauria. In: Nesbitt, S.J., Desojo, J.B., Irmis, R.B. (Eds.), Anatomy, Phylogeny, and Palaeobiology of Early Archosaurs and their Kin. Geological Society Special Publication, vol. 379, Geological Society of London, London, pp. 91–117.

Stovall, J.W., Wharton Jr., J.B., 1936. A new species of phytosaur from Big Spring, Texas. J. Geol. 44, 183–192.

Strauss, D., Sadler, P.M., 1989. Classical confidence intervals and Bayesian probability estimates for ends of local taxon ranges. Math. Geol. 21, 411–427.

Sues, H.-D., Fraser, N.C., 2010. Triassic Life on Land: The Great Transition. Columbia University Press, New York, NY. 236 p.

Sues, H.-D., Shubin, N.H., Olsen, P.E., Amaral, W.W., 1996. On the cranial structure of a new protosuchid (Archosauria: Crocodyliformes) from the McCoy Brook Formation (Lower Jurassic) of Nova Scotia, Canada. J. Vertebr. Paleontol. 16 (1), 34–41.

Sulej, T., 2002. Species discrimination of the Late Triassic temnospondyl amphibian *Metoposaurus diagnosticus*. Acta Palaeontol. Pol. 47 (3), 535–546.

Sulej, T., 2007. Osteology, variability, and evolution of *Metoposaurus*, a temnospondyl from the Late Triassic of Poland. Palaeontol. Pol. 64, 29–139.

Tedford, R.H., 1970. Principles and practices of mammalian geochronology in North America. In: Proceedings of the North American Paleontological Convention, Part F, pp. 666–703.

Walsh, S.L., 1998. Fossil datum and paleobiological event terms, paleontostratigraphy, chronostratigraphy, and the definition of Land Mammal "Age" boundaries. J. Vertebr. Paleontol. 18 (1), 150–179.

Walsh, S.L., 2001. Eubiostratigraphic units, quasibiostratigraphic units, and "assemblage zones." J. Vertebr. Paleontol. 20 (4), 761–775.

Walsh, S.L., 2004. Time and time-rock again: an essay on the (over)simplification of stratigraphy. Palaeontol. Newsl. 57, 19–25.

Wang, S.C., Zimmerman, A.E., Mcveigh, B.S., Everson, P.J., Wong, H., 2012. Confidence intervals for the duration of a mass extinction. Paleobiology 38 (2), 265–277.

Westphal, F., 1976. Phytosauria. In: Kuhn, O. (Ed.), Handbuch der Paläoherpetologie: Thecodontia, Gustav-Fischer-Verlag, München, pp. 99–120.

Whiteside, J.H., Olsen, P.E., Kent, D.V., Fowell, S.J., Et-Touhami, M., 2007. Synchrony between the Central Atlantic magmatic province and the Triassic-Jurassic mass-extinction event? Palaeogeogr. Palaeoclimatol. Palaeoecol. 244, 345–367.

Whiteside, J.H., Grogan, D.S., Olsen, P.E., Kent, D.V., 2011. Climatically driven biogeographic provinces of Late Triassic tropical Pangea. PNAS 108 (22), 8972–8977.

Williston, S.W., 1904. Notice of some new reptiles from the Upper Trias of Wyoming. J. Geol. 12 (8), 688–697.

Wood II, H.E., Chaney, R.W., Clark, J., Colbert, E.H., Jepsen, G.L., Reedside Jr., J.B., Stock, C., 1941. Nomenclature and correlation of the North American continental Tertiary. Geol. Soc. Am. Bull. 52, 1–48.

Woodburne, M.O., 1977. Definition and characterization in mammalian chronostratigraphy. J. Paleontol. 51 (2), 220–234.

Woodburne, M.O., 1987. Mammal ages, stages, and zones. In: Woodburne, M.O. (Ed.), Cenozoic Mammals of North America: Geochronology and Biostratigraphy. University of California Press, Berkeley, CA, pp. 18–23.

Woodburne, M.O., 1989. Hipparion horses: a pattern of endemic evolution and intercontinental dispersal. In: Prothero, D.R., Schoch, R.M. (Eds.), The Evolution of Perissodactyls. Oxford University Press, New York, NY, pp. 197–233.

Woodburne, M.O., 1996. Precision and resolution in mammalian chronostratigraphy: principles, practices, examples. J. Vertebr. Paleontol. 16 (3), 531–555.

Woodburne, M.O., 2004. Principles and procedures. In: Woodburne, M.O. (Ed.), Late Cretaceous and Cenozoic Mammals of North America. Columbia University Press, New York, NY, pp. 1–20.

Woodburne, M.O., 2006. Mammal ages. Stratigraphy 3 (4), 229–261.

Woody, D.T., 2006. Revised stratigraphy of the Lower Chinle Formation (Upper Triassic) of Petrified Forest National Park, Arizona. Museum of Northern Arizona Bulletin, vol. 62, Museum of Northern Arizona, Flagstaff, pp. 17–45.

Woody, D.T., Parker, W.G., 2004. Evidence for a transitional fauna within the Sonsela Member of the Chinle Formation, Petrified Forest National Park, Arizona. J. Vertebr. Paleontol. 24 (Suppl. 3), 132A.

Zawiskie, J.M., 1986. Triassic faunal succession. In: Padian, K. (Ed.), The Beginning of the Age of Dinosaurs: Faunal Change across the Triassic-Jurassic Boundary. Cambridge University Press, Cambridge, pp. 161–169.

Zeigler, K.E., Lucas, S.G., Heckert, A.B., 2003 (imprint 2002). A phytosaur skull from the upper Triassic Snyder quarry (Petrified Forest Formation, Chinle Group) of north-central New Mexico. N. M. Mus. Nat. Hist. Sci. Bull. 21, 171–177.

Zeigler, K.E., Lucas, S.G., Morgan, V.L., 2005. Vertebrate fauna of the Upper Triassic Mesa Montosa Member (Petrified Forest Formation, Chinle Group), Chama Basin, northern New Mexico. New Mexico Geological Society Field Conference Guidebook, vol. 56, New Mexico Geological Society, Albuquerque, pp. 335–340.

Zeigler, K.E., Kelley, S., Geissman, J.W., 2008. Revisions to stratigraphic nomenclature of the Upper Triassic Chinle Group in New Mexico: new insights from geologic mapping, sedimentology, and magnetostatigraphic/paleomagnetic data. Rocky Mt. Geol. 43 (2), 121–141.

Jeffrey W. Martz is assistant professor of geology in the Department of Natural Sciences at the University of Houston-Downtown. His research focuses on the evolution and ecology of archosaurs during the Late Triassic, particularly in western North America. In addition to studying the systematics of early dinosauromorphs and pseudosuchian archosaurs, he is currently developing detailed lithostratigraphic and biostratigraphic models for the Upper Triassic Chinle Formation in western North America that will be used to track the relationship between paleoenvironmental (especially paleoclimatic) change and paleoecology. Much of his work has focused on increasing the methodological rigor of Late Triassic paleontological and stratigraphic research.

Methods in Paleopalynology and Palynostratigraphy: An Application to the K-Pg Boundary

A. Bercovici*, J. Vellekoop[†]
*Smithsonian Institution, Washington, DC, United States
[†]KU Leuven, Leuven, Belgium

Contents

WHAT IS PALYNOLOGY?

The study of microfossils, or micropaleontology, is a major field of research in the geological sciences with numerous applications (Armstrong and Brasier, 2005). Because of their

Terrestrial Depositional Systems
http://dx.doi.org/10.1016/B978-0-12-803243-5.00003-0

small sizes, usually less than 1 mm, microfossils can be recovered from various sedimentary rocks, both marine and terrestrial, in great numbers and diversity. A small sample can reveal an entire ecosystem of organisms that can be used to reconstruct paleoenvironmental conditions (e.g., climate, water depth, and salinity), evolutionary history of a group (paleobiogeography, origination, and extinction of species), and for relative dating (biostratigraphy and correlation). For these reasons, micropaleontology is used routinely for the study of core samples, just as macrofossils such as mollusk shells, vertebrates, or plant fossils have been studied when examining exposed surfaces in cores. The different branches of micropaleontology focus on specific fossil groups, characterized by their constitutive material, whether inorganic walled (calcareous microfossils such as foraminifera and coccolith, siliceous such as radiolarians, diatoms, and phytoliths, or phosphatic such as conodonts), or organic walled, the latter being specifically relevant to palynology.

At first glance, one would think that the term *palynology* refers to the branch of botanical sciences that specifically studies pollen. It is certainly true that Hyde and Williams (1944) had this in mind when they coined the term "palynology" for the first time, as a substitute for "pollen science" or "pollen analysis" (the latter having a distinct use in Quaternary science). Deriving from the greek παλύνω "I sprinkle", palynology is the study of palynomorphs, a general term for organic-walled microfossils typically ranging from 5 to 500 μm, consisting at least partly of very resistant organic molecules such as sporopollenin or chitin, and collectively found in the residue of palynological samples after dissolution of the mineral matrix by strong inorganic acids. Thus palynology encompasses the study of an array of different microfossil groups, both plant and animal, marine and terrestrial, in which palynologists specialize. Palynologists all rely on a common extraction method that involves the use of inorganic acids and reagents to recover palynomorphs from rocks and sediments.

THE MAJOR GROUPS OF PALYNOMORPHS AND THEIR APPLICATION

The purpose of this chapter is to present a quick review of the major groups of palynomorphs (Fig. 2) with emphasis on terrestrial palynomorphs, specifically pollen and spores. A more detailed review is given in the three-volume palynology encyclopedia of Jansonius and McGregor (1996) and Traverse (2007). Each of these groups offers a plethora of applications in the study of the evolution of life, relative dating (biostratigraphy), paleobiology, paleoenvironment, paleogeography, paleoclimatology, and geological resources (hydrocarbon production).

Because acid dissolution of a rock or sediment sample is a chemically selective process, the resulting organic residue is usually composed of a broad variety of material of plant and animal origin within the same sample. It is not uncommon to recover marine palynomorphs from terrestrial settings by erosion and redeposition of older marine rocks, and terrestrial palynomorphs can be transported by air and water into nearshore marine settings.

Acritarchs

Acritarchs are organic-walled microfossils of unknown or uncertain affinity, though most are derived from unicellular eukaryotes. They range from mid-Precambrian to recent times (Fig. 1). Many acritarchs still have a dubious biological affinity. Whereas many probably represent the resting stage (cyst) of marine phytoplanktonic algae, various forms have now been assigned to the division Prasinophyta of the green algae. Morphology has been used to define informal groupings (Downie, 1973; Fensome et al., 1990; Martin, 1993; Colbath and Grenfell, 1995), which are useful for Precambrian and Paleozoic biostratigraphy (Martin, 1993; Vidal and Knoll, 1993; Molyneux et al., 1996) and paleobiogeography (Le Hérissé et al., 1997; Servais et al., 2003, 2004). Acritarchs reached their acme in the Paleozoic and are largely but not exclusively marine.

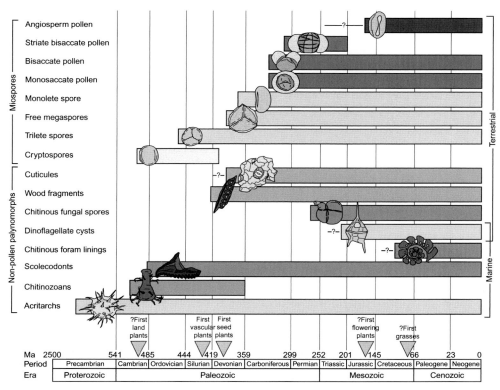

Fig. 1 Stratigraphic range chart of the major groups of palynomorphs. *(Modified after Traverse, A., 2007. Paleopalynology. Springer, New York, 814 pp).*

Chitinozoans

Chitinozoans are a group of exclusively marine, bottle-shaped microfossils of uncertain affinity (Cashman, 1990; Jaglin and Paris, 1992; Paris et al., 2004), although their pseudochitin wall suggests animal origin. They radiated during the Paleozoic (Fig. 1) with peak diversity during the Middle Ordovician and Silurian, although some early occurrences in the Cambrian have been documented (Bloeser et al., 1977; Shen et al., 2013). The group disappeared at the end of the Devonian period. Like acritarchs, chitinozoans evolved into a diversity of shapes and forms (Jansonius, 1970; Miller, 1996), making them a major useful group for Paleozoic biostratigraphy (Grahn, 1978; Paris, 1990, 1996; Nôlvak and Grahn, 1993; Verniers et al., 1995).

Scolecodonts

Another common group of Paleozoic fossils includes scolecodonts, which occurs associated with acritarchs and chitinozoans in early Paleozoic marine shales. Scolecodonts are the chitinous mouthparts of polychaetes (marine annelid worms), and have a spotty stratigraphic range from early Ordovician to recent (Fig. 1). Scolecodonts have a limited biostratigraphic utility and are more useful for paleoenvironmental and thermal maturation studies (Szaniawski, 1996).

Dinoflagellates

Dinoflagellates are motile unicellular algae characterized by a pair of flagellae. Many dinoflagellates are photosynthetic, whereas others are mixotrophic. Dinoflagellates have been an important part of the marine phytoplankton since the mid-Mesozoic (Fig. 1) and represent the second largest primary producers in the world's oceans today, next to the diatoms. Whereas most are strictly marine, some dinoflagellates occupy brackish and freshwater environments. Dinoflagellates also exhibit remarkable traits: In addition to chlorophyll, some possess carotenoid pigments (dinoxanthin and peridinin), giving them a flamboyant red coloration, whereas others are bioluminescent. Some species form blooms in the oceans, a phenomenon called "red tide" due to coloration of the water resulting from the intense concentration of algal cells. These blooms have important economic implications, as many dinoflagellates also produce toxins that can kill great numbers of fish and invertebrates, and can accumulate in many filter-feeders, rendering them unsafe for human consumption. Dinoflagellates have a complex life cycle during which the motile planktonic cell can form a resting cyst (encystment) that remains dormant on the sea floor during unfavorable conditions such as winter (Dale, 1983). Incubation studies have shown that in this resting cyst, dinoflagellates can remain dormant for decades (Ribeiro et al., 2011). With ameliorating conditions, the dinoflagellate cell can escape the cyst through an opening (the archaeopyle) to become motile again (excystment). The empty cyst is left behind on the sea floor. Although almost all fossil dinoflagellates are preserved as cysts, only ∼10%–20% of living dinoflagellate species are known to encyst following sexual reproduction (Dale, 1983; Evitt, 1985). Many of these dinoflagellate taxa produce a cyst that is made of very resistant material (dinosporin, an organic compound related to sporopollenin), which can be potentially preserved in the fossil record. These organic-walled dinoflagellate cysts are commonly referred to as "dinocysts." Other dinoflagellates produce nonpreservable (i.e., dinosporin-free) resting cysts, which quickly decay after excystment. While in biological studies the taxonomy of dinoflagellates is based on living, motile cells, the fossil record is represented by resting cysts only. Consequently, the species that formed a particular dinocyst morphology is often unknown. Therefore, a separate taxonomic system, based on form taxa (using morphological traits to define generic and specific systematics), has been developed for dinocysts (Kofoid, 1907; Evitt, 1985). Dinoflagellate cysts are commonly identified based on morphological features such as size, shape, wall structure, surface features, and the archeopyle. Living dinoflagellates often comprise overlapping cellulose plates, to create a sort of armor called the "theca." Resting cysts often bear ridges or linear elements, which simulate thecal sutures and divide the surface into more or less polygonal areas that resemble the thecal plates. The arrangement of these sutural features on the resting cyst is referred to as the "paratabulation." This paratabulation is an important feature used in dinoflagellate cyst taxonomy. Kofoid (1907) developed a classification of dinoflagellate cysts based on this paratabulation, which is referred to as the Kofoid system.

In the fossil record, over 4400 fossil dinocyst species have been identified (Williams et al., 2017). It is no surprise that with such abundance and diversity in the more recent fossil record (Evitt, 1985), dinoflagellates represent one of the major branches of study by palynologists and are especially useful in describing the dynamics of Mesozoic and Cenozoic oceans (Fensome et al., 1996). Dinoflagellate cysts have been extensively used for biostratigraphy and correlation (Stover et al., 1996), but because they are sensitive to water temperature, salinity, and depth, they are also valuable as paleoenvironmental indicators (Dale, 1996; Sluijs et al., 2005; Vellekoop et al., 2015). For example, the abundance ratio between two major dinoflagellates groups, the Gonyaulacaceae and Peridiniaceae, is routinely used as an indication for productivity and/or nutrient availability in surface waters (Dale, 1996; Sluijs et al., 2005).

Pollen and Spores

Pollen and spores represent the most important group of microfossils in terrestrial settings. Beside scolecodonts, they differ from any other palynomorph group described within this chapter as they do not represent individual organisms, but rather reproductive structures produced during the life cycle of plants. Spores (Playford and Dettemann, 1996) are produced by cryptogams (the so-called lower plants), which include bryophytes (mosses, liverworts, and hornworts) and pteridophytic vascular plants (ferns, lycopsids, and horsetails). Pollen (Jarzen and Nichols, 1996), on the other hand, is produced by seed plants, the gymnosperms (conifers and older relatives) and angiosperms (flowering plants). Because terrestrial plants are affixed to their substratum, their reproduction and dispersion relies on the propagation of pollen and spores, which are produced in vast numbers and released into the environment, traveling widely and rapidly in wind and/or water, and eventually settling on the surface of the land and on water bodies, including places where preservation is most likely, such as at the bottom of ponds, lakes, rivers, and oceans. The fact that they are very resistant, microscopic, produced in large numbers, and disseminated over large areas makes them an ideal biostratigraphic group. The outer wall of pollen and spores (the exine), is made of sporopollenin, one of the most resistant and chemically inert biopolymers (Fig. 2). Whereas pollen and spores are mostly produced by land plants (embryophytes), it is not uncommon to recover them from nearshore marine deposits where they may be transported by river systems into estuaries and coastal lagoons. The broad range of dispersal of pollen and spores make terrestrial to marine correlation possible, and is one of their most valuable biostratigraphic functions. Pollen and spores are also useful paleoecological and paleoenvironmental markers when they can be linked to a parent plant, and when the ecology of that parent plant is known. Collectively,

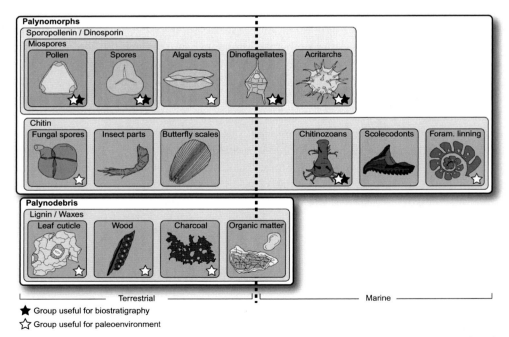

Fig. 2 The composition of the major groups of palynomorphs and their origin (terrestrial and/or marine).

pollen and spores are referred to using the term miospores (Fig. 2), sporomorphs, or sometimes sporopollen in the Chinese literature.

Spores appear very early in the fossil record, associated with the colonization of land by the first terrestrial plants. The oldest known forms of sporelike bodies are called cryptospores, from the Middle Cambrian (~500 million years old, Fig. 1, Strother, 2016). The first trilete spores appear in the late Ordovician (Fig. 1). Whereas palynological evidence is the oldest indication for the existence of true land plants, the first fossils of macroscopic land plants, the Cooksoniales, is recorded later during the Middle Silurian (~425 million years, Fig. 1). The temporal disconnect or delay between the first appearance of a plant group in the palynological record and its later appearance as macrofossils is a recurrent observation in the rock record. This delay can be simply explained by a taphonomical bias, as pollen and spores offer a better preservational potential than the presumably small and delicate parent plants. Although some cryptospores resemble the spores of modern bryophytes (Gray, 1985; Richardson, 1992; Renzaglia et al., 2015), their botanical affinity is still highly debated. Nonetheless, they remain of great use for biostratigraphy and correlation of early Paleozoic terrestrial and nearshore marine deposits (Richardson, 1996; Streel and Loboziak, 1996; Loboziak et al., 2004). Just as paleontologists have subdivided the Phanerozoic into three major eras, the Paleozoic, Mesozoic, and Cenozoic, each representative of revolutions in faunal assemblages, introducing the next revolution initiated by the emergence of the gymnosperms (conifers, cycadophytes, pteridospermaphytes, Caytoniales, etc.) and their associated cortège of pollen grains during the late Carboniferous (Clayton, 1996; Owens, 1996; Fig. 1). One of the notable characteristics of the late Paleozoic floras is the appearance of taeniate (= striated) pollen near the end of the Carboniferous, which persists up to the end of the Triassic (Warrington, 1996a,b; Playford and Dino, 2004). The last floral revolution is marked by the first significant appearance of angiosperms (flowering plants) in the early Cretaceous and a rapid diversification from the mid-Cretaceous (Batten and Koppelhus, 1996; Batten, 1996a; Frederiksen, 1996; Fig. 1).

Pollen and spores are retrieved from and isolated in sediments, disconnected from their parent plants, meaning that their natural affinities are often obscure. The principle of anatomical connection (e.g., extracting pollen from a fossilized cone or flower), or the use of modern analogues (comparison with extant plants), can help to elucidate the botanical relationship of specific pollen and spores taxa. As is common in paleobotany, where organs are preserved individually and disconnected rather than in the form of entire organisms, a paranomenclatural system based on form taxa has been developed, similar to the way that vertebrate paleontology has its own naming conventions for ichnotaxa and ootaxa. Potonié developed an informal, artificial classification of spores and pollen based on morphological traits, called the turmal system. The problem of identifying parent plants has led to major differences in methods employed for Quaternary (palynology) and pre-Neogene (paleopalynology). In the former, almost all taxa recorded are from extant species, whereas in the latter, the taxa are almost invariably extinct. Thus palynologists use natural genera and species whereas paleopalynologists use form taxa, and many use the turmal system, especially for Paleozoic and Mesozoic taxa. In a similar way, palynologists use a very different approach to palynostratigraphy. Quaternary dating and correlation strongly rely on relative abundance counts (formally called pollen analysis, and represented in a pollen diagram) and the recognition of changing climate indicated by shifts in marker species, whereas studies in deep time traditionally rely on the description of the first and last occurrence of taxa (originations and extinctions) in the stratigraphic record (albeit the benefit of using relative abundance counts is being acknowledged; Phillips et al., 1974; Phillips and Peppers, 1984; Willard, 1993; Eble and Grady, 1990; Willard and Phillips, 1993; Eble et al., 1994; Bercovici et al., 2009b; Dolby et al., 2011).

Cuticles, Wood Fragments, and Palynodebris

Fragments of leaf cuticles are also a common component of terrestrial palynological assemblages. They represent the waxy coating covering the epidermal cells of leaves. Anatomical features such as epidermal cell walls are very often clearly visible, as well as stomata, guard cells, and hair bases. The appearance of cuticles in the fossil record coincides with the evolution of the first vascular plants (tracheophytes, Fig. 1). Plant taxonomy based on cuticles remains challenging as a lot of variability in epidermal cell morphology is expressed through environmental constraints. Nevertheless, general botanical affinities can be deduced (Upchurch, 1984a,b, 1995; Cleal and Zodrow, 1989; Cleal and Shute, 1992, 2003; Cleal et al., 1990; Kerp and Barthel, 1993; Barthel, 1997), and efforts are being made to build a catalog to help identify dispersed cuticular remains (Barclay et al., 2007). Cuticles are, however, extremely valuable for paleoenvironment as stomatal density is a function of atmospheric pCO_2, and can thus provide a record of atmospheric composition through the last $\sim$300 Ma of Earth's history (Beerling et al., 1998; Royer, 2001; Retallack, 2001; Beerling et al., 1998, 2002; Beerling and Royer, 2002a,b; Montañez et al., 2007, 2016). Recent developments are also investigating the use of epidermal cells for characterizing diverse paleoecosystem functions such as canopy density, as epidermal cells vary in area and shape as a function of incident light. Calibration of the method has been conducted on phytholiths (Strömberg, 2004; Strömberg and McInerney, 2011; Dunn et al., 2015a,b), which are silica infilling of epidermal cells (and thus not palynomorphs as they are siliceous, and thus destroyed by the acid extraction method used by palynologists), but the method is readily transferable to dispersed leaf cuticles.

The appearance of wood fragments in the palynological record coincides with the evolution of the first vascular plants (tracheophytes), which possess lignified tissues (xylem) for conduction. Because lignin is a hard and resistant material, wood fragments survive well during transport over long distances, especially as sedimentary particles in streams, and are commonly found in palynological preparations. Several distinctions are made depending on whether the fragments are translucent (brown wood) or opaque (black wood). The shape and size of the wood particles can be used as an indicator of transport and erosion, as is commonly done for siliciclastic sedimentary particles like quartz and feldspars in thin sections. Wood can also be preserved as charcoal fragments, which are carbonized material resulting from wildfire (Scott, 2001, 2009; Belcher et al., 2013; Rimmer et al., 2015). Wildfire is one of the best preservational agents for plant remains; fire instantly removes water from the organic tissues while retaining their three-dimensional shapes (Scott et al., 2000). In addition, heating removes most of the volatiles from the organic matter (OM), leaving charcoal as a nearly inert material. Charcoal fragments over ~200 μm are usually out of the scope of palynology and are considered as "mesofossils/mesoflora"; they have been of considerable help in the study of early flowers (Friis et al., 2011), among other contributions to paleobotany.

Other "Nonpollen" Palynomorphs

Members of the algal-grade green plants also produce microscopic sporopollenin-walled structures (Colbath and Grenfell, 1995; Zippi, 1998; Van Geel, 2001; Worobiec, 2014). For example, the Zygnemataceae are well represented in the fossil record; they produce thick-walled, resistant structures (zygospore) of varying sizes and shapes depending on the species, but which display an equatorial line of rupture into two more or less equal parts. The Zygnemataceae are unicellular green algae restricted to freshwater streams and lakes, and are thus useful as paleoenvironmental indicators. Other common freshwater algae are colonial, such as *Pediastrum*, *Palambages*, and *Botryococcus*.

Fungal spores and hyphae are another form of chitinous microfossil commonly recovered in palynological preparations; these are most generally from the Ascomycetes (Traverse, 2007). While usually not abundant, several exceptional and notable examples exist such as those spread out across the Permian-Triassic transition. The proliferation of fungal spores specifically attributed to *Reduviasporonites*, coined the end-Permian "fungal spike" event, has been observed at many localities worldwide (Eshet et al., 1995; Visscher et al., 1996), and represents the signature of decomposers thriving in a devastated landscape (Looy et al., 1999, 2001; Visscher et al., 2004, 2011; Sephton et al., 2009). Such a short-timed event and global occurrence would represent a unique correlation potential for many terrestrial Permian-Triassic sections. There is still, however, significant debate over the affinity and ecological significance of *Reduviasporonites* due to its close morphological resemblance to the zygnematalean green algae *Tympanicysta* (Afonin et al., 2001; Foster et al., 2002; Spina et al., 2015). Ultimately, geochemical signature (Sephton et al., 2009; Visscher et al., 2011) and preservation potential (Visscher et al., 2011; Bercovici et al., 2015; Hochuli, 2016; Bercovici and Vajda, 2016) could help resolve this issue.

Last, most common in coastal and nearshore marine environments are chitinous inner linings of foraminifera, which are almost always of planispiral forms. Unlike for the calcareous foraminiferal tests, chitinous linings are not useful for taxonomy, but their occurrence can be a useful paleoenvironmental indicator of marine-influenced environments. Chitinous remains of insect and arthropod skeleton parts, butterfly wing scales, and insect and copepod eggs can also be found in palynological residues, but are rare and of very limited use.

METHODS IN PALEOPALYNOLOGY

Palynologists have developed a series of standard methods that are applied to collect, extract, identify, and describe organic-walled microfossils. This chapter presents a quick overview of this workflow.

Sample Collection

Because palynomorphs are microfossils generally in the 5–200 μm range, they are too small to be seen with the naked eye. Palynological sampling thus requires a good knowledge of sedimentology in order to identify those lithologies most likely to preserve OM particles of microscopic size. Ideal facies usually correspond to fine-grained mudstone or silty mudstone, indicating deposition in a calm aqueous environment where palynomorphs could settle and get buried. Coarser lithology like fine sandstones may also produce, but usually the palynological recovery is orders of magnitude lower. Color of the rock is another very important indicator, as usually rocks of dark gray/black color are rich in OM. On the other hand, OM and palynomorphs are destroyed by oxidizing conditions; therefore, paleosols are not good candidates, whereas red, orange, and white rocks are always barren. Local tectonic activity can affect the OM by thermal maturation. Palynomorphs can get "cooked" and change color according to thermal stress, from transparent yellow to orange, dark brown, and opaque/black. Whereas opaque palynomorphs are useless for light microscopy (though they can still be studied using scanning electron microscopy (SEM)), color is a good indicator for evaluating thermal maturation of the OM due to increasing subsurface temperatures (Goodhue and Clayton, 2010).

Whether collected in the field from outcrop or cores, or from material in museum collections, cross-contamination by surrounding matrix or dirty tools should always be of major concern. On outcrop, contamination can easily come from the modern soil, cracks in the rocks, and modern erosion and surface weathering (which can be quite deep). This is especially true for soft rocks that require the excavation of a trench to reach unweathered sediments (Fig. 3A). For core samples, the coring mud is an additional source of potential contamination. Collection of the sample itself is usually straightforward and consists of the extraction of ~10 g of sediment, or more for facies with poor recovery (e.g., sandstones and shallow marine limestones). This is usually done using a

Fig. 3 (A) A typical stratigraphic section in the soft rocks of the lowermost part of the Fort Union Formation (southwestern North Dakota). The *red* stratigraphic markers indicate the contact between different lithological units. Palynological samples were collected in stratigraphic sequence from the clean surface of the highwall using a sharp knife. (B) Palynologist Doug Nichols holds a freshly collected sample from the Hell Creek Formation. The ~50 g of sediment contained within the ziplock bag are enough to make many dozens of palynological slides. (C) Dyed palynological residues stored in vials. The residue is used to mount microscope slides or SEM stubs. (D) Palynological slides for optical microscopy observation. The dyed residue is mounted in a permanent medium for long-term preservation.

rock hammer or a knife in case of soft rocks. Samples can conveniently be stored in airtight ziplock bags with an acid-free paper label inside (Fig. 3B).

Palynological Extraction

Back in the laboratory, samples are processed chemically to remove the mineral fraction of the rocks, typically using strong inorganic acids. Many of the chemicals used in palynological preparation are very hazardous and should be handled by a trained lab technician with appropriate protection and under a fume hood specifically designed to handle hydrofluoric acid (HF). The preparation techniques usually consist of minor variations around a similar theme (Wood et al., 1996). Siliciclastic rocks such as clays, silts, mudstones, sandstones, shales, and carbonates (limestones) are ground down into small fragments of ~5 mm to improve the surface contact with the acid and speed up the chemical reaction. Samples are first covered in hydrochloric acid (HCl) to remove carbonates, then thoroughly washed and rinsed three or four times. The second acid treatment removes silicates using 40%–70% HF for ~24 h in plastic beakers. At the end of the reaction, the organic residue should be apparent as a fine black sludge, which is an indication of a positive sample. Samples are then cleaned of acid by a series of decantations and rinses, and a last treatment using boiling HCl is used to remove the insoluble fluorosilicate crystals that formed during the previous step and that would greatly impair observation. The residue can be checked immediately in a wet mount, and optional steps can be taken if deemed necessary. Sieving is commonly done in order to remove the largest fragments as well as the smallest amorphous organic particles. The standard fraction usually of interest for palynology is 200–10 μm, although some angiosperm pollen grains and fern spores can be <10 μm, and some megaspore can exceed 1 mm. The residue can be checked immediately in a wet mount, and optional steps can be taken if deemed necessary. Centrifugation of macerate in heavy liquid using $ZnCl_2$ or $ZnBr_2$ (density of 1.6–2.5) can be applied to further clean the residue, separating the remaining minerals from organic material, which float to the surface. Some laboratories apply an ultrasound treatment to break up agglutinated particles in the residue. Other treatments using oxidizers can also be applied, including sodium hydroxide (NaOH) or potassium hydroxide (KOH) to remove humic acids, and nitric acid or Schultze's reagent to reduce the opacity of the OM. These steps should be carefully controlled in order to avoid completely oxidizing, and thus destroying, the organic residue. Staining of samples (with Safranin Red dye, which gives a pinkish color to palynomorphs) may be used if palynomorphs are clear/translucent to improve contrast. The residue that is obtained can then be stored in vials, either dried or in alcohol (Fig. 3C) for future use.

Because modern pollen is likely to be carried in surrounding air and tap water, it is not uncommon to have contaminants within samples. This is not a big concern when working with ancient rocks as modern pollen and spores appear three-dimensional and glassy, often with cellular content in contrast to palynomorphs, which have been compressed by sediments. Contamination can be a serious problem when working on Quaternary samples as modern pollen can be indistinguishable from the study sample. Processing needs to be conducted in a hermetically sealed lab environment with positive pressure gradient.

Some exceptions to the acid treatment workflow are worth noting. Carbonaceous shale and coal are often devoid of mineral matter, so that the acid treatments can be omitted and oxidation is the only step required, though treatment with bases or Schultze's solution may be necessary to separate palynomorphs from encasing organic debris. Pollen and spores can also be recovered from amber, with their cellular content (De Franceschi et al., 2000). Finally, extraction of in situ pollen and spores preserved in fossilized plant reproductive structures (sporangia, cones, and flowers; Balme, 1995; Eklund et al., 2004; Friis et al., 2004; Kerp et al., 2013; Zavialova and Karasev, 2015) or from insects (Peñalver et al., 2012; Wappler et al., 2015; Labandeira et al., 2016) and coprolites (Prasad et al., 2005; Vajda et al., 2016) is an important source of behavioral and taxonomic information.

Palynological Sample Mounting and Observation

Traditional transmitted light microscopy, using a compound microscope, is the preferred method of observation for palynological slides. Because of the palynomorph size-range, observation is best done at 100×, 200×, and 400× for slide scanning, palynofacies analysis and counts, while 1000× under oil immersion is more suited for morphological studies and photomicrography. Contrast enhancement techniques reveal details of the surface and ornamentation of palynomorphs, and consist of adjusting the aperture diaphragm, and using differential interference contrast (DIC, also called Nomarski Interference Contrast) on microscopes with special optics and polarizing prisms. Because of the shallow depth of field of optical microscopes, especially at high magnification, focus has to be constantly adjusted to observe the entire grain. This focus-through process allows the palynologist to observe the surface ornamentation, details of the germination aperture, and thickness and layering of the spore/pollen wall, all important taxonomic characteristics. Image postprocessing methods such as focus stacking can be used to reconstruct depth of field on photographs (Bercovici et al., 2009a).

For optical microscopy, the material is prepared and mounted for observation on microscope glass slides (Fig. 3D). The simplest way is to make a wet mount slide by trapping a drop of residue between a microscope slide and a coverslip. Wet mounts are performed throughout the palynological extraction process in order to evaluate the preservational condition and OM recovery and to decide whether extra steps such as sieving or oxidation are necessary. Wet mounts are, however, not suitable for thorough observation as the material can easily move under the coverslip, and water induces a refractive index mismatch for observation at higher magnifications (the refractive index of water is ~1.33, and a desirable mounting medium should have a refractive index similar to glass, ~1.46). Temporary mounts can be made using glycerin jelly as a mounting medium. The biggest advantage of temporary mounts is the possibility of remelting the glycerin jelly to release specimens for observation using other techniques like electron microscopy, or the possibility to roll palynomorphs like tiny ball bearings by sliding the coverslip on top of the slide. Glycerin jelly mounts can have a shelf life of many years if the jelly is spiked with phenol and the coverslip edge is sealed with transparent nail polish to avoid development of mold growth. Ensuring that specimens can be precisely relocated and preserved permanently is instrumental to the verification and replicability of palynological research; thus permanent mounting mediums are used for storage in collection. Canada balsam used to be a traditional mounting medium, but has a tendency to crack and yellow as it ages. Now, modern polymers like epoxy and polyester resin display very good stability, optimum refractive index, and are very easy to use. Several drops of the palynological residue are spread on a coverslip with a drop of polyvinyl alcohol. The polyvinyl alcohol attaches the residue to the coverslip, ensuring that the material lies on the same optical plane. The coverslip is then flipped onto the slide bearing several drops of the mounting medium, and left to cure. Position of important and photographed slides for archiving and taxonomic description can be noted with the use of a gridded micrometer slide called an England Finder.

Electron microscopy can be beneficial in resolving details of the surface and ornamental structure of palynomorphs at much higher magnification than optical microscopy.

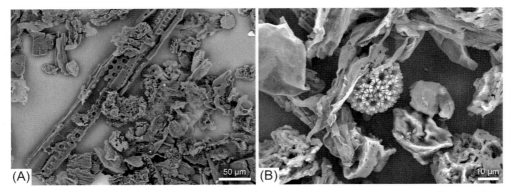

Fig. 4 (A) SEM image of palynological residue (Hell Creek Formation) showing a variety of palynodebris including a large tracheid with pitting (wood fragment). (B) SEM image of *Erdtmanipollis cretaceus*, an uppermost Maastrichtian angiosperm pollen species from the Hell Creek Formation among palynodebris.

SEM provides three-dimensional images of the outer surface of imaged objects (Fig. 4A and B), whereas transmitted electron microscopy (TEM) is used to elucidate the internal structure through ultrathin sections roughly 100 nm thick. Because electron microscopy cannot be performed on material already mounted on optical microscopy slides, it is often desirable to keep palynological residue in vials for mounting on metal stubs and gold plated for SEM, or for embedding and ultramicrotomy sectioning for TEM. As an alternative to TEM, Villanueva-Amadoz et al. (2012) investigated the use of focused ion beam (FIB) to microsection palynological material in a controlled manner, directly under SEM. This method enables definition of a specific sectioning path, which is not possible with TEM. Confocal laser scanning microscopy (Feist-Burkhardt and Pross, 1999; Peyrot et al., 2007), a hybrid system between optical microscopy and SEM, uses a scanned laser beam to provide successive fluorescence images of optical slices that can be stacked using dedicated software, to reconstruct three-dimensional images.

Palynological Analysis

Several methods exist to assess the content of palynological slides. A typical biostratigraphic approach would rely on presence/absence data conducted on a series of samples collected in stratigraphic sequence. The identification of the first appearance datum, last appearance datum, and/or association zones of key taxa is used to recognize events and biostratigraphic boundaries that can be correlated. In addition to presence/absence, relative abundance count data can provide extra information, especially for paleoenvironmental characterization. Presence/absence data usually rely on scanning the entire content of a palynological slide, while relative abundance counts rely on the accurate identification of at least 200 palynomorphs for statistical robustness. The identification and counting process is usually done manually by a palynologist, which is often very time consuming. Tools based on computer image recognition have been developed to automate this process (Weller et al., 2006), although they only provide satisfactory results for identifying and sorting palynomorphs into broad categories. In addition to stratigraphic range charts, various multivariate analysis such as principal component analysis or cluster analysis can be employed to explore large palynological datasets and extract information relevant to facies dependency, climate change, and paleobiogeography.

Palynological slides can be characterized though palynofacies analysis in a manner similar to that used for description of sedimentary facies from rock thin sections, in which the proportion, shape, and size of the constitutive minerals are derived. The concept of palynofacies analysis was introduced by Batten (1973, 1996b, 1982) and Batten and Stead (2005) and aims at describing the proportion, shape, color, and size of every piece of organic debris (palynodebris) present on a slide. General palynofacies categories defined by Batten include pollen, spores, dinoflagellates, structured OM including various phytoclasts such as brown wood, black wood, charcoal and cuticles, and amorphous OM. Palynofacies data is highly relevant in the characterization of depositional environments, as the proportion, preservation, and shape of palynodebris is a function of taphonomy. Elements more resistant to mechanical abrasion like wood fragments tend to be more abundant in energetic environments such as river deposits, while more delicate material like palynomorphs and cuticles are more abundant in calm depositional environments such as lacustrine and floodplain deposits. Because depositional settings are responsible for variation in the composition of palynologial assemblages at the first order, it is instrumental for palynologists to record the detailed sedimentological context from which samples are taken and to confront the complementary information given by palynofacies and sedimentary facies data. The insoluble macromolecular OM dispersed in sedimentary rocks, called kerogen, is a major source for oil and natural gas (Batten, 1981a). Thus palynofacies analysis is not only critical for characterizing depositional environments and taphonomy, but is also of prime importance for assessing potential source rocks for petroleum (Vandenbroucke and Largeau, 2007).

A CASE STUDY IN PALYNOSTRATIGRAPHY: THE CRETACEOUS-PALEOGENE TRANSITION

Reconstructing the history of the Earth and the evolution of life ultimately requires that a broad range of geological and paleontological data gathered from different countries and depositional settings can be accurately compared, dated, and correlated. The Cretaceous-Paleogene (K–Pg) boundary represents one of the major transitions in life's history for which palynology can provide crucial correlation data, both on land (using pollen and spores) and in coastal and marine environments (using dinoflagellates). The K–Pg boundary is best known for being the latest mass extinction event (excluding the ongoing "sixth mass extinction"), responsible for the demise of many key groups 66 million years ago, including the nonavian dinosaurs, pterosaurs, (Buffetaut, 1984, 1990; Smit and van der Kaars, 1984; Sheehan et al., 1991; Sheehan and Fastovsky, 1992; Pearson et al., 2001, 2002; Fastovsky and Sheehan, 2005; Lyson et al., 2011; Longrich et al., 2013; Fastovsky and Weishampel, 2017), marine reptiles (Gallager et al., 2012), ammonites, and various marine protists (Smit, 2012).

Death From Above

Prior to the 1980s, paleontologists proposed several theories to explain this extensive die-off. These included gradual scenarios as a response to global climate and sea-level changes, or strictly dinosaur-centric models, which failed to explain the demise of entire marine and terrestrial ecosystems. The gradualistic ideas were challenged by Alvarez et al. (1980), who reported the discovery of a sharp iridium anomaly that they detected in Gubbio (Italy) and Stevns Klint in Denmark (Fig. 5A), two outcrops exposing marine rocks that span the K-Pg boundary. Such an anomalously high concentration of iridium could only be linked to an extraterrestrial origin, and Alvarez et al. (1980) proposed that a large impact from an extraterrestrial body occurred precisely at the K-Pg boundary and was responsible for the observed mass extinctions (Alvarez et al., 1982, 1984; Alvarez, 1983). The Alvarez publication spawned a breadth of research spanning every discipline of geology and paleontology in attempts to test this new hypothesis (Smit and Hertogen, 1980; Alvarez, 1997; Nichols and Johnson, 2008).

The results of Alvarez et al. (1980) have been independently replicated over the years with the identification of a threefold impact-induced stratigraphic sequence comprising (1) an ~1-cm-thick kaolinitic "boundary" clay layer representing the low angle ejecta, very fine material blasted at supersonic speeds from the impact site (Izett, 1990; Schulte et al., 2010); (2) an ~1–2-cm-thick layer containing millimeter-sized spherules called microtectites representing the high angle ejecta, or incandescent material blasted from the impact site, which solidifies into glass as it rains down from the atmosphere (Bohor et al., 1987a; Izett, 1990; Ocampo et al., 1996; Schulte et al., 2010); and (3) an ~1–2-cm-thick fining upward sequence containing shocked quartz fractured by intense pressure at the impact site (Bohor et al., 1984, 1987b; Bohor and Izett, 1986; Izett, 1990; Alvarez et al., 1995; Nichols and Johnson, 2002), Ni spinels (Robin et al., 1992; Rocchia et al., 1996) and the iridium anomaly (and associated metals such as Fe, Cr, Ni, and Au) from the blasted impactor material (Orth et al., 1981, 1982; Izett, 1990; Sweet et al., 1999; Lerbekmo et al., 1999; Sweet and Braman, 2001; Vajda et al., 2001, 2004; Wigforss-Lange et al., 2007; Schulte et al., 2010; Ferrow et al., 2011).

Today, about 350 K-Pg boundary sections have been identified (Fig. 6, Schulte et al., 2010), among which 105 are from terrestrial deposits (Nichols and Johnson, 2008). Along these localities, the K-Pg boundary impact sequences increase in thickness approaching the potential impact site where the thickest ejecta is recorded (Claeys et al., 2002;

Fig. 5 (A) The K-Pg boundary in the nearshore marine deposits of Stevns Klint (Denmark), located within the ~5-cm-thick Fish Clay Member. (B) The K-Pg boundary in the terrestrial deposits of Mud Buttes (North Dakota, USA). (C) Latest Cretaceous red floodplain and paleosol sequences in Provence (southern France).

Schulte et al., 2010). Hildebrand et al. (1991) finally identified an ~180-km-diameter crater of compatible size and age in the Yucatán Peninsula (Mexico) using gravity anomaly maps, under the village of Chicxulub. The thickest deposits occur in Belize and southern Mexico, proximal to the crater, where the successions reach up to ~45 m of impact breccia and spherules (Ocampo et al., 1996). Farther away in northern Mexico and Texas, spherules and ejecta are mixed within an ~10-m sequence of tsunami deposits (Schulte et al., 2006, 2011). In North America and Europe, the K–Pg impact sequence is usually 1–5 cm

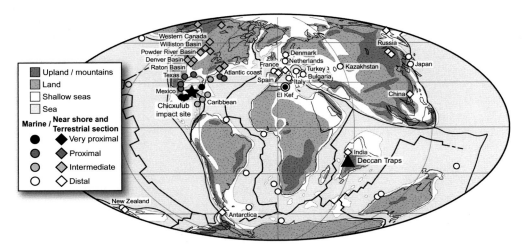

Fig. 6 Location of the known marine (*circles*), coastal and terrestrial (*diamonds*) K-Pg boundary sections, color coded depending on the type and thickness of impact ejecta deposits (see Schulte et al., 2010). The Global Boundary Stratotype Section and Point (GSSP) for the K-Pg is located in the shallow marine deposits of El Kef, Tunisia.

thick. The thinnest deposits constitute of only a few millimeters in antipodal regions such as New Zealand (Moody Creek Mine, Vajda et al., 2001). The impact sequence has been dated with a compatible age of 66.0 Ma at many locations (Schulte et al., 2010; Kamo et al., 2011; Renne et al., 2013; Sprain et al., 2015; Clyde et al., 2016).

Independently of the implications of such a large asteroid impact, the simple fact that ejected material rained down in a matter of hours to days to form a uniform blanket on the surface of the Earth, across continents and oceans, represents a unique opportunity for global correlation. Thus the K-Pg boundary and the base of the Danian stage is set to coincide with the base of the boundary claystone and the presence of the iridium anomaly, as it has been officially defined at the El Kef Global Stratotype Section and Point in Tunisia (Molina et al., 2006). However, because of the incompleteness of the sedimentary record, the critical K-Pg boundary impact sequence and its suite of indicators can be missing altogether because of nonpreservation, nondeposition, or erosion, which are common phenomena within terrestrial deposits.

Blowin' in the Wind

The asteroid impact theory of mass extinction proposed by Alvarez et al. (1980) revolutionized the understanding of the K-Pg boundary. However, evidence pointing to the indication of an abrupt and catastrophic event had existed for more than a decade. The answer was blowing in the wind in the form of pollen and spores: Palynologists had observed firsthand a profound and synchronous change in terrestrial palynofloras at the K-Pg boundary from several North American localities in the Western Interior (Norton and Hall, 1969; Oltz, 1969; Leffingwell, 1970; Tschudy, 1970, 1984). Contrasting with the apparent gradual image given by dinosaurs and other fossil vertebrate data of very coarse resolution and often lacking the appropriate stratigraphic framework at the time, palynologists used a rigorous biostratigraphic approach, helped by the high temporal resolving power and large amount of microfossils found in the Western Interior K–Pg deposits.

Today, this change in palynoflora has been reported from over a hundred localities worldwide (Nichols and Johnson, 2008; Bercovici et al., 2012a; Vajda and Bercovici, 2012, 2014), and palynology represents the second best option to identify the K–Pg and correlate terrestrial deposits when the geochemical and mineralogical evidence of the Chicxulub impact are absent. These localities are mostly within North America, including the Raton Basin in Colorado and New Mexico (Pillmore et al., 1984, 1999; Tschudy et al., 1984; Nichols and Fleming, 1990), the Denver Basin in Colorado (Nichols and Fleming, 2002; Barclay et al., 2003), the Powder River Basin of Wyoming and Montana (Lance Formation, Leffingwell, 1970; Wolfe, 1991; Nichols et al., 1992), the Williston Basin in Montana and North Dakota (Hell Creek Formation, Norton and Hall, 1969; Tschudy, 1984; Nichols et al., 2000; Hotton, 2002; Johnson et al., 2002; Fastovsky and Bercovici, 2016), extending farther north into Canada (Saskatchewan and Alberta, Lerbekmo et al., 1987; Nichols et al., 1986; Sweet and Braman, 1992, 2001; Sweet et al., 1999), and all the way up to the Canadian Northwest Territories (Sweet et al., 1990, 1999; Sweet and Braman, 2001). Palynological evidence also has been adduced on other continents (Saito et al., 1986; Brinkhuis and Schiøler, 1996), most specifically in New Zealand (Vajda et al., 2001, 2004; Vajda and Raine, 2003). The North American record has proved to be the most valuable for documenting the evolution of the latest Cretaceous terrestrial ecosystems leading up to the K–Pg boundary, because of the existence of numerous excellent exposures of upper Maastrichtian to lower Danian strata spanning the K–Pg boundary (Fig. 5B). These outcrops were located on vast deltaic floodplains on the shore of the Western Interior Seaway, allowing for fast sedimentation rates and burial and preservation of abundant fossils of all kinds (vertebrates, invertebrates, plant macrofossils, and pollen). The presence of gray, organic-rich deposits (Fig. 5B) ensures abundant recovery of fossil pollen and spores; in contrast, many of the terrestrial outcrops of the K–Pg boundary outside of North America consist of red beds (Fig. 5C). Such outcrops are devoid of palynomorphs and other OM, are usually difficult to assess using other means, and thus remain stratigraphically unconstrained. For this reason, we will use the palynological record of the Hell Creek Formation of southwestern North Dakota (Fastovsky and Bercovici, 2016) as an example for the case study.

The Hell Creek Palynoflora and the End of the Cretaceous

By the end of the Cretaceous, flowering plants already dominated the global vegetation, supplanting the ferns, conifers, Bennettitales, and seed ferns that had dominated earlier Mesozoic floras (Heimhofer et al., 2007; Friis et al., 2011). The latest Maastrichtian is divided into four main palynological provinces (Fig. 7) defined by the restricted geographic occurrence of specific species of pollen (Herngreen and Chlonova, 1981; Herngren et al., 1996; Vajda and Bercovici, 2014). The Western Interior of North America is contained within the *Aquilapollenites* province (Fig. 7), which is characterized by the occurrence of triprojectate pollen, a morphological group of angiosperm pollen that features a set of three distinctive arm-like projections, each bearing a pore at the distal end. The *Aquilapollenites-Triprojectacites* complex is of uncertain affinity, but shares morphological resemblance to some pollen grains produced by the modern Santalales (Funkhouser, 1961; Jarzen, 1977; Muller, 1984), and reached its maximum distribution and diversity during the Maastrichtian, but most representatives became extinct at the K-Pg boundary, with rare occurrences up to the Oligocene (Farabee, 1991). The palynology of the upper Maastrichtian successions is well constrained, and the Hell Creek assemblages correspond to the *Wodehouseia spinata* Assemblage Zone, the first occurrence of this species marking the base of the upper Maastrichtian (Nichols, 1990, 1994, 2007; Bercovici et al., 2012a; Braman and Sweet, 2012). The *Wodehouseia spinata* assemblage of North America features diverse and abundant fern spores and angiosperm pollen, including several species of *Aquilapollenites* (Nichols and Johnson, 2002).

The palynological assemblage of the Hell Creek Formation is extremely diverse (Nichols, 2002; Hotton, 2002; Bercovici et al., 2009b) and paints the image of a lush, subtropical to warm-temperate evergreen forest with a mix of deciduous and evergreen elements. The understory vegetation was represented by many hydrophilous ground-

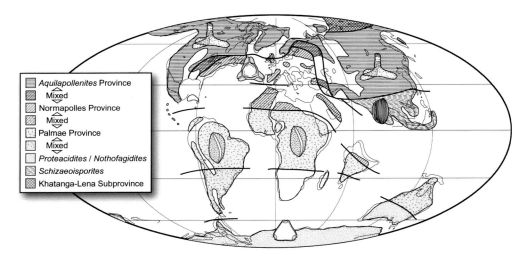

Fig. 7 The main palynological provinces of the Maastricthtian, with representative pollen morphologies. *From Vajda, V., Bercovici, A., 2014. The global vegetation pattern across the Cretaceous-Paleogene mass extinction interval; a template for other extinction events. Glob. Planet. Chang. 122, pp. 29–49; (Modified from Herngren, G.F.W., Kevdes, M., Rovnina, L.V., Smirnova, S.B., 1996. Cretaceous palynofloral provinces: a review. In: Jansonius, J., McGregor, D.C. (Eds.), Palynology: Principles and Applications, vol. 3. American Association of Stratigraphic Palynologists, pp. 1157–1188).*

dwelling plants including the several groups of lycopsids (clubmosses [Lycopodiaceae], spikemosses [Selaginellaceae], and quillworts [Isoetaceae]), and elements of the major groups of bryophytes (hornworts [Anthocerotales], liverworts [Marchantiales], and mosses [sphagnum]). Most of the understory vegetation was probably dominated by ferns (Nichols, 2002; Vajda et al., 2013), which are represented by many spore species, and generally account for ~50% of Hell Creek palynological assemblages. Heterosporous aquatic fern spores are also present, including the free-floating fern *Azolla*, which formed mats on standing water surfaces. Shrubby gymnosperms are represented, including cycads (*Cycadopites* sp., which can also correspond to *Ginkgo*), which had mostly replaced bennettitaleans by the end of the Cretaceous (Nichols and Johnson, 2008).

The Hell Creek canopy comprised evergreen conifers, especially the Pinaceae (pine family). Bisaccate pollen grains with large sacci (*Podocarpidites* sp.) suggest the presence of podocarps, a group today mainly endemic in the Southern Hemisphere. Other relict plant groups may be present, such as Cheirolepidaceae (*Classopollis* sp.), pteridosperms, and Caytoniales (*Alisporites* sp., *Vitreisporites* sp.) (Vajda et al., 2013). Araucariaceae were very marginal components of the flora if present at all (Johnson, 2002). Angiosperms are the most diverse group, as they likely occupied various habitats, from herbs to understory tree to the canopy. They included monocots such as Arecaceae (palms) and various other freshwater aquatic species. Woody dicots represented most of the diversity recorded, which matches well with the very diverse leaf floras recovered from the Hell Creek Formation (Johnson et al., 1989; Johnson and Hickey, 1990; Johnson, 2002; Wilf and Johnson, 2004).

Floral Turnover at the K-Pg Boundary

The mass extinction and floral turnover that occurred at the K-Pg boundary are readily identifiable using presence/absence and relative abundance counts thanks to the disappearance of many pollen species. Thus, the palynologically defined K-Pg boundary is based on the extinction of Cretaceous taxa (Hotton, 1988, 2002; Nichols, 2002), which is synchronous with the Chicxulub impact but independent of other factors such as local paleoenvironmental variations (Nichols and Fleming, 1990). This change is also recorded synchronously in the leaf floras, which show an even greater level of extinction (Wolfe and Upchurch, 1986, 1987; Johnson, 1989, 1992; Upchurch, 1989; Johnson, 2002; Barclay et al., 2003; Barclay and Johnson, 2004; Wilf and Johnson, 2004; Blonder et al., 2014). The Cretaceous taxa that go extinct at the K-Pg boundary have been termed "K-taxa" (Nichols, 2002) or "K species" (Hotton, 2002), and comprise 32 angiospermous taxa and one gymnosperm (Nichols, 2002; see Fig. 8 for a depiction of some representative K-taxa). K-taxa are usually quite abundant up to the K-Pg boundary, with an average relative abundance of ~15% (min: ~5%, max: ~50%, Bercovici et al., 2009b; Fig. 9A). The K-taxa comprise almost all species of *Aquilapollenites* and many other angiosperm taxa that are very distinctively shaped; thus recognition of their disappearance from palynological assemblages is generally straightforward, contributing to the

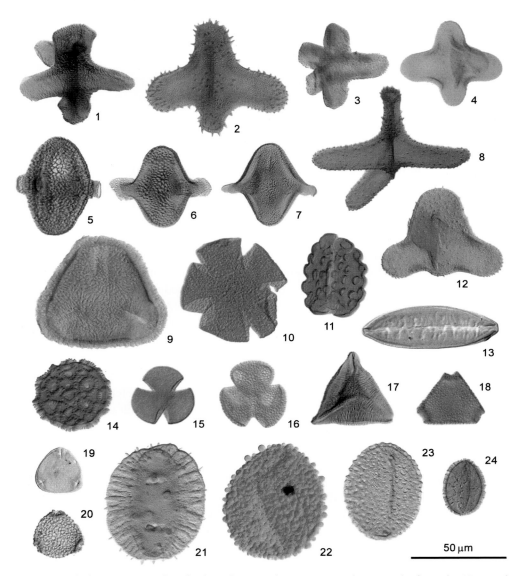

Fig. 8 Optical photomicrographs of selected terminal Cretaceous palynomorphs from the Marmarth area (Williston Basin, North Dakota). 1: *Aquilapollenites conatus* Norton 1965; 2: *Aquilapollenites attenuatus* Funkhouser 1961; 3: *Aquilapollenites quadricretaeus* Chlonova 1961; 4: *Aquilapollenites marmarthensis* Nichols 2002; 5: *Aquilapollenites reductus* Norton 1965; 6: *Aquilapollenites oblatus* Srivastava 1968; 7: *Aquilapollenites mtchedlishvili* (Mtchedlishvili, 1961) Tschudy and Leopold 1971; 8: *Aquilapollenites collaris* (Tschudy and Leopold, 1971) Nichols 1994; 9: *Styxpollenites calamitas* Nichols 2002; 10: *Leptopecopites pocockii* (Srivastava, 1967) Srivastava 1978; 11: *Marsypiletes cretaceus* Robertson 1973 emend. Robertson and Elsik 1978; 12: *Aquilapollenites quadrilobus* Srivastava & Rouse, 1970; 13: *Ephedripites multipartitus* (Chlonova, 1961) Yu et al., 1981; 14: *Erdtmanipollis* sp.; 15: *Tricolpites microreticulatus* Belsky, Boltenhagen and Potonié 1965; 16: *Libopollis jarzenii* Farabee et al., 1984; 17: *Striatellipollis striatellus* (Mtchedlishvili *in* Samoilovitch et al., 1961) Krutzsch 1969; 18: *Tschudypollis retusus* (Anderson, 1960) Nichols 2002; 19: *Myrtipites scabratus* Norton *in* Norton and Hall 1969; 20: *Retibrevitricolporites beccus* Sweet 1986; 21: *Wodehouseia spinata* Stanley 1961; 22: *Racemonocolpites formosus* Sweet 1986; 23: *Liliacidites complexus* (Stanley, 1965) Leffingwell 1971; 24: *Liliacidites altimurus* Leffingwell 1971. All taxa with the exception of *Wodehouseia spinata* are considered to be restricted to the Cretaceous and can be used as markers of the K-Pg mass extinction event.

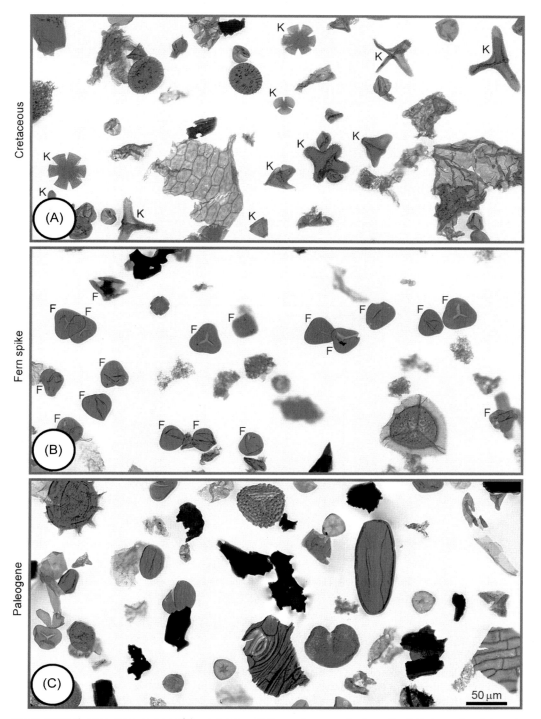

Fig. 9 Optical microscope image of three representative pollen samples from the K-Pg boundary interval. (A) ~50 cm below the K-Pg boundary, showing a diverse assemblage of angiosperm pollen including K-taxa (marked with "K"); (B) 1 cm above the K-Pg boundary, showing abundance of the trilete fern spore *Cyathidites* sp. (marked with "F"); (C) ~50 cm above the K-Pg boundary, showing a diverse assemblage of pollen and spores, but no K-taxa. Stratigraphic position of each image is indicated in Fig. 10.

speed and precision of palynological identification of the K–Pg boundary. Sweet and Braman (2001) proposed that these pollen taxa, featuring a thick and structurally complex exine, could be zoophilous (more especially entomophilous, or insect pollinated), indicating a possible relationship with the extinction of insects at the K–Pg boundary (Labandeira et al., 2002a,b; Wilf et al., 2006; Rehan et al., 2013; Donovan et al., 2014).

In addition to simply tracking the disappearance of the K-taxa using a simple presence/absence approach, a relative abundance count can greatly help in providing a more accurate placement for the K–Pg boundary (Bercovici et al., 2009b; Vajda and Bercovici, 2014). This is mainly due to the fact that K-taxa can persist in very low abundance (<0.5%), for a short period of time after the K–Pg boundary (Fig. 10). The palynofloral turnover is clear, from assemblages with many K-taxa to assemblages with almost none, and with a general change in the abundance and distribution of boundary-crossing taxa (Bercovici et al., 2009b). This synchronous and abrupt change is very desirable for correlation, and is used throughout the many localities of the Western Interior (Nichols and

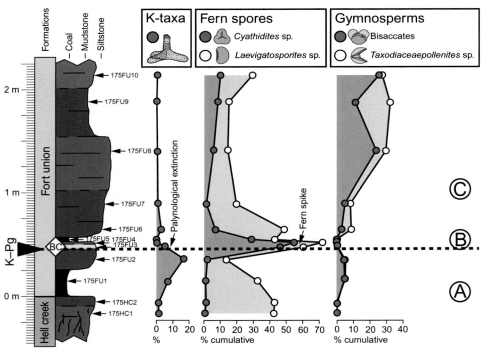

John's Nose section (Bercovici et al., 2012b)

Fig. 10 Relative abundance data for 12 palynological samples from the John's Nose K–Pg boundary section in southwestern North Dakota (Bercovici et al., 2012b). The relative abundance data show a major drop and near extinction of the K-taxa at the K–Pg boundary, an abundance spike of the spore taxa *Cyathidites* sp. followed by *Laevigatosporites* sp. immediately after the boundary, and an increased abundance of gymnosperms in the earliest Paleocene. A, B, and C refer to pollen assemblages shown in Fig. 9.

Johnson, 2008). However, because the vast majority of the K-taxa are bound to the *Aqui-lapollenites* province (Fig. 7), identification of the K-Pg boundary and correlation with other regions of the world can be problematic. In these other provinces, the disappearance of different suites of taxa is used (Batten, 1981b; Vajda and Raine, 2003; Ferrow et al., 2011; Bercovici et al., 2012a; Vajda and Bercovici, 2012).

The K-Pg Boundary Fern Spike Event

Compositionally, pollen samples recovered from ~1 cm immediately above the K-Pg boundary are virtually devoid of K-taxa, but are also characterized by an unusually high abundance of fern spores, representing >70% of the assemblage (Figs. 9B and 10, Orth et al., 1981; Tschudy et al., 1984; Fleming and Nichols, 1988, 1990; Nichols and Fleming, 1990; Wolfe, 1991). The taxonomic signature is also very specific, corresponding to a succession of two monospecific spikes, the first composed of the trilete spore *Cyathidites* sp., and the second composed of the monolete spore *Laevigatosporites* sp. (Fig. 10). This event, called the "fern spike," has been detected in many K-Pg boundary localities across the world (Vajda and Bercovici, 2014), and is generally interpreted as an indication of plant cover devastation (Fleming and Nichols, 1988, 1990; Wolfe, 1991; Sweet et al., 1990, 1999; Sweet and Braman, 2001; Vajda et al., 2001, 2004; Vajda and McLoughlin, 2004, 2007; Vajda and Bercovici, 2014).

In this model, the blast induced by the Chicxulub impact would have destroyed a vast proportion of the plant cover, leaving extensive areas of denuded land. Ferns are pioneer recolonizers of devastated landscapes, rapidly germinating from spores, and are classic examples of so-called disaster floras. Recent examples of fern spikes include the localized early recolonization of the denuded ground by ferns after the massive landslides induced by the 1980 eruption of Mount St. Helens, or the short-lived dominance of ferns in the Krakatau floras following the 1883 eruption (Tschudy et al., 1984). In a similar way, ferns are the first colonizers of freshly deposited lava flows in Hawaii, a biotic response to destabilized landscapes. Further evidence of the loss of the plant cover is given from the New Zealand K-Pg localities, showing proliferation of fungi (a "fungal spike"), increased microbial activity at Stevns Klint (Denmark, Sepúlveda et al., 2009), and worm burrows at Mud Buttes (North Dakota, Chin et al., 2013), all indicating proliferation of saprotrophs thriving on decomposing OM during an early phase postimpact when sunlight was suppressed by soot ejected into the atmosphere (Vajda and McLoughlin, 2004; Vajda, 2012).

Because both *Cyathidites* sp. and *Laevigatosporites* sp. are ubiquitous components of any palynological assemblage in the late Cretaceous, caution should be taken not to overinterpret their presence as a representation of the K-Pg boundary fern spike. Moreover, the very narrow stratigraphic interval in which the fern spike is recorded, only 1–2 cm thick, can easily be missed if the sampling resolution is too coarse. Coupled with the disappearance of K-taxa, a general increase in the relative abundance of fern spores can be used as a supporting evidence of floral turnover associated with the K-Pg boundary.

Recolonization and Appearance of New Species

Whereas the changes observed at the K–Pg within palynological assemblages are drastic and abrupt, the earliest Paleocene flora does not give the sense of an entirely new world. Since the extinction involved the 33 K-taxa (~30% of the species), the immediate recovery was sourced from the surviving boundary crossers (Fig. 9C). Ultimately, the earliest Paleocene palynoflora is less diverse, and assemblages have a tendency to be dominated by a few pollen species (Bercovici et al., 2009b). Palynological assemblages in the Western Interior usually show increased abundance of freshwater aquatic taxa (Bercovici et al., 2009b; Fastovsky and Bercovici, 2016), and an increased relative abundance of pollen from conifers (*Taxodiaceae* and *Pinaceae*, Fig. 10), which are long-lived plants of climax vegetation, and were high pollen producers (Sweet et al., 1999; Sweet and Braman, 2001; Bercovici et al., 2009b).

New Paleocene-restricted species such as *Momipites inaequalis* appear; however, these new Paleocene taxa remain relatively rare and hard to identify (Nichols, 2003; Braman and Sweet, 2012; Bercovici et al., 2012b; Vajda and Bercovici, 2014). More importantly, the first occurrence of the new Paleocene taxa is delayed. Consequently, they cannot be used as reliable stratigraphic markers for the K–Pg boundary.

Palynological Correlation of the Terrestrial K–Pg Boundary With the Marine Realm

Comparable to the terrestrial K–Pg boundary sites, palynology also provides a powerful tool for biostratigraphy and correlation of the K–Pg boundary transition in the marine realm. In the marine realm, dinoflagellates enable a very detailed biostratigraphic zonation, allowing for a detailed age assessment of uppermost Cretaceous and lowermost Paleogene deposits (Brinkhuis et al., 1998; Williams et al., 2004; Vellekoop et al., 2015). Shallow marine records, like estuaries and coastal lagoons, are especially well suited for correlating terrestrial to marine records, as these generally comprise both terrestrial palynomorphs and dinoflagellate cysts. As an example, the shallow marine mid-Waipara River section in New Zealand comprises a diverse dinoflagellate assemblage (Willumsen, 2003), as well as abundant terrestrial palynomorphs, recording the K–Pg boundary fern spike event (Vajda et al., 2001). Permitting a direct correlation between the terrestrial and marine realm, these sites are crucial in assessing the timing and extent of environmental changes across the K–Pg boundary interval.

CONCLUSIONS

Whether used for biostratigraphy, paleoenvironmental analysis, or characterization of organic content of rocks, palynology has proven to be an essential tool for the study of unoxidized terrestrial sediments and sedimentary rocks. Because palynomorphs are very small and abundant, they are ideally suited for core sample study and for high-resolution biostratigraphic analysis, and have been instrumental in the establishment of biozones for the correlation of terrestrial and nearshore marine deposits. Fossil pollen and spores can be used to reconstruct a picture of past vegetation and can provide information on ancient climates. Furthermore, palynological processing and analysis is cost effective and provides a fast turnaround, in comparison with other analytical techniques.

ACKNOWLEDGMENTS

Antoine Bercovici is currently supported by a Deep Time Peter Buck Postdoctoral Fellowship from the NMNH Smithsonian Institution. Assistance in the field was kindly provided by Dean A. Pearson at the Pioneer Trails Regional Museum in Bowman, North Dakota, and by Tyler R. Lyson at the Marmarth Research Foundation and Denver Museum of Nature and Science. Carol Hotton and Bill DiMichele (NMNH Smithsonian Institution) are thanked for their review and helped to improve the English style of this manuscript.

REFERENCES

Afonin, S.A., Barinova, S.S., Krassilov, V.A., 2001. A bloom of *Tympanicysta* Balme (green algae of zygnematalean affinities) at the Permian–Triassic boundary. Geodiversitas 23, 481–487.

Alvarez, L.W., 1983. Experimental evidence that an asteroid impact led to the extinction of many species 65 million years ago. Proc. Natl. Acad. Sci. U. S. A. 80, 627–642.

Alvarez, W., 1997. T. rex and the Crater of Doom. Princeton University Press, Princeton, NJ. 208 p.

Alvarez, L.W., Alvarez, W., Asaro, F., Michel, H.V., 1980. Extraterrestrial cause for the Cretaceous-Tertiary extinction. Science 208, 1095–1108. http://dx.doi.org/10.1126/science208.4448.1095.

Alvarez, W., Alvarez, L.W., Asaro, F., Michel, H.V., 1982. Current status of the impact theory for the terminal Cretaceous extinction. In: Silver, L.T., Schultz, P.H. (Eds.), Geological Implications of Impacts of Large Asteroids and Comets on the Earth. Geological Society of America, Boulder, CO, pp. 305–315. Geological Society of America Special Paper 190.

Alvarez, W., Kauffman, E.G., Surlyk, F., Alvarez, L.W., Asaro, F., Michel, H.V., 1984. Impact theory of mass extinctions and the invertebrate fossil record. Science 223, 1135–1141.

Alvarez, W., Claeys, P., Kieffer, S.W., 1995. Emplacement of Cretaceous–Tertiary boundary shocked quartz from Chicxulub crater. Science 269, 930–935.

Armstrong, A., Brasier, M.D., 2005. Micropaleontology, second ed. Blackwell, Malden, MA. 296 p.

Balme, B.E., 1995. Fossil in site spores and pollen grains: an annotated catalogue. Rev. Palaeobot. Palynol. 87, 81–323.

Barclay, R.S., Johnson, K.R., 2004. West Bijou site Cretaceous–Tertiary boundary, Denver Basin, Colorado. Geol. Soc. Am. Field Guide 5, 59–68.

Barclay, R.S., Johnson, K.R., Betterton, W.J., Dilcher, D.L., 2003. Stratigraphy and megaflora of a K–T boundary section in the eastern Denver Basin, Colorado. Rocky Mt Geol. 38, 45–71.

Barclay, R., McElwain, J.C., Dilcher, D., Sageman, B., 2007. The Cuticle Database: developing an interactive tool for taxonomic and paleoenvironmental study of the fossil cuticle record. Cour. Forsch. Inst. Senckenburg 258, 39–55.

Barthel, M., 1997. Epidermal structures of sphenophylls. Rev. Palaeobot. Palynol. 95, 115–127.

Batten, D.J., 1973. Use of palynologic assemblage-types in Wealden correlation. Palaeontology 16, 1–40.

Batten, D.J., 1981a. Palynofacies, organic maturation and source potential for petroleum. In: Brook, J. (Ed.), Organic Maturation Studies and Fossil Fuel Exploration. Academic Press, London, pp. 201–223.

Batten, D.J., 1981b. Stratigraphic, palaeogeographic and evolutionary significance of Late Cretaceous and Early Tertiary *Normapolles* pollen. Rev. Palaeobot. Palynol. 35, 125–137.

Batten, D.J., 1982. Palynofacies, palaeoenvironments and petroleum. J. Micropalaeontol. l, 107–114.

Batten, D.J., 1996a. Upper Jurassic and Cretaceous miospores. In: Jansonius, J., McGregor, D.C. (Eds.), In: Palynology: Principles and Applications, vol. 2. American Association of Stratigraphic Palynologists, pp. 807–831.

Batten, D.J., 1996b. Palynofacies and palaeoenvironmental interpretation. In: Jansonius, J., McGregor, D.C. (Eds.), In: Palynology: Principles and Applications, vol. 2. American Association of Stratigraphic Palynologists, pp. 1011–1064.

Batten, D.J., Koppelhus, E.B., 1996. Biostratigraphic significance of Uppermost Triassic and Jurassic miospores in Northwestern Europe. In: Jansonius, J., McGregor, D.C. (Eds.), In: Palynology: Principles and Applications, vol. 2. American Association of Stratigraphic Palynologists, Dallas, TX, pp. 795–806.

Batten, D.J., Stead, D.T., 2005. Palynofacies analysis and its stratigraphic application. In: Koutsokos, E.A.M. (Ed.), Applied Stratigraphy. Springer, Dordrecht, pp. 203–226.

Beerling, D.J., Royer, D.L., 2002a. Reading a CO_2 signal from fossil stomata. New Phytol. 153, 387–397.

Beerling, D.J., Royer, D.L., 2002b. Fossil plants as indicators of the phanerozoic global carbon cycle. Annu. Rev. Earth Planet. Sci. 30, 527–556.

Beerling, D.J., McElwain, J.C., Osborne, C.P., 1998. Stomatal responses of the 'living fossil' *Ginkgo biloba* L. to changes in atmospheric CO_2 concentrations. J. Exp. Bot. 49, 1603–1607.

Beerling, D.J., Lomax, B.H., Royer, D.L., Upchurch Jr., G.R., Kump, L.R., 2002. An atmospheric pCO_2 reconstruction across the Cretaceous-Tertiary boundary from leaf megafossils. Proc. Natl. Acad. Sci. U. S. A. 99, 7836–7840.

Belcher, C.M., Collinson, M.E., Scott, A.C., 2013. A 450-million-year history of fire. In: Belcher, C.M. (Ed.), Fire Phenomena and the Earth System: An Interdisciplinary Guide to Fire Science. John Wiley & Sons, Oxford. http://dx.doi.org/10.1002/9781118529539.ch12.

Bercovici, A., Vajda, V., 2016. Terrestrial Permian–Triassic boundary sections in South China. Glob. Planet. Chang. 143, 31–33.

Bercovici, A., Pearson, D., Nichols, D.J., Wood, J., 2009a. Biostratigraphy of selected K/T boundary sections in southwestern North Dakota, USA: toward a refinement of palynological identification criteria. Cretac. Res. 30, 632–658.

Bercovici, A., Hadley, A., Villanueva-Amadoz, U., 2009b. Improving depth of field resolution for palynological photomicrography. Paleontol. Electron. 12. PE Article Number: 12.2.5T.

Bercovici, A., Vajda, V., Sweet, A., 2012a. Pollen and spore stratigraphy of the Cretaceous–Paleogene mass-extinction interval in the Northern Hemisphere. J. Stratigr. 36, 165–178.

Bercovici, A., Vajda, V., Pearson, D., Villanueva-Amadoz, U., Kline, D., 2012b. Palynostratigraphy of John's Nose, a new Cretaceous–Paleogene boundary section in southwestern North Dakota, USA. Palynology 36, 36–47.

Bercovici, A., Cui, Y., Forel, M.-B., Yu, J., Vajda, V., 2015. Terrestrial paleoenvironment characterization across the Permian–Triassic boundary in South China. J. Asia Earth Sci. 98, 225–246.

Bloeser, B., Schopf, J.W., Horodyski, R.J., Breed, J.W., 1977. Chitinozoans from the Late Precambrian Chuar Group of the Grand Canyon, Arizona. Science 195, 67–69.

Blonder, B., Royer, D.L., Johnson, K.R., Miller, I., Enquist, B.J., 2014. Plant ecological strategies shift across the Cretaceous–Paleogene boundary. PLoS Biol. 12, e1001949. http://dx.doi.org/10.1371/journal.pbio.1001949.

Bohor, B.F., Izett, G., 1986. World-wide size distribution of shocked quartz at the K/T boundary: evidence for a North American impact site. Lunar Planet. Sci. 52, 68–69.

Bohor, B.F., Foord, E.E., Modreski, P.J., Triplehorn, D.M., 1984. Mineralogic evidence for an extraterrestrial impact event at the Cretaceous–Tertiary boundary. Science 224, 867–869.

Bohor, B.F., Modreski, P.J., Foord, E.E., 1987a. Shocked quartz in the Cretaceous–Tertiary boundary clays—evidence for a global disruption. Science 236, 705–709.

Bohor, B.F., Triplehorn, D.M., Nichols, D.J., Millard, H.T., 1987b. Dinosaurs, spherules and the "magic layer"; a new K–T boundary clay site in Wyoming. Geology 15, 896–899.

Braman, D.R., Sweet, A.R., 2012. Biostratigraphically useful Late Cretaceous–Paleocene Terrestrial palynomorphs from the Canadian Western Interior Sedimentary Basin. Palynology 36, 8–35.

Brinkhuis, H., Schiøler, P., 1996. Palynology of the Geulhemmerberg, Cretaceous/Tertiary boundary section (Limburg, SE Netherlands). Geol. Mijnb. 75, 193–213.

Brinkhuis, H., Bujak, J.P., Smit, J., Versteegh, G.J.M., Visscher, H., 1998. Dinoflagellate-based sea surface temperature reconstructions across the Cretaceous–Tertiary boundary. Palaeogeogr. Palaeoclimatol. Palaeoecol. 141, 67–83.

Buffetaut, E., 1984. Selective extinctions and terminal Cretaceous events. Nature 310, 276.

Buffetaut, E., 1990. Vertebrate extinctions and survival across the Cretaceous–Tertiary boundary. Tectonophysics 171, 337–345.

Cashman, P.B., 1990. The affinity of the chitinozoans: new evidence. Mod. Geol. 5, 59–69.

Chin, K., Pearson, D.A., Ekdale, A.A., 2013. Fossil worm burrows reveal very early terrestrial animal activity and shed light on trophic resources after the end- Cretaceous mass extinction. PLoS One. 8(8)8 e70920. http://dx.doi.org/10.1371/journal.pone.0070920.

Claeys, P., Kiessling, W., Alvarez, W., 2002. Distribution of Chicxulub ejecta at the Cretaceous-Tertiary boundary. In: Koeberl, C., MacLeos, K.G. (Eds.), Catastrophic Events and Mass Extinctions: Impacts and Beyond. Geological Society of America, Boulder, CO, pp. 55–68. Geological Society of America Special Paper 356.

Clayton, G., 1996. Mississipian miospores. In: Jansonius, J., McGregor, D.C. (Eds.), In: Palynology: Principles and Applications, vol. 2. American Association of Stratigraphic Palynologists, Dallas, TX, pp. 589–596.

Cleal, C.J., Shute, C.H., 1992. Epidermal features of some Carboniferous neuropteroid fronds. Rev. Palaeobot. Palynol. 71, 191–206.

Cleal, C.J., Shute, C.H., 2003. Systematics of the Late Carboniferous medullosalean pteridosperm *Laveineopteris* and its associated *Cyclopteris* leaves. Palaeontology 46, 353–411.

Cleal, C.J., Zodrow, E.L., 1989. Epidermal structure of some medullosan *Neuropteris* foliage from the Middle and Upper Carboniferous of Canada and Germany. Palaeontology 32, 837–882.

Cleal, C.J., Shute, C.H., Zodrow, E.L., 1990. A revised taxonomy for Palaeozoic neuropterid foliage. Taxon 39, 486–492.

Clyde, W.C., Ramezani, J., Johnson, K.R., Bowring, S.A., Jones, M.M., 2016. Direct high-precision U-Pb geochronology of the end-Cretaceous extinction and calibration of Paleocene astronomical timescales. Earth Planet. Sci. Lett. 452, 272–280.

Colbath, G.K., Grenfell, H.R., 1995. Review of biological affinities of Paleozoic acid-resistant, organic-walled eukaryotic algal microfossils (including "acritarchs"). Rev. Palaeobot. Palynol. 86, 287–314.

Dale, B., 1983. Dinoflagellate resting cysts: "benthic plankton" In: Fryxell, G.A. (Ed.), Survival Strategies of the Algae. Cambridge University Press, Cambridge, pp. 69–136.

Dale, B., 1996. Dinoflagellate cyst ecology: modelling and geological applications. In: Jansonius, J., McGregor, D.C. (Eds.), In: Palynology: Principles and Applications, vol. 2. American Association of Stratigraphic Palynologists, Dallas, TX, pp. 1249–1276.

De Franceschi, D., Dejax, J., De-Ploeg, G., 2000. Pollen extraction from amber (Sparnacian of Le Quesnoy, Paris Basin): towards a new speciality in palaeo-palynology. Comptes Rendus de l'Academie des Sciences Serie II A Sciences de la Terre et des Planetes 330, 227–233.

Dolby, G., Falcon-Lang, H.J., Gibling, M.R., 2011. A conifer-dominated palynological assemblage from Pennsylvanian (late Moscovian) alluvial drylands in Atlantic Canada: implications for the vegetation of tropical lowlands during glacial phases. J. Geol. Soc. 168, 571–584.

Donovan, M.P., Wilf, P., Labandeira, C.C., Johnson, K.R., Peppe, D.J., 2014. Novel insect leaf-mining after the end-Cretaceous extinction and the demise of Cretaceous leaf miners, Great Plains, USA. PLoS One. 9e103542. http://dx.doi.org/10.1371/journal.pone.0103542.

Downie, C., 1973. Observations on the nature of the acritarchs. Palaeontology 16, 239–259.

Dunn, R.E., Le, T., Strömberg, C.A.E., 2015a. Light environment and epidermal cell morphology in grasses. Int. J. Plant Sci. 176, 832–847. http://dx.doi.org/10.1086/683278.

Dunn, R.E., Strömberg, C.A.E., Madden, R.H., Kohn, M.J., Carlini, A.A., 2015b. Linked canopy, climate, and faunal change in the Cenozoic of Patagonia. Science 347, 258–261.

Eble, C.F., Grady, W.C., 1990. Paleoecological interpretation of a Middle Pennsylvanian coal bed in the Central Appalachian Basin, USA. Int. J. Coal Geol. 16, 255–286.

Eble, C.F., Hower, J.C., Andrews, W.M., 1994. Paleoecology of the Fire Clay coal bed in a portion of the Eastern Kentucky coal field. Palaeogeogr. Palaeoclimatol. Palaeoecol. 106, 287–305.

Eklund, H., Doyle, J.A., Herendeen, P., 2004. Morphological phylogenetic analysis of living and fossil chlobantaceae. Int. J. Plant Sci. 165, 107–151.

Eshet, Y., Rampino, M.R., Visscher, H., 1995. Fungal event and palynological record of ecological crisis and recovery across the Permian-Triassic boundary. Geology 23, 967–980.

Evitt, W.R., 1985. Sporopollenin Dinoflagellate Cysts: Their Morphology and Interpretation. American Association of Stratigraphic Palynologists Foundation, Dallas, TX. 300 pp.

Farabee, M.J., 1991. Botanical affinities of some Triprojectacites fossil pollen. Am. J. Bot. 78, 1172–1181.

Fastovsky, D.E., Bercovici, A., 2016. The Hell Creek Formation and its contribution to the Cretaceous–Paleogene extinction: a short primer. Cretac. Res. 57, 368–390.

Fastovsky, D.E., Sheehan, P.M., 2005. The extinction of the dinosaurs in North America. GSA Today 15, 4–10.

Fastovsky, D.E., Weishampel, D.B., 2017. The Evolution and Extinction of the Dinosaurs, third ed. Cambridge University Press, Cambridge. 485 p.

Feist-Burkhardt, S., Pross, J., 1999. Morphological analysis and description of Middle Jurassic dinoflagellate cyst marker species using confocal laser scanning microscopy, digital optical microscopy and conventional light microscopy. Bull. Centres Rech. Explor. Prod. Elf-Aquitaine 22, 103–145.

Fensome, R.A., Williams, G.L., Barrs, M.S., Freeman, J.E., Hill, J.M., 1990. Acritarchs and fossil prasinophytes: an index to genera, species and intraspecific taxa. AASP Contrib Ser 25, 1–771.

Fensome, R.A., Riding, J.B., Taylor, F.R.J., 1996. Dinoflagellates. In: Jansonious, J., McGregor, D.C. (Eds.), In: Palynology: Principles and Applications, vol. 1. American Association of Stratigraphic Palynologists Foundation, Dallas, TX, pp. 107–169. Chapter 6.

Ferrow, E., Vajda, V., Bender Koch, C., Peucker-Ehrenbrink, B., Suhr-Willumsen, P., 2011. Multiproxy analysis of a new terrestrial and a marine Cretaceous–Paleogene (K–Pg) boundary site from New Zealand. Geochim. Cosmochim. Acta 75, 657–672.

Fleming, R.F., Nichols, D.J., 1988. The "Tschudy effect": fern-spore abundance anomaly at the Cretaceous-Tertiary boundary. Palynology 12, 238.

Fleming, R.F., Nichols, D.J., 1990. The fern-spore abundance anomaly at the Cretaceous–Tertiary Boundary: a regional bio-event in western North America. In: Kauffman, E.G., Walliser, O.H. (Eds.), Extinction Events in Earth History. Lectures Notes in Earth Sciences, 30. Springer Verlag, New York, pp. 351–364.

Foster, C.B., Stephenson, M.H., Marshall, C., Logan, G.A., Greenwood, P.F., 2002. A revision of Reduviasporonites Wilson 1962: description, illustration, comparison and biological affinities. Palynology 26, 35–58.

Frederiksen, N.O., 1996. Uppermost Cretaceous and Tertiary spore/pollen biostratigraphy. In: Jansonius, J., McGregor, D.C. (Eds.), Palynology: Principles and Applications, vol. 2. American Association of Stratigraphic Palynologists, Dallas, TX, pp. 831–841.

Friis, E.M., Pedersen, K.R., Crane, P.R., 2004. Araceae from the Early Cretaceous of Portugal: evidence on the emergence of monocotyledons. Proc. Natl. Acad. Sci. U. S. A. 101, 16565–16570.

Friis, E.M., Crane, P.R., Pedersen, K.R., 2011. Early Flowers and Angiosperm Evolution. Cambridge University Press, Cambridge.

Funkhouser, J.W., 1961. Pollen of the genus *Aquilapollenites*. Micropaleontology 7, 193–198.

Gallager, W.B., Miller, K.G., Sherrell, R.M., Browning, J.V., Field, P.M., Olsson, R.K., Sugarman, P.J., Wahyudi, H., 2012. On the last mosasaurs: late Maastrichtian mosasaurs and the Cretaceous–Paleogene boundary in New Jersey. Bull. Soc. Geol. Fr. 183, 145–150.

Goodhue, R., Clayton, G., 2010. Palynomorph darkness index (PDI)—a new technique for assessing thermal maturity. Palynology 34, 147–156.

Grahn, Y., 1978. Chitinozoan stratigraphy and palaeoecology at the Ordovician–Silurian boundary in Skåne, southern-most Sweden. Sver. Geol. Undersökning, Ser. C 744, 1–16.

Gray, J., 1985. The microfossil record of early land plants: advances in understanding of early terrestrialization, 1970–1984. Philos. Trans. R. Soc. Lond. B309, 167–195.

Heimhofer, U., Hochuli, P.A., Burla, S., Weissert, H., 2007. New records of Early Cretaceous angiosperm pollen from Portuguese coastal deposits: implications for the timing of the early angiosperm radiation. Rev. Palaeobot. Palynol. 144, 39–76.

Herngreen, G.F.W., Chlonova, A.F., 1981. Cretaceous microfloral provinces. Pollen Spores 23, 441–555.

Herngren, G.F.W., Kevdes, M., Rovnina, L.V., Smirnova, S.B., 1996. Cretaceous palynofloral provinces: a review. In: Jansonius, J., McGregor, D.C. (Eds.), In: Palynology: Principles and Applications, vol. 3. American Association of Stratigraphic Palynologists, Dallas, TX, pp. 1157–1188.

Hildebrand, A.R., Penfield, G.T., Kring, D.A., Pilkington, M., Camargo, Z.A., Jacobsen, S.B., Boynton, W.V., 1991. Chicxulub crater: a possible Cretaceous/Tertiary boundary impact crater on the Yucatán Peninsula, Mexico. Geology 19, 867–871.

Hochuli, P.A., 2016. Interpretation of "fungal spikes" in Permian-Triassic boundary sections. Glob. Planet. Chang. 144, 48–50.

Hotton, C.L., 1988. Palynology of the Cretaceous–Tertiary Boundary in Central Montana, U.S.A., and Its Implication for Extraterrestrial Impact. University of California, Davis, CA. Ph.D. dissertation.

Hotton, C.L., 2002. Palynology of the Cretaceous–Tertiary boundary in central Montana: evidence for extraterrestrial impact as a cause for the terminal Cretaceous extinction. In: Hartman, J.H., Johnson, K.R., Nichols, D.J. (Eds.), The Hell Creek Formation and the Cretaceous–Tertiary Boundary in the Northern Great Plains: An Integrated Continental Record of the End of the Cretaceous. Geological Society of America, Boulder, CO, pp. 191–216. Geological Society of America Special Paper 361.

Hyde, H.A., Williams, D.W., 1944. The Right word. Pollen Anal Circ 8, 6.

Izett, G.A., 1990. The Cretaceous/Tertiary Boundary Interval, Raton Basin, Colorado and New Mexico, and Its Content of Shock-Metamorphosed Minerals: Evidence Relevant to the K–T Boundary Impact-Extinction Theory. Geological Society of America, Boulder, CO. Geological Society of America Special Paper 249.

Jaglin, J.C., Paris, F., 1992. Examples of Teratology in the Chitinozoa from the Pridoli of Libya and implications for biological significance of this group. Lethaia 25, 151–164.

Jansonius, J., 1970. Classification and stratigraphic application of Chitinozoa. Proceedings of the North American Paleontological Convention 1969 Part G, 789–808.

Jansonius, J., McGregor, D.C., 1996. Palynology: Principles and Applications. American Association of Stratigraphic Palynologists, Salt Lake City, UT. 3 volumes.

Jarzen, D.M., 1977. *Aquilapollenites* and some Santalalean genera. A botanical comparison. Grana 16, 29–39.

Jarzen, D.M., Nichols, D.J., 1996. Pollen. In: Jansonius, J., McGregor, D.C. (Eds.), Palynology: Principles and Applications, vol. 1. American Association of Stratigraphic Palynologists, Dallas, TX, pp. 261–291.

Johnson, K.R., 1989. A High-Resolution Megafloral Biostratigraphy Spanning the Cretaceous-Tertiary in the Northern Great Plains. Yale University, New Haven, CT. unpubl. Ph.D. thesis, 556 p.

Johnson, K.R., 1992. Leaf-fossil evidence for extensive floral extinction at the Cretaceous-Tertiary boundary, North Dakota, USA. Cretac. Res. 13, 91–117. http://dx.doi.org/10.1016/0195-6671(92)90029-P.

Johnson, K.R., 2002. Megafloral of the Hell Creek and lower fort Union Formations in the western Dakotas: vegetational response to climate change, the Cretaceous-Tertiary boundary event, and rapid marine transgression. In: Hartman, J., Johnson, K.R., Nichols, D.J. (Eds.), The Hell Creek Formation and the Cretaceous-Tertiary Boundary in the Northern Great Plains. Geological Society of America, Boulder, CO, pp. 329–391. Geological Society of America Special Paper 361.

Johnson, K.R., Hickey, L.J., 1990. Megafloral change across the Cretaceous/Tertiary boundary in the northern Great Plains and Rocky Mountains, USA. In: Shapton, V.L., Ward, P.D. (Eds.), Global Catastrophes in Earth History: An Interdisciplinary Conference on Impacts, Volcanism, and Mass Mortality. Geological Society of America, Boulder, CO, pp. 433–444. Geological Society of America Special Paper 247.

Johnson, K.R., Nichols, D.J., Attrep Jr., M., Orth, C.J., 1989. High-resolution leaf-fossil record spanning the Cretaceous/Tertiary boundary. Nature 307, 224–228.

Johnson, K.R., Nichols, D.J., Hartman, H.H., 2002. Hell Creek Formation: a 2002 synthesis. In: Hartman, J., Johnson, K.R., Nichols, D.J. (Eds.), The Hell Creek Formation and the Cretaceous-Tertiary Boundary in the Northern Great Plains. Geological Society of America, Boulder, CO, pp. 503–510. Geological Society of America Special Paper 361.

Kamo, S.L., Lana, C., Morgan, J.V., 2011. U–Pb ages of shocked zircon grains link distal K–Pg boundary sites in Spain and Italy with the Chicxulub impact. Earth Planet. Sci. Lett. 310, 401–408.

Kerp, H., Barthel, M., 1993. Problems of cuticular analysis of pteridosperms. Rev. Palaeobot. Palynol. 78, 1–18.

Kerp, H., Welleman, C.H., Krings, M., Kearney, P., Hass, H., 2013. Reproductive organs and in situ spores of *Asteroxylon* mackiei Kidston & Lang, the most complex plant from the Lower Devonian Rhynie Chert. Int. J. Plant Sci. 174, 293–308.

Kofoid, C.A., 1907. The plats of Ceratium with a note on the unity of the genus. Zool. Anz. 32, 177–183.

Labandeira, C.C., Johnson, K.R., Lang, P., 2002a. Preliminary assessment of insect herbivory across the Cretaceous-Tertiary boundary: major extinction and minimum rebound. In: Hartman, J.H., Johnson, K.R., Nichols, D.J. (Eds.), The Hell Creek Formation and the Cretaceous–Tertiary Boundary in the Northern Great Plains: An Integrated Continental Record of the End of the Cretaceous. Geological Society of America, Boulder, CO, pp. 297–318. Geological Society of America Special Paper 361.

Labandeira, C.C., Johnson, K.R., Wilf, P., 2002b. Impact of the terminal Cretaceous event on plant-insect associations. Proc. Natl. Acad. Sci. U. S. A. 99, 2061–2166.

Labandeira, C.C., Yang, Q., Santiago-Blay, J.A., Hotton, C.L., Monteiro, A., Wang, Y.J., Goreva, Y., Shih, C.K., Silijeström, S., Rose, T.R., Dilcher, D.L., Dong, R., 2016. The evolutionary convergence of mid-Mesozoic lacewings and Cenozoic butterflies. Proc. R. Soc. B 283, 9.

Le Hérissé, A., Gourvennec, R., Wicander, R., 1997. Biogeography of Late Silurian and Devonian acritarchs and prasionphytes. Rev. Palaeobot. Palynol. 98, 105–124.

Leffingwell, H.A., 1970. Palynology of Lance (Late Cretaceous) and Fort Union (Paleocene) formations of the type Lance area, Wyoming. In: Kosanke, R.M., Cross, A.T. (Eds.), Symposium on Palynology of the Late Cretaceous and Early Tertiary. Geological Society of America, Boulder, CO, pp. 1–64. Geological Society of America Special Paper 127.

Lerbekmo, J.F., Sweet, A.R., St. Louis, R.M., 1987. The relationship between the iridium anomaly and palynological floral events at three Cretaceous–Tertiary boundary localities in western Canada. Geol. Soc. Am. Bull. 99, 325–330.

Lerbekmo, J.F., Sweet, A.R., Davidson, R.A., 1999. Geochemistry of the Cretaceous–Tertiary (K–T) boundary interval south-central Saskatchewan and Montana. Can. J. Earth Sci. 36, 717–724.

Loboziak, S., Melo, J.H.G., Streel, M., 2004. Devonian palynostatigraphy in Western Gondwana. In: Koustoukos, E.A.M. (Ed.), Applied Stratigraphy. Springer Verlag, New York, pp. 73–100.

Longrich, N.R., Bhullar, B.-A.S., Gauthier, J.A., 2013. Mass extinction of lizards and snakes at the Cretaceous–Paleogene boundary. Proc. Natl. Acad. Sci. U. S. A. 109, 21396–21401. http://dx.doi.org/10.1073/pnas.1211526110.

Looy, C.V., Brugman, W.A., Dilcher, D.L., Visscher, H., 1999. The delayed resurgence of equatorial forests after the Permian-Triassic ecologic crisis. Proc. Natl. Acad. Sci. U. S. A. 96, 13857–13862.

Looy, C.V., Twitchett, R.J., Dilcher, D.L., Van Konijnenburg-van Cittert, J.H.A., Visscher, H., 2001. Life in the end-Permian dead zone. Proc. Natl. Acad. Sci. U. S. A. 98, 7879–7883.

Lyson, T.R., Bercovici, A., Chester, S.G.B., Sargis, E.J., Pearson, D., Joyce, W.G., 2011. Dinosaur extinction: closing the "3 m gap" Biol. Lett. 7, 925–928. http://dx.doi.org/10.1098/rsbl.2011.0470.

Martin, F., 1993. Acritarchs—a review. Biol. Rev. 69, 475–539.

Miller, M.A., 1996. Chitinozoa. In: Jansonius, J., McGregor, D.C. (Eds.), In: Palynology: Principles and Applications, vol. 1. American Association of Stratigraphic Palynologists, pp. 307–336.

Molina, E., Alegret, L., Arenillas, I., Arz, J.A., Gallala, N., Hardenbol, J., Luterbacher, H., Steurbaut, E., Vandenberghe, N., Zaghbib-Turki, D., 2006. The Global Boundary Stratotype Section and Point for the base of the Danian Stage (Paleocene, Paleogene, "Tertiary", Cenozoic) at El Kef, Tunisia: original definition and revision. Episodes 29, 263–273.

Molyneux, S.G., Le Hérissé, A., Wicander, R., 1996. Paleozoic phytoplankton. In: Jansonious, J., McGregor, D.C. (Eds.), In: Palynology: Principles and Applications, vol. 2. American Association of Stratigraphic Palynologists Foundation, Dallas, TX, pp. 493–529.

Montañez, I.P., Tabor, N.J., Niemeier, D., DiMichele, W.A., Frank, T.D., Fielding, C.R., Isbell, J.L., Birgenheier, L.P., Rygel, M.C., 2007. CO2-forced climate and vegetation instability during Late Paleozoic deglaciation. Science 315, 87–91.

Montañez, I.P., McElwain, J.C., Poulsen, C.J., White, J.D., DiMichele, W.A., Wilson, J.P., Griggs, G., Hren, M.T., 2016. Climate, pCO_2 and terrestrial carbon cycle linkages during late Palaeozoic glacial-interglacial cycles. Nat. Geosci. 9, 824–828.

Muller, J., 1984. Significance of fossil pollen for angiosperm history. Ann. Mo. Bot. Gard. 71, 419–443.

Nichols, D.J., 1990. Geologic and biostratigraphic framework of the non-marine Cretaceous–Tertiary boundary interval in western North America. Rev. Palaeobot. Palynol. 70, 77–88.

Nichols, D.J., 1994. A revised palynostratigraphic zonation of the nonmarine Upper Cretaceous, Rocky Mountain region, United States. In: Caputo, M.V., Peterson, J.A., Franczyk, K.J. (Eds.), Mesozoic Systems of the Rocky Mountain Region, USA. Society of Economic Paleontologists and Mineralogists, Denver, CO, Rocky Mountain Section.

Nichols, D.J., 2002. Palynology and palynostratigraphy of the Hell Creek Formation in North Dakota: a microfossil record of plants at the end of Cretaceous time. In: Hartman, J.H., Johnson, K.R., Nichols, D.J. (Eds.), The Hell Creek Formation and the Cretaceous–Tertiary Boundary in the Northern Great Plains: An Integrated Continental Record of the End of the Cretaceous. Geological Society of America, Boulder, CO, pp. 393–456. Geological Society of America Special Paper 361.

Nichols, D.J., 2003. Palynostratigraphic framework for age determination and correlation of the nonmarine lower Cenozoic of the Rocky Mountains and Great Plains region. In: Raynolds, R.G., Flores, R.M. (Eds.), Cenozoic Systems of the Rocky Mountain Region. Rocky Mountain Section of the Society for Sedimentary Geology (SEPM), Denver, CO, pp. 107–134.

Nichols, D.J., 2007. Selected plant-microfossil records of the terminal Cretaceous event in terrestrial rocks, western North America. Palaeogeogr. Palaeoclimatol. Palaeoecol. 255, 22–34.

Nichols, D.J., Fleming, R.F., 1990. Plant microfossil record of the terminal Cretaceous event in the western United States and Canada. In: Shapton, V.L., Ward, P.B. (Eds.), Global Catastrophes in Earth History: An Interdisciplinary Conference on Impacts, Volcanism, and Mass Mortality. Geological Society of America, Boulder, CO, pp. 445–455. Geological Society of America Special Paper 247.

Nichols, D.J., Fleming, R.F., 2002. Palynology and palynostratigraphy of Maastrichtian, Paleocene and Eocene strata in the Denver Basin, Colorado. Rocky Mt Geol. 37, 135–163.

Nichols, D.J., Johnson, K.R., 2002. Palynology and microstratigraphy of Cretaceous-tertiary boundary sections in southwester North Dakota. In: Hartman, J., Johnson, K.R., Nichols, D.J. (Eds.), The Hell Creek Formation and the Cretaceous-Tertiary Boundary in the Northern Great Plains. Geological Society of America, Boulder, CO, pp. 95–143. Geological Society of America Special Paper 361.

Nichols, D.J., Johnson, K.R., 2008. Plants and the K–T Boundary. Cambridge University Press, Cambridge. 280 p.

Nichols, D.J., Jarzen, D.M., Orth, C.J., Oliver, P.Q., 1986. Palynological and iridium anomalies at Cretaceous-Tertiary boundary, south-central Saskatchewan. Science 231, 714–717.

Nichols, D.J., Brown, J.L., Attrep Jr., M., Orth, C.J., 1992. A new Cretaceous-Tertiary boundary locality in the western Powder River Basin, Wyoming: biological and geological implications. Cretac. Res. 13, 3–30.

Nichols, D.J., Murphy, E.C., Johnson, K.R., Betterton, W.J., 2000. A second K-T boundary locality in North Dakota verified by palynostratigraphy and shocked quartz. Geol. Soc. Am. Abstr. Programs 32 (7), 130.

Nôlvak, J., Grahn, Y., 1993. Ordovician chitinozoan zones from Baltoscandinavia. Rev. Palaeobot. Palynol. 79, 245–269.

Norton, N.J., Hall, J.W., 1969. Palynology of Upper Cretaceous and Lower Tertiary in the type locality of the Hell Creek Formation, Montana, U.S.A. Palaeontogr. Abt. B 125, 1–64.

Ocampo, A.C., Pope, K.O., Fischer, A.G., 1996. Ejecta blanket deposits of the Chicxulub crater from Albion Island, Belize. In: Ryder, G., Fastovsky, D., Gartner, S. (Eds.), The Cretaceous–Tertiary Event and Other Catastrophes in Earth History. Geological Society of America, Boulder, CO, pp. 75–88. Geological Society of America Special Paper 307.

Oltz, D.F., 1969. Numerical analyses of palynological data from Cretaceous and early Tertiary sediments in east central Montana. Palaeontogr. Abt. B 128, 90–166.

Orth, C.J., Gilmore, J.S., Knight, J.D., Pillmore, C.L., Tschudy, R.H., Fassett, J.E., 1981. An iridium anomaly at the Cretaceous–Tertiary boundary in northern New Mexico. Science 214, 1341–1343.

Orth, C.J., Gilmore, J.S., Knight, J.D., Pillmore, C.L., Tschudy, R.H., Fassett, J.E., 1982. Iridium abundance measurements across the Cretaceous/Tertiary boundary in the San Juan and Raton Basins of northern New Mexico. In: Silver, L.T., Schultz, P.H. (Eds.), Geological Implications of Impacts of Large Asteroids and Comets on the Earth. Geological society of America, Boulder, CO, pp. 423–433. Geological Society of America Special Paper 190.

Owens, B., 1996. Upper Carboniferous spores and pollen. In: Jansonius, J., McGregor, D.C. (Eds.), In: Palynology: Principles and Applications, vol. 2. American Association of Stratigraphic Palynologists, Dallas, TX, pp. 597–606.

Paris, F., 1990. The Ordovician chitinozoan biozones of the Northern Gondwana Domain. Rev. Palaeobot. Palynol. 66, 181–209.

Paris, F., 1996. Chitinozoan biostratigraphy and palaeoecology. In: Jansonius, J., McGregor, D.C. (Eds.), In: Palynology: Principles and Applications, vol. 2. American Association of Stratigraphic Palynologists, Dallas, TX, pp. 531–552.

Paris, F., Achab, A., Asselin, E., Chen, X., Grahn, Y., Nölvak, J., Obut, O., Samuelsson, J., Sennikov, N., Vecoli, M., Verniers, J., Wang, X., Winchester-Seeto, T., 2004. Chitinozoans. In: Webby, D.B., Paris, F., Droser, M.L., Percival, G. (Eds.), The Great Ordovician Biodiversification Event. Columbia University Press, New York, pp. 294–311.

Pearson, D.A., Schaefer, T., Johnson, K.R., Nichols, D.J., 2001. Palynologically calibrated vertebrate record from North Dakota consistent with abrupt dinosaur extinction at the Cretaceous–Tertiary boundary. Geology 29, 39–42.

Pearson, D.A., Schaefer, T., Johnson, K.R., Nichols, D.J., Hunter, J.P., 2002. Vertebrate biostratigraphy of the Hell Creek Formation in southwestern North Dakota and northwestern South Dakota. In: Hartman, J., Johnson, K.R., Nichols, D.J. (Eds.), The Hell Creek Formation and the Cretaceous-Tertiary Boundary in the Northern Great Plains. Geological society of America, Boulder, CO, pp. 145–167. Geological Society of America Special Paper 361.

Peñalver, E., Labandeira, C.C., Barrón, E., Delclòs, X., Nel, P., Nel, A., Tafforeau, P., Soriano, C., 2012. Thrips pollination of Mesozoic gymnosperms. Proc. Natl. Acad. Sci. U. S. A. 109, 8623–8628.

Peyrot, D., Barrón, E., Comas-Rengifo, M.J., Thouand, E., Tafforeau, P., 2007. A confocal laser scanning and conventional wide field light microscopy study of *Classopollis* from the Toarcian-Aalenian of the Fuentelsaz section (Spain). Grana 46, 217–226.

Phillips, T.L., Peppers, R.A., 1984. Changing patterns of Pennsylvanian coal-swamp vegetation and implications of climatic control on coal occurrence. Int. J. Coal Geol. 3, 205–255.

Phillips, T.L., Peppers, R.A., Avcin, M.J., Laughnan, P.F., 1974. Fossil plants and coal: patterns of change in Pennsylvanian coal swamps of the Illinois Basin. Science 184, 1367–1369.

Pillmore, C.L., Tschudy, R.H., Orth, C.J., Gilmore, J.S., Knight, J.D., 1984. Geologic framework of non-marine Cretaceous-Tertiary boundary sites, Raton Basin, New Mexico and Colorado. Science 223, 1180–1183.

Pillmore, C.L., Nichols, D.J., Fleming, R.F., 1999. Field guide to the continental Cretaceous-Tertiary boundary in the Raton Basin, Colorado and New Mexico. In: Lagerson, D.R., Lester, A.P., Trudgill, B.D. (Eds.), Colorado and Adjacent Areas. Geological Society of America, Boulder, CO, pp. 135–155. Geological Society of America Field Guide 1.

Playford, G., Dettemann, M.E., 1996. Spores. In: Jansonius, J., McGregor, D.C. (Eds.), Palynology: Principles and Applications, vol. 1. American Association of Stratigraphic Palynologists, Dallas, TX, pp. 227–260.

Playford, G., Dino, R., 2004. Carboniferous and Permian palynostratigraphy. In: Koustoukos, E.A.M. (Ed.), Applied Stratigraphy. Springer Verlag, New York, pp. 101–122.

Prasad, V., Strömberg, C.A.E., Alimohammadian, H., Sahini, A., 2005. Dinosaur coprolites and the early evolution of grasses and grazers. Science 310, 1177–1180.

Rehan, S.M., Leys, R., Schwarz, M.P., 2013. First evidence for a massive extinction event affecting bees close to the K-T boundary. PLoS One. 8e76683. http://dx.doi.org/10.1371/journal.pone.0076683.

Renne, P.R., Deino, A.L., Hilgen, F.J., Kuiper, K.F., Mark, D.F., Mitchell III, W.S., Morgan, L.E., Mundil, R., Smit, J., 2013. Time scales of critical events around the Cretaceous-Paleogene boundary. Science 339, 684–687.

Renzaglia, K.S., Crandall-Stotler, B., Pressel, S., Duckett, J.G., Schuette, S., Strother, P.K., 2015. Permanent spore dyads are not 'a thing of the past': on their occurrence in the liverwort *H. aplomitrium* (Haplomitriopsida). Botanical J. 179, 658–669.

Retallack, G.J., 2001. A 300-million-year record of atmospheric carbon dioxide from fossil plant cuticles. Nature 411, 287–290.

Ribeiro, S., Berge, T., Lundholm, M., Andersen, T.J., Abrantes, F., Ellegaard, M., 2011. Phytoplankton growth after a century of dormancy illuminates past resilience to catastrophic darkness. Nat. Commun. 2, 311. http://dx.doi.org/10.1038/ncomms1314.

Richardson, J.B., 1992. Origin and evolution of the earliest land plants. In: Scopf, J.W. (Ed.), Major Events in the History of Life. Jones and Bartlett, Boston, MA, pp. 95–118.

Richardson, J.B., 1996. Lower and Middle Palaeozoic records of terrestrial palynomorphs. In: Jansonius, J., McGregor, D.C. (Eds.), Palynology: Principles and Applications, vol. 2. American Association of Stratigraphic Palynologists, Dallas, TX, pp. 555–574.

Rimmer, S.M., Hawkins, S.J., Scott, A.C., Cressler, W.L., 2015. The rise of fire: Fossil charcoal in late Devonian marine shales as an indicator of expanding terrestrial ecosystems, fire, and atmospheric change. Am. J. Sci. 315, 713–733. http://dx.doi.org/10.2475/08.2015.01.

Robin, E., Bonté, P., Froget, L., Jéhanno, C., Rocchia, R., 1992. Formation of spinels in cosmic objects during atmospheric entry: a clue to the Cretaceous–Tertiary boundary event. Earth Planet. Sci. Lett. 108, 181–190.

Rocchia, R., Robin, E., Froget, L., Gayraud, J., 1996. Stratigraphic Distribution of Extraterrestrial Markers at the Cretaceous–Tertiary Boundary in the Gulf of Mexico Area: Implications for the Temporal Complexity of the Event. Geological society of America, Boulder, CO. Geological Society of America Special Paper 307, pp. 279–286.

Royer, D.L., 2001. Stomatal density and stomatal index as indicators of paleoatmospheric CO_2 concentration. Rev. Palaeobot. Palynol. 114, 1–28.

Saito, T., Yamanoi, T., Kaiho, K., 1986. End–Cretaceous devastation of terrestrial flora in the boreal Far East. Nature 323, 253–255.

Schulte, P., Speijer, R.P., Mai, H., Kontny, A., 2006. The Cretaceous–Paleogene (K–P) boundary at Brazos, Texas: sequence stratigraphy, depositional events and the Chicxulub impact. Sediment. Geol. 184, 77–109.

Schulte, P., Alegret, L., Arenillas, I., Arz, J.A., Barton, P.J., Bown, P.R., Bralower, T.J., Christeson, G.L., Claeys, P., Cockell, C.S., Collins, G.S., Deutsch, A., Goldin, T.J., Goto, K., Grajales-Nishimura, J.M., Grieve, R.A.F., Gulick, S.P.S., Johnson, K.R., Kiessling, W., Koeberl, C., Kring, D.A., MacLeod, K.G., Matsui, T., Melosh, J., Montanari, A., Morgan, J.V., Neal, C.R., Nichols, D.J., Norris, R.D., Pierazzo, E., Ravizza, G., Rebolledo-Vieyra, M., Reimold, W.U., Robin, E., Salge, T., Speijer, R.P., Sweet, A.R., Urrutia-Fucugauchi, J., Vajda, V., Whalen, M.T., Willumsen, P.S., 2010. The Chicxulub asteroid impact and mass extinction at the Cretaceous–Paleogene boundary. Science 327, 1214–1218.

Schulte, P., Smit, J., Deutsch, A., Salge, T., Friese, A., Beichel, K., 2011. Tsunami backwash deposits with Chicxulub impact ejecta and dinosaur remains from the Cretaceous–Palaeogene boundary in the La Popa Basin, Mexico. Sedimentology 59, 20–45.

Scott, A.C., 2001. Preservation by fire. In: Briggs, D.E.G., Crowther, P.J. (Eds.), Palaeobiology II. Blackwells, Oxford, pp. 277–280.

Scott, A.C., 2009. Forest fire in the fossil record. In: Cerdà, A., Robichaud, P. (Eds.), Fire Effects on Soils and Restoration Strategies. Science Publishers, New Hampshire, pp. 1–37.

Scott, A.C., Moore, J., Brayshay, B., 2000. Fire and the palaeoenvironment. Palaeogeogr. Palaeoclimatol. Palaeoecol. 164, 1–412.

Sephton, M.A., Visscher, H., Looy, C.V., Verchovsky, A.B., Watsin, J.S., 2009. Chemical constitution of a Permian-Triassic disaster species. Geology 37, 875–878.

Sepúlveda, J., Wendler, J.E., Summons, R.E., Hinrichs, K., 2009. Rapid resurgence of marine productivity after the Cretaceous–Paleogene mass extinction. Science 326, 129–132.

Servais, T., Li, J., Molyneux, S., Raevsaya, E., 2003. Ordovician organic-walled microphytoplankton (acritarch) distribution: the global scenario. Palaeogeogr. Palaeoclimatol. Palaeoecol. 195, 149–172.

Servais, T., Li, J., Stricanne, L., Vecoli, M., Wicander, R., 2004. Acritarchs. In: Webby, D.B., Paris, F., Droser, M.L., Percival, G. (Eds.), The Great Ordovician Biodiversification Event. Columbia University Press, New York, pp. 348–360.

Sheehan, P.M., Fastovsky, D.E., 1992. Major extinctions of land-dwelling vertebrates at the Cretaceous-Tertiary boundary, eastern Montana. Geology 20, 556–560.

Sheehan, P.M., Fastovsky, D.E., Hoffmann, R.G., Berghaus, C.B., Gabriel, D.L., 1991. Sudden extinction of the dinosaurs. Latest Cretaceous, upper Great Plains, USA. Science 254, 835–839.

Shen, C., Aldridge, R.J., Williams, M., Vandenbroucke, T.R.A., Zhang, X.-G., 2013. The earliest chitinozoans discovered in the Cambrian Duyun fauna of China. Geology 41, 191–194. http://dx.doi.org/10.1130/G33763.1.

Sluijs, A., Pross, J., Brinkhuis, H., 2005. From greenhouse to icehouse; organic-walled dinoflagellate cysts as paleoenvironmental indicators in the Paleogene. Earth Sci. Rev. 68, 281–315. http://dx.doi.org/10.1016/j.earscirev.2004.06.001.

Smit, J., 2012. A flashback on the dawn of the meteorite impact/extinction theory. Acta Palaeontol. Pol. 57, 677–679.

Smit, J., Hertogen, J., 1980. An extraterrestrial event at the Cretaceous-Tertiary boundary. Nature 285, 198–200.

Smit, J., van der Kaars, W.A., 1984. Terminal Cretaceous extinctions in the Hell Creek area, Montana: compatible with catastrophic extinctions. Science 223, 1177–1179.

Spina, A., Cirilli, S., Utting, J., Jansonius, J., 2015. Palynology of the Permian and Triassic of the Tesero and Bulla sections (Western Dolomites, Italy) and consideration about the enigmatic species *Reduviasporonites chalastus*. Rev. Palaeobot. Palynol. 218, 3–14.

Sprain, C., Renne, P.R., Wilson, G.P., Clemens Jr., W.A., 2015. High-resolution chronostratigraphy of the terrestrial Cretaceous-Paleogene transition and recovery interval in the Hell Creek region, Montana. Geol. Soc. Am. Bull. 127, 393–409. http://dx.doi.org/10.1130/B31076.1.

Stover, L.E., Brinkhuis, H., Damassa, S.P., de Verteuil, L., Helby, R.J., Monteil, E., Partridge, A.D., Powell, A.J., Riding, J.B., Smelror, M., Williams, G.L., 1996. Mesozoic-Tertiary dinoflagellates, acritarchs and prasinophytes. In: Jansonious, J., McGregor, D.C. (Eds.), In: Palynology: Principles and Applications, vol. 2. American Association of Stratigraphic Palynologists Foundation, pp. 641–750.

Streel, M., Loboziak, S., 1996. Middle and Upper Devonian miospores. In: Jansonius, J., McGregor, D.C. (Eds.), Palynology:Principles and Applications, vol. 2. American Association of Stratigraphic Palynologists, Dallas, TX, pp. 575–587.

Strömberg, C.A., 2004. Using phytolith assemblages to reconstruct the origin and spread of grass-dominated habitats in the great plains of North America during the late Eocene to early Miocene. Palaeogeogr. Palaeoclimatol. Palaeoecol. 207, 239–275.

Strömberg, C.A., McInerney, F.A., 2011. The Neogene transition from C3 to C4 grasslands in North America: assemblage analysis of fossil phytoliths. Paleobiology 37, 50–71.

Strother, P.K., 2016. Systematics and evolutionary significance of some new cryptospores from the Cambrian of easter Tennessee, USA. Rev. Palaeobot. Palynol. 227, 28–41.

Sweet, A.R., Braman, D.R., 1992. The K–T boundary and the contiguous strata in western Canada: interactions between paleoenvironments and palynological assemblages. Cretac. Res. 13, 31–79.

Sweet, A.R., Braman, D.R., 2001. Cretaceous–Tertiary palynofloral perturbations and extinctions within the *Aquilapollenites* phytogeographic province. Can. J. Earth Sci. 38, 249–269.

Sweet, A.R., Braman, D.R., Lerbekmo, J.F., 1990. Palynofloral response to the K/T boundary events: a transitory interruption within a dynamic system. In: Sharpton, V.L., Ward, P.D. (Eds.), Global Catastrophes in Earth history: An Interdisciplinary Conference on Impacts, Volcanism, and Mass Mortality. Geological society of America, Boulder, CO, pp. 457–469. Geological Society of America Special Paper 247.

Sweet, A.R., Braman, D.R., Lerbekmo, J.F., 1999. Sequential palynological changes across the composite Cretaceous–Tertiary boundary claystone and contiguous strata, western Canada and Montana, U.S.A. Can. J. Earth Sci. 36, 743–768.

Szaniawski, H., 1996. Scolecodonts. In: Jansonius, J., McGregor, D.C. (Eds.), Palynology: Principles and Applications, vol. 1. American Association of Stratigraphic Palynologists, Dallas, TX, pp. 337–354.

Traverse, A., 2007. Paleopalynology. Springer, New York. 814 pp.

Tschudy, R.H., 1970. Palynology of the Cretaceous-Tertiary boundary in the northern Rocky Mountain and Mississippi Embayment regions. In: Kosanke, R.M., Cross, A.T. (Eds.), Symposium on Palynology of the Late Cretaceous and Early Tertiary. Geological society of America, Boulder, CO, pp. 65–111. Geological Society of America Special Paper 127.

Tschudy, R.H., 1984. Palynological evidence for change in continental floras at the Cretaceous Tertiary boundary. In: Berggen, W.A., Van Couvering, J.A. (Eds.), Catastrophes in Earth History: The New Uniformitarianism. Princeton University Press, Princeton, NJ, pp. 315–337.

Tschudy, R.H., Pillmore, C.L., Orth, C.J., Gilmore, J.S., Knight, J.D., 1984. Disruption of the terrestrial plant ecosystem at the Cretaceous-Tertiary boundary, Western Interior. Science 225, 1030–1032.

Upchurch Jr., G.R., 1984a. Cuticular evolution in Early Cretaceous angiosperms from the Potomac Group of Virginia and Maryland. Ann. Mo. Bot. Gard. 71, 518–546.

Upchurch Jr., G.R., 1984b. The cuticular anatomy of early angiosperm leaves from the Lower Cretaceous Potomac Group of Virginia and Maryland, Part 1, Zone 1 leaves. Am. J. Bot. 71, 192–202.

Upchurch Jr., G.R., 1989. Terrestrial environmental changes and extinction patterns at the Cretaceous-Tertiary boundary, North America. In: Donovan, S.K. (Ed.), Mass Extinction: Processes and Evidence. Belhaven Press, London, pp. 195–216.

Upchurch Jr., G.R., 1995. Dispersed angiosperm cuticles: their history, preparation, and application to the rise of angiosperms in Cretaceous to Paleocene coals, southern Western Interior of North America. Int. J. Coal Geol. 28, 161–227.

Vajda, V., 2012. Fungi, a driving force in normalization of the terrestrial carbon cycle following the end–Cretaceous extinction. In: Talent, J.A. (Ed.), Earth and Life. Global Biodiversity, Extinction Intervals and Biogeographic Perturbations Through Time. Springer, Science, Dordrecht, pp. 132–144.

Vajda, V., Bercovici, A., 2012. Pollen and spore stratigraphy of the Cretaceous-Paleogene mass-extinction interval in the Southern Hemisphere. J. Stratigr. 36, 154–165.

Vajda, V., Bercovici, A., 2014. The global vegetation pattern across the Cretaceous–Paleogene mass extinction interval: a template for other extinction events. Glob. Planet. Chang. 122, 29–49.

Vajda, V., McLoughlin, S., 2004. Fungal proliferation at the Cretaceous–Tertiary boundary. Science 303, 1489.

Vajda, V., McLoughlin, S., 2007. Extinction and recovery patterns of the vegetation across Cretaceous–Palaeogene boundary—a tool for unravelling the causes of the end–Permian mass–extinction. Rev. Palaeobot. Palynol. 144, 99–112.

Vajda, V., Raine, J.I., 2003. Pollen and spores in marine Cretaceous/Tertiary boundary sediments at mid–Waipara River, North Canterbury, New Zealand. N. Z. J. Geol. Geophys. 46, 255–273.

Vajda, V., Raine, J.I., Hollis, C.J., 2001. Indication of global deforestation at the Cretaceous–Tertiary boundary by New Zealand fern spike. Science 294, 1700–1702.

Vajda, V., Raine, J.I., Hollis, C.J., Strong, C.P., 2004. Global effects of the Chicxulub impact on terrestrial vegetation—review of the palynological record from New Zealand Cretaceous/Tertiary boundary. In: Dypvik, H., Clayes, P. (Eds.), Cratering in Marine Environments and on Ice. Springer Verlag, New York, pp. 57–74.

Vajda, V., Lyson, T.R., Bercovici, A., Doman, J., Pearson, D.A., 2013. A snapshot into the terrestrial ecosystem of an exceptionally well-preserved dinosaur (Hadrosauridae) from the Upper Cretaceous of North Dakota, USA. Cretac. Res. 46, 114–122.

Vajda, V., Pesquero Fernández, M.D., Villanueva-Amadoz, U., Lehsten, V., Alcalá, L., 2016. Dietary and environmental implications of Early Cretaceous predatory dinosaur coprolites from Teruel, Spain. Palaeogeogr. Palaeoclimatol. Palaeoecol. 464, 134–142.

van Geel, B., 2001. Non-pollen palynomorphs. In: Smol, J.P., Birks, H.J.B., Last, W.M. (Eds.), Tracking Environmental Change Using Lake Sediments. In: Terrestrial, Algal and Siliceous Indicators, vol. 3. Kluwer, Dordrecht, pp. 99–119.

Vandenbroucke, M., Largeau, C., 2007. Kerogen origin, evolution and structure. Org. Geochem. 38, 719–833.

Vellekoop, J., Smit, J., van de Schootbrugge, B., Weijers, J.W.H., Galeotti, S., Sinninghe Damsté, J.S., Brinkhuis, H., 2015. Palynological evidence for prolonged cooling along the Tunisian continental shelf following the K–Pg boundary impact. Palaeogeogr. Palaeoclimatol. Palaeoecol. 426, 216–228.

Verniers, J., Nestor, V., Paris, F., Dufka, P., Sutherland, S., Vangrootel, G., 1995. Global chitinozoa biozonation for the Silurian. Geol. Mag. 132, 651–666.

Vidal, G., Knoll, A.H., 1993. Proterozoic plankton. Mem. Geol. Soc. Am. 161, 265–267.

Villanueva-Amadoz, U., Benedetti, A., Méndez, J., Sender, L.M., Diez, J.B., 2012. Focused ion beam nano-sectioning and imaging: a new method in characterisation of paleopalynological remains. Grana 51, 1–9.

Visscher, H., Brinkhuis, H., Dilcher, D.L., Elsik, W.C., Eshet, Y., Looy, C.V., Rampino, M.R., Traverse, A., 1996. The terminal Paleozoic fungal event: evidence of terrestrial ecosystem destabilization and collapse. Proc. Natl. Acad. Sci. U. S. A. 93, 2155–2158.

Visscher, H., Looy, C.V., Collinson, M.E., Brinkhuis, H., Van Konijnenburg-Van Cittert, J.H.A., Küschner, W.M., Sephton, M.A., 2004. Environmental mutagenesis during the end-Permian ecological crisis. Proc. Natl. Acad. Sci. U. S. A. 101, 12952–12956.

Visscher, H., Sephton, M.A., Looy, C.V., 2011. Fungal virulence at the time of the end-Permian biosphere crisis? Geology 39, 883–886.

Wappler, T., Labandeira, C.C., Engel, M.S., Zetter, R., Grímsson, F., 2015. Specialized and generalized pollen-collection strategies in an ancient bee lineage. Curr. Biol. 25, 3092–3098. http://dx.doi.org/10.1016/j.cub.2015.09.021.

Warrington, G., 1996a. Permian spores and pollen. In: Jansonius, J., McGregor, D.C. (Eds.), In: Palynology: Principles and Applications, vol. 2. American Association of Stratigraphic Palynologists, pp. 607–619.

Warrington, G., 1996b. Triassic spores and pollen. In: Jansonius, J., McGregor, D.C. (Eds.), In: Palynology: Principles and Applications, vol. 2. American Association of Stratigraphic Palynologists, pp. 755–766.

Weller, A.F., Harris, A.J., Ware, J.A., 2006. Artificial neural networks as potential classification tools for dinoflagellate cyst images: a case using the self-organizing map clustering algorithm. Rev. Palaeobot. Palynol. 141, 287–302.

Wigforss-Lange, J., Vajda, V., Ocampo, A., 2007. Trace element concentrations in the Mexico-Belize ejecta layer: a link between the Chicxulub impact and the global Cretaceous–Paleogene boundary. Meteorit. Planet. Sci. 42, 1871–1882.

Wilf, P., Johnson, K.R., 2004. Land plant extinction at the end of the Cretaceous: a quantitative analysis of the North Dakota megafloral record. Paleobiology 30, 347–368.

Wilf, P., Labandeira, C.C., Johnson, K.R., Ellis, B., 2006. Decoupled plant and insect diversity after the end-Cretaceous extinction. Science 313, 1112–1115.

Willard, D.A., 1993. Vegetational patterns in the Springfield Coal (Middle Pennsylvanian, Illinois Basin): comparison of miospore and coal-ball records. Geological society of America, Boulder, CO, pp. 139–152. Geological society of America Special Papers 286.

Willard, D.A., Phillips, T.L., 1993. Paleobotany and palynology of the Bristol Hill Coal Member (Bond Formation) and Friendsville Coal Member (Mattoon Formation) of the Illinois Basin (Upper Pennsylvanian). Palaios 8, 574–586.

Williams, G.L., Brinkhuis, H., Pearce, M.A., Fensome, R.A., Weegink, J.W., 2004. Southern Ocean and global dinoflagellate cyst events compared: index events for the Late Cretaceous–Neogene. In: Exon, N.F., Kennett, J.P., Malone, M.J. (Eds.), In: Proc. ODP Sci. Results, 189, Ocean Drilling Program, College Station, TX, pp. 1–98. http://dx.doi.org/10.2973/odp.proc.sr.189.107.2004.

Williams, G.L., Fensome, R.A., MacRae, R.A., 2017. The Lentin and Williams index of fossil dinoflagellates, 2017 ed. AASP Contrib Ser0160-884348, American Association of Stratigraphic Palynologists Foundation. January 2017.

Willumsen, P.S., 2003. Marine Palynology Across the Cretaceous-Tertiary Boundary in New Zealand. Victoria University of Wellington, Wellington (PhD dissertation).

Wolfe, J.A., 1991. Palaeobotanical evidence for a June 'impact winter' at the Cretaceous-Tertiary boundary. Nature 352, 420–423.

Wolfe, J.A., Upchurch Jr., G.R., 1986. Vegetation, climatic and floral changes at the Cretaceous-Tertiary boundary. Nature 324, 148–152.

Wolfe, J.A., Upchurch Jr., G.R., 1987. Leaf assemblages across the Cretaceous-Tertiary boundary in the Raton Basin, New Mexico and Colorado. Proc. Natl. Acad. Sci. U. S. A. 84, 5096–5100.

Wood, G.D., Gabriel, A.M., Lawson, J.C., 1996. Palynological techniques—processing and microscopy. In: Jansonius, J., McGregor, D.C. (Eds.), In: Palynology: Principles and Applications, vol. 1. American Association of Stratigraphic Palynologists, Dallas, TX, pp. 29–50.

Worobiec, E., 2014. Fossil zygospores of Zygnemataceae and other micro remains of freshwater algae from two Miocene palaeosinkholes in the Pole region, SW Poland. Acta Palaeobotanica 54, 113–157.

Zavialova, N., Karasev, E., 2015. Exine ultrastructure of in situ Protohaploxypinus from a Permian peltasperm pollen organ, Russian Platform. Rev. Palaeobot. Palynol. 213, 27–41.

Zippi, P.A., 1998. Freshwater algae from the Mattagami Formation (Albian), Ontario: paleoecology, botanical affinities, and systematic taxonomy. Micropaleontology 44, 1–78.

FURTHER READING

Traverse, A., 1988. Plant evolution dances to a different beat. Plant and animal evolutionary mechanisms compared. Hist. Biol. 1, 227–301.

Sedimentologist's Guide for Recognition, Description, and Classification of Paleosols

N.J. Tabor*, T.S. Myers*, L.A. Michel*†
*Southern Methodist University, Dallas, TX, United States
†Tennessee Technological University, Cookeville, TN, United States

Contents

INTRODUCTION

Soils form near the Earth's surface, at the interface of the lithosphere, hydrosphere, atmosphere, and biosphere. They record physical and chemical interactions among Earth's materials and biota, mediated by climate. Occasionally, soils are buried, lithified, and incorporated into the geological record. These fossilized soils are called paleosols, whereas Quaternary soils that are no longer developing on the surface of the Earth, but are not yet lithified, are termed buried soils. Paleosols preserve valuable data related to Earth's past environments, climates, and geomorphologies that may be unlocked if the conditions affecting their formative processes are adequately understood. Because

Terrestrial Depositional Systems
http://dx.doi.org/10.1016/B978-0-12-803243-5.00004-2

paleosols compose a major part of the terrestrial sedimentary record (Retallack, 1986), they represent a vast archive of information about Earth history.

Soils and paleosols share many similarities, yet they are fundamentally different entities. Soils are dynamic mixtures of minerals and organic materials, whereas paleosols are inert sedimentary layers that have been incorporated into the rock record. Nevertheless, many researchers have adopted an approach to the study of paleosols (paleopedology) that is based principally on the study of modern soils (pedology). Other paleopedologists use classification schemes and analytical methods specifically tailored to paleosols. The difference between these two approaches to paleopedology is *not* merely semantic. Analytical methods and classification systems developed for modern soils are not necessarily appropriate for paleosols and vice versa. When applied to paleosols in the deep-time sedimentary record, modern soil taxonomy and analytical techniques do not necessarily provide reliable information about original conditions of soil formation. Conversely, paleosol classifications and analytical methods cannot provide detailed data equivalent to those reported in modern soil science studies.

This chapter seeks to facilitate an understanding of soil formation, how soils become paleosols, and particular aspects of these deep-time weathering profiles that are useful for understanding Earth surface processes. The chapter summarizes methods for documentation and characterization of modern soils that are relevant to paleosols and considers the applicability of modern soil science techniques to deep-time (i.e., pre-Pleistocene) paleosol profiles. This chapter is designed as a beginner's guide to understanding the differences between soils and paleosols, and how to recognize and describe paleosol properties that reflect important processes in ancient terrestrial environments. It highlights, in particular, the limitations and potential pitfalls associated with indiscriminate application of modern soil science methods in paleoenvironmental and paleoclimatological research. This contribution is not intended as a review of the paleoenvironmental and paleoclimatological information recorded in paleosols (e.g., Sheldon and Tabor, 2009; Tabor and Myers, 2015).

RECOGNITION AND DESCRIPTION

Early research on paleosols was primarily chronostratigraphic in nature, attempting to use hiatuses associated with soil formation as a stratigraphic correlation tool for large parts of the Quaternary in North America and Europe (Gile et al., 1981; Ruhe, 1956, 1969). Subsequent recognition of the abundance of paleosols in pre-Quaternary sedimentary strata and the utility of paleosol profiles for paleoenvironmental and paleoclimate reconstruction has fueled the growth of paleopedology as an increasingly prominent discipline within the geosciences. Deep-time paleosol research has provided insights into paleoclimate, sediment accumulation rates, stream behavior, terrestrial ecosystems, and sequence stratigraphy. Therefore, it is appropriate that sedimentologists, stratigraphers, and paleoclimatologists who work in terrestrial sedimentary strata possess a working knowledge of paleosols and a fundamental understanding of their significance.

Soil-forming processes produce a variety of macroscopic and microscopic features that may survive in the rock record. These physical features are recognizable in outcrop or thin sections and provide important information concerning soil formation (pedogenesis) and transformation of soils into paleosols. These features include changes in the discrete layers (horizons) within a paleosol profile, textures, soil structural elements, colors and color mottling, nodules and concretions, and biological structures.

Time in the field is limited, so it is necessary to prioritize observations that must be made in the field over those that may be addressed adequately later in a laboratory setting. The most important procedures involve observations of pedogenic features for which identification and interpretation are dependent on context. Field techniques advocated by Retallack (1988) for paleosol workers were drawn from an earlier version of the United States Department of Agriculture (USDA) Soil Survey Manual, with only minor modifications. Although scientists are often inclined to collect as many data as possible, constraints related to fieldwork dictate that observations should be limited to those that are necessary for paleosol identification and classification or that provide valuable information about paleoclimate, paleoenvironment, genetic processes, or diagenetic alteration. This guide to paleosol description emphasizes practical and useful observations rather than an exhaustive compilation of properties. It is tempting to apply the same descriptors developed for modern soils by the Soil Survey Staff (1999) to paleosols, but paleosols have undergone physical and chemical alterations and therefore are not directly equivalent to modern soils. There should be no expectation that appropriate methods of paleosol description mirror those devised for modern soils, except in a general sense. We advocate minimal usage of descriptive categories and terminology from modern soil science unless it is used in a strictly interpretive manner. This guide focuses on common paleosol features and morphologies rather than exotic examples that are rarely encountered.

Soil Horizon Definitions and Description

Soil formation is a progression of integrated events. First, substrate is emplaced at or near the surface of the Earth (Fig. 1). The initial substrate may vary in terms of its origin (sediments or sedimentary, metamorphic, or igneous rock), chemical composition (e.g., quartz sandstone, granite, phyllite), and texture. As time passes, a weathering zone forms at the substrate surface and begins to penetrate downward. This barely weathered substrate is called parent material. As the weathering zone continues to develop, characteristic properties emerge that are related to local environmental factors, the most important of which are climate (precipitation and temperature) and vegetation (Birkeland, 1999; Buol et al., 2003). The parent material within the weathering zone is slowly modified and reorganized, developing into discrete layers called horizons, which together compose the solum. The characteristics and properties of the soil profile, comprising the solum and

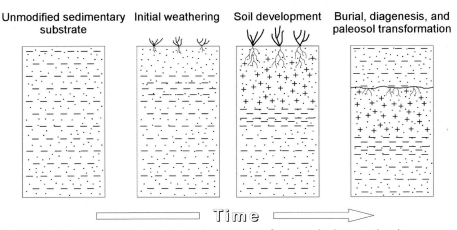

Fig. 1 Schematic diagrams showing idealized progression from newly deposited sedimentary strata, through intervening stages of pedogenesis, to final preservation as a paleosol. Important processes represented here include growth of plants and preservation of root structures, translocation and accumulation of clays, and precipitation of pedogenic carbonate *(crosses)*.

the underlying parent material, are referred to as soil morphology. The style, process, and magnitude of parent material alteration may be used to recognize and name soil horizons. Unlike the rule of "original horizontality" in sedimentary geology, the modified zones and horizons of soils tend to parallel the land surface, which need not be horizontal. In other words, soil horizons conform to slopes and gradients of the landscape.

The uppermost part of a soil profile usually accumulates organic matter (OM) due to plant growth and microbial activity, and loses some mineral and organic constituents through the process of eluviation, the physical and chemical transport of material downward through a soil profile. In this way, the upper horizons act like a physical sieve and chemical chromatograph that separates fine, soluble materials below from coarse, chemically recalcitrant materials above. Zones of eluviation and OM accumulation occur in both E and A horizons. Eluviated material accumulates in underlying horizons, a process termed illuviation, and may become organized in soil structural units called peds. Zones within a soil profile that exhibit evidence of illuviation and/or development of ped structure are called B horizons. Beneath the zone of illuviation, small amounts of soluble materials may accumulate, and soil materials may undergo weak, in situ physical and chemical alteration, or no alteration at all. These lowermost parts of a soil profile are called the C or R horizon depending on their composition and texture, although R horizon designations are seldom used in paleosol descriptions.

Soil horizons are labeled using a series of uppercase and lowercase letters and numbers in order to convey detailed information about a soil profile to someone *in absentia*. Master horizons are designated with capital letters: O, A, E, B, C, R, and sometimes K (Table 1). Lowercase letters called soil horizon descriptors are used in conjunction with master

Table 1 Distinguishing characteristics of soil master horizons and commonly encountered variations of these horizons

O Horizons

These unique horizons are composed primarily of OM deposited in situ, in a soil-forming environment. O horizons often contain some minor amount of mineral material. Minimum amounts of OM required in order to name an O horizon are

- ~30% OM if the mineral material is ~50% clay
- ~20% OM if the mineral material has no clay
- Proportional relationships between OM and clay when mineral material in the horizon contains somewhere between 0% and 50% clay.

O horizons typically form under the following conditions:

(1) Anoxic conditions created by prolonged periods of soil saturation inhibit normal oxidation of plant matter and allow OM to accumulate on the surface of the soil profile.

(2) Extremely acidic conditions created by accumulation of pine bark and pine needles inhibit activity of soil bacteria such that OM accumulation rates exceed rates of fungal-mediated decomposition.

A Horizons

The top horizon of most soils is usually the A horizon. Although A horizons are zones of OM accumulation, they contain less than the minimum amount of OM required for O horizons. A horizon OM is primarily derived from subsurface decomposition of roots and other plant material. This relatively small amount of OM imparts a significantly darker color to the soil because it is well mixed with mineral matter and coats each soil particle.

A horizons lose material through eluviation of organic colloids, clays, iron and aluminum oxides, and water-soluble bases that are washed downward through the profile and accumulate in lower horizons. The A horizon is usually the most leached part of a profile, unless an E horizon is present.

Several kinds of A horizons are recognized:

(1) A horizon—In natural, uncultivated soils the A horizon has relatively dark coloration and moderate to strong granular structure.

(2) AB horizon—Transitional to the underlying B horizon (when present), it is somewhat lighter in color, but does not possess the major characteristics of a B horizon. More closely resembles the A than the B horizon and has no significant clay content. Commonly has subangular blocky structure.

(3) A/B horizon—Transitional B horizon that contains salients or pockets of an A horizon. If material from A horizon is dominant, it is a B/A horizon.

(4) AC horizon—Transition horizon between the A and C horizon when B horizon is absent. Similar in appearance to an AB horizon.

E Horizons

In E horizons, intense eluviation has removed coatings of iron and/or OM from soil particles, creating gray coloration and platy structure. They contain very little clay and OM relative to adjacent horizons. E horizons are common in soils developed in forests or basin areas where ponding occurs.

Continued

Table 1 Distinguishing characteristics of soil master horizons and commonly encountered variations of these horizons—cont'd

---

B Horizons

B horizons are zones where materials eluviated from A or E horizons accumulate and
development of soil structure obliterates the original structure of the parent material.
Materials that often accumulate in B horizons, a process called illuviation, include: silicate
clays, colloidal humus, and oxides of iron and aluminum. Zones with accumulations of salts
or carbonates are not B horizons unless soil structure is also present.
Several types of B horizons exist:
(1) BA horizon—Transitional horizon that is more like the B than the A horizon. Typically has
subangular blocky structure, relatively dark coloration, and clay content intermediate
between that of the B horizon and the A horizon.
(2) B horizon—May contain accumulations of illuvial clay or calcium carbonate and possess
relatively dark coloration and well-developed blocky structure. Often divided into
subhorizons (e.g., Bt1, Bt2).
(3) BC horizon—A transitional horizon between the B horizon and parent material of the
B horizon. If the parent material of the B horizon is like the material in the C horizon,
the BC may be said to be transitional to the C horizon. However, it is usually recognized by a
decrease in the clay content, by a weakening of the structure, and/or by a lighter color than
the B above it.

C Horizons

C horizons are often incorrectly described as parent material for A and B horizons. The
partially weathered C horizon may or may not bear some resemblance to the original
parent material from which the solum formed. The C horizon consists of minimally
weathered material and usually underlies the B horizon (if present). It often has large,
weakly developed structural units or calcium carbonate accumulations. C horizons are
typically light colored and usually contain little clay relative to adjacent horizons. Cr
horizons are saprolite formed from weathered bedrock where rock structure is preserved.

R Horizons

The R horizon consists of consolidated bedrock, and the weathered residuum from this
horizon is the parent material for the soil. The designation 2R is sometimes used when the
bedrock lithology differs from that of the overlying, unconsolidated material (e.g., granite
bedrock beneath mud-rich glacial deposits).

These master horizon designations, derived from the USDA Soil Taxonomy (Soil Survey Staff, 1999), are appropriate for
use in paleosol studies.

horizon designations to connote specific physical or chemical characteristics (Table 2). If
multiple master horizons within a single soil profile share the same descriptor(s), but differ
in other properties (e.g., texture, color), the horizons are numbered sequentially beginning with the uppermost layer. Horizons that combine features of two different master
horizons are called transitional horizons, and are labeled by combining the two master
horizon designations. For more detailed information on terminology used to describe
modern soil horizons, see Birkeland (1999) or Schaetzl and Anderson (2005). Descriptive

Table 2 Horizon descriptors frequently used with master horizon designations for paleosol profiles

Horizon descriptors	Description
a	Highly decomposed plant OM; few, if any, recognizable plant parts; O horizon modifier
b	Buried soil horizon; not typically used in paleosol profiles associated with ancient stratigraphic successions
c	Nodules or concretions present; B horizon modifier
e	Moderately decomposed plant OM; no more than half recognizable plant parts; O horizon modifier
f	Evidence of ice in soil profile at the time of diagenesis; rare
g	Strong gleying, including redox accumulations, depletions, and mottles; may modify any horizon except for Bw
h	Illuvial accumulation of OM (e.g., humus); often occurs with sesquioxides (e.g., Bhs); B horizon modifier; rare
i	Slightly decomposed plant OM; more than half recognizable plant parts; O horizon modifier
j	Acid-sulfate minerals (e.g., alunite, jarosite); A, B, or C horizon modifier; rare
k	Diffuse, nodular, tubular, concretionary, or ped-coating cements of carbonate, abundant
m	Induration via pedogenic cementation during pedogenesis; cements may include calcite, gypsum; B horizon modifier; uncommon
n	Accumulation of sodium salts (natric conditions); columnar pedogenic structure common; B horizon modifier; rare
o	Fe-oxide concentration due to in situ chemical weathering; B horizon modifier; rare
p	Plow layer; typically has weak structure, abrupt lower boundary, and relatively light coloration; not used for paleosols
q	Diffuse, nodular, concretionary or ped-coating cements of silica; B, BC, or C horizon modifier; common in strata dominated by volcanic ash, otherwise rare
r	Weathered bedrock (e.g., saprolite); C horizon modifier
s	Illuvial accumulation of sesquioxides (e.g., Fe_2O_3, Al_2O_3); B horizon modifier; often used in conjunction with "h" descriptor (e.g., Bhs); rare)
ss	Pedogenic slickenplanes; B or C horizon modifier; often used in conjunction with "k" or "t" descriptors (e.g., Bkss); abundant
t	Illuvial accumulation of clay-sized materials; B horizon modifier; often used in conjunction with "t" descriptor (e.g., Bkt); common
v	Iron-cemented zone (may or may not be laterally continuous); B or C horizon modifier; equivalent to plinthite; uncommon
w	Development of soil structure and/or reddening (rubifaction); a basic B horizon modifier; common to abundant
x	Induration or semilithification at the time of pedogenesis; source of hardening is often related to accumulation of silica cements and clays; evidence for such conditions at the time of pedogenesis typically include evidence for lack of root penetration; B or C horizon modifier; rare

Continued

Table 2 Horizon descriptors frequently used with master horizon designations for paleosol profiles—cont'd

Horizon descriptors	Description
y	Diffuse, nodular, tubular, concretionary or ped-coating cements of gypsum or pseudomorphs after gypsum; B horizon modifier; may be used in conjunction with "k" or "n" descriptors (e.g., Bky, Byn); uncommon
z	Accumulation of salts and/or pseudomorphs of salts that more soluble than gypsum; rare

These descriptors are derived from the USDA Soil Taxonomy (Soil Survey Staff, 1999).

labels designed for use with modern soil horizons may be applied to paleosols because the horizon designators and descriptors are based on features related to the formative processes or sedimentary context associated with the profile, and these features are resistant to eradication or significant alteration during the transition from soil to paleosol.

Paleosol Horizon Description

Paleosol horizons may be labeled in the same manner as modern soil horizons. Paleosol O, A, E, B, C, and R horizons are frequently identified and reported (Kraus, 1999; Mack et al., 1993; Retallack, 1988). Any layers composed of primarily OM accumulated in situ on a paleo-surface, even those that have undergone substantial amounts of compaction, are designated O horizons (Mack et al., 1993). A horizons have been reported in paleosol profiles (Retallack, 1988, 1994a), but dark soil coloration related to elevated OM content, the distinguishing attribute of modern A horizons, is infrequently preserved in paleosols due to oxidation in the burial environment. Furthermore, erosion of soil surfaces prior to burial often removes all or part of A horizons before soils become paleosols (Bown and Kraus, 1987; Tabor and Montañez, 2004). AC horizons, however, are quite common in poorly developed soils associated with sandstone- and siltstone-rich strata deposited adjacent to fluvial channels (Tabor et al., 2006; Tabor and Montañez, 2004). E horizons, chemically leached horizons composed of mostly recalcitrant minerals such as quartz, are also relatively rare but have been described in Upper Paleozoic deposits where they are referred to historically as sinters or ganisters (see reviews by Besly and Fielding, 1989; Retallack, 1988). B horizons are one of the most abundant types of horizon in the paleosol record, and play a major role in paleosol classification. Although certain types of B horizons are particularly abundant (Table 2), other B horizons are relatively rare, probably due to paleoclimatic, geological, and research biases of the sedimentary rock record. C horizons are also quite common due to the abundance of sediment-hosted paleosol profiles in the geological record. R horizons are rarely reported but have been identified in paleosols formed atop limestone, granite, and basalt (Capo, 1993; Maynard et al., 1995; Sheldon, 2006; Tabor et al., 2004).

The morphology of the upper and lower boundaries of a profile and the horizon boundaries within a profile provide clues about processes operating during pedogenesis (Birkeland, 1999; Kraus, 1999; Retallack, 1988). The spatial relationship of paleosol profiles with underlying and overlying strata and relationships among horizons within profiles have been covered extensively in previous studies, and no new concepts are offered here. Paleosol profile and horizon boundary relationships are described in terms of the vertical thickness of gradients and the general morphology of the transition (Table 3).

Three processes in particular complicate recognition and interpretation of paleosols: cumulative deposition, polygenesis, and erosion. In standard models of pedogenesis, soil formation occurs during depositional hiatuses and is interrupted periodically by brief depositional events that add new sedimentary material. Cumulate soils form when gradual, continuous sedimentation occurs during pedogenesis, creating a thick, cumulate profile (Gerard, 1987; Marriott and Wright, 1993). The soil profile, in effect, slowly extends upward as sediments accumulate with ongoing generation of accommodation space. Polygenesis involves the creation of a composite soil profile in which original soil features are overprinted by later phases of pedogenesis. Polygenetic soil profiles may be created when weathering penetrates through a thin cover of sedimentary material into a buried soil (similar to soil welding described by Ruhe and Olson, 1980), or when climatic conditions change significantly during soil development (Gile et al., 1966; Miller et al., 1996). These processes may alter or mask the original characteristics of the overprinted horizons. It is also common in paleosols for the upper part of the profile to be truncated by erosion prior to final burial, or for the A horizon to be obliterated and incorporated into

Table 3 Paleosol horizon boundary descriptions modified from Retallack (1988)

Component	Descriptor term	Characteristics
Vertical transition	Scoured	Transition equivalent to a knife blade's thickness. Relatively rare within profiles, but common at the upper boundary of a paleosol where it contacts overlying strata
	Abrupt	Transition between horizons is <2 cm
	Clear	Transition between horizons is 2–5 cm
	Gradual	Transition between horizons is 5–15 cm
	Diffuse	Transition between horizons is gradual/gradational over a distance >15 cm
Lateral continuity	Smooth	Planar and usually conformable/parallel with master bedding planes in surrounding strata
	Wavy	Sinusoidal (regular or irregular) with wavelengths greater than amplitude
	Irregular	Sinusoidal (regular or irregular) with amplitudes greater than wavelengths
	Broken	Includes components of underlying and overlying horizons due to pedogenic or erosive processes

another horizon of a succeeding pedogenic event. Because of erosion, most paleosols do not appear to preserve A horizons (Blodgett, 1988; Holliday, 1989). The influence of all three processes (i.e., cumulative deposition, polygenesis, and erosion) may be assessed by careful observations of morphological features and analysis of mineralogical compositions within a paleosol profile.

Recognizing horizons and their boundaries is the first step in describing a paleosol profile. Once the upper and lower bounds of the profile are identified, one should define individual horizons based on color, texture, mineralogy, and other characteristics. Next, one should describe the types of horizons present, their thicknesses, and their spatial relationships, including the morphology of boundary contacts and boundary-zone thicknesses. If describing a paleosol profile in isolation from the surrounding strata, it may be easiest to begin with the uppermost horizon and work downwards. However, if paleosols are being described as part of a larger stratigraphic section that includes unmodified sedimentary strata, it is more appropriate to begin with the lowermost paleosol horizons and work upward in order to preserve the continuity of the stratigraphic section and maintain stratigraphic order. One should always observe lateral changes in paleosol horizons to determine if upper horizons have been scoured, but in some cases lateral observations may not be possible (e.g., description from sedimentary core samples: Rosenau et al., 2013a,b).

Soil Texture

Soil texture refers to the relative percentage of sand, silt, and clay within a soil layer. Only particles <2 mm in equivalent spherical diameter (e.s.d.) are included in soil texture descriptions because most physicochemical activity occurs in this fine-size fraction, although assignment of soil textural classes involves consideration of particles >2 mm e.s.d.—termed "skeletal grains" because of their low water-holding capacity. Soil texture and textural classes are an especially important aspect of modern soil research because they affect water-holding capacity and base saturation, which relate to agronomic productivity. However, modern soil texture and textural class assignments are not appropriate for paleosols, which are composed of sediments and sedimentary rocks that are either partially or completely lithified. The textural contrast between soils and paleosols is reflected in the different grain-size definitions used for soil texture (Soil Survey Staff, 1999) and sedimentary texture (Pettijohn, 1975). Modern soil science defines the clay-size fraction with a 2 μm upper bound because of the biochemical importance of this threshold, whereas sedimentology uses a 4 μm cutoff that coincides with a change in hydrodynamic behavior. In our experience, the sedimentary textural classification defined by Pettijohn (1975) is more practical for description of paleosol texture than the textural classes utilized in modern soils research because a paleosol profile is a type of sediment or sedimentary rock that no longer has all of the textural attributes that are appropriate for the operational definitions employed in soil science.

The sedimentary approach advocated here may be used to describe the size, shape, and composition of the particles that compose paleosol horizons. One should begin by characterizing grain size using the Wentworth scale (Wentworth, 1922), and note any gradational changes in grain size within a horizon. The Wentworth scale does not compensate for increases in apparent crystal and grain size that are the result of postburial cementation and diagenesis and are unrelated to the original soil textural class. Sorting and shape (roundness and sphericity) of macroscopic grains should be recorded, just as they are for any sedimentary deposit. Dilute hydrochloric acid may be used to test for effervescent reaction and determine if the matrix is calcareous. If desired, detailed textural and compositional (e.g., weight percent carbonate) observations may be conducted in the laboratory. In most cases, preliminary field observations will be adequate to determine how much weathering a paleosol has undergone and to characterize the relative maturity of a profile. Revisiting samples in a controlled laboratory setting is always recommended in order to support field observations.

Soil Structure

Although soil texture may undergo substantial alteration during burial and lithification, soil structural units—called aggregates or peds—are often preserved in paleosol profiles. Soil structure descriptions include observations of the shape, size, and strength of aggregates or peds within a profile. Peds form as soil particles that are forced together by root growth, shrink-swell processes related to wetting and drying, freeze-thaw cycles, and other physicochemical dynamics. OM and its decomposition products may act as a cement or adhesive, promoting development of soil structure in the upper horizons of a soil profile.

Peds in a soil or paleosol may be observed by crumbling pieces of material and noting the shape and size of the natural breakage planes associated with the smallest divisible units (Fig. 2). Soil structural classification depends on the shape (structural type), size (structural class), and physical strength (structural grade) of the structural unit, all of which are subject to change during burial.

Structural grade is determined by the visibility of the peds. In soils with strong ped structure, peds are clearly visible in an exposed profile. In profiles with moderate structure, peds are observable only after manipulation of a hand sample. Weak structure describes peds that are subtle and very difficult to discern. If peds are not detectable, the soil horizon is called structureless. Horizons in which there is no visible difference between the interior and exterior morphology of natural breakage surfaces, and the surrounding matrix is coherent (sticks together), are deemed massive and structureless. More coarsely grained soil horizons are described as possessing single-grained structure.

Soil structure plays an important role in pedogenesis. Coherent, clay-rich soil material with little or no structure inhibits penetration of plant roots and water, and typically

Fig. 2 Examples of paleosol ped structures. (A) Subangular blocky ped with clay coatings from an Argillisol in the Upper Triassic Ischigualasto Formation, Argentina. (B) Wedge-shaped aggregate from a gleyed Vertisol in the Upper Jurassic Lourinhã Formation, Portugal. (C) Subangular blocky structure in the Bk horizon of a Calcisol in the Lower Permian strata of the New Red Sandstone, Scotland. (D) Columnar to prismatic ped structure with secondary angular blocky structure in a spodic Gleysol/gleyed Spodosol from Upper Permian strata of the Wutongguo Formation, Xinjiang, western China. (E) Prismatic structure in a Bw horizon of a gleyed Vertisol in upper Oligocene strata, Chilga, western Ethiopia.

precludes development of a healthy and diverse soil biota. In contrast, shrinking, swelling, and cracking of soil materials promote structural development and permit roots and moisture to penetrate to significant depths. Therefore, ped structure in paleosol profiles provides important information about biologic and hydrologic processes within the original soil.

In some paleosol profiles, ped structure is readily identifiable, but it may be more difficult to detect in indurated paleosols that initially possessed weakly developed structure. Paleosol structure should be assessed using only freshly exposed rock that has not been altered by modern weathering; this typically requires digging into the outcrop at least 30 cm. Paleosol profiles with well-developed ped structure will readily crumble when dislodged with a sharp implement or digging tool. Paleosol descriptions use the same ped structure classification and terminology as modern soils (Soil Survey Staff, 1999). The most commonly encountered ped shapes in paleosols are angular blocky and wedge shaped. Platy, granular, columnar, and prismatic structures are less common.

Angular blocky peds are roughly cubic or slightly rectangular in shape, with sharp angles at the intersections of the ped faces. When the edges and corners of blocky peds are rounded rather than sharp, the structure is called subangular blocky. Wedge-shaped aggregates, sometimes called lentils (Blokhuis, 1982), are trapezoidal peds with tapering faces that meet to form a wedge shape. The ped faces of wedge-shaped aggregates often exhibit slickensides and may constitute part of large-scale slickenplanes (Fig. 2). Wedge-shaped peds form only in soil profiles where seasonal precipitation and abundant fine clay combine to induce shrink-swell (vertic) processes. Granular structure in modern soils, which is sometimes attributable to earthworm cast production (e.g., Zhang and Schrader, 1993), frequently occurs near the soil surface in A horizons where biological activity and organic content is typically high (Birkeland, 1999). Granular structure is relatively uncommon in paleosols because A horizons are frequently eroded, leaving only B and C horizons. Platy structure, commonly found in organic-depleted E horizons of modern soils (Birkeland, 1999), is also encountered infrequently in paleosols due to erosion of near-surface horizons. However, disintegration of stratified materials during development of BC- and some AC horizons often creates platy structure. Prismatic and columnar structures—oblong, vertically oriented peds—often form deeper within soil profiles (i.e., B and C horizons) where ped-forming processes are less dominant and structural features are less abundant (Schaetzl and Anderson, 2005). Because the intensity of pedogenic processes decreases downprofile, ped sizes generally increase with depth, and the lowermost horizons in a profile are usually structureless or massive. Weakly developed paleosol profiles recognized on the basis of rooting alone are also often structureless.

Strength of soil structure directly relates to soil maturity and parent material texture. Well-developed structure is characteristic of mature paleosol profiles formed in fine-grained parent material, over longer intervals of time, in stable landscapes, or under intense weathering conditions (Kraus, 1999; Marriott and Wright, 1993). The depth and abundance of wedge-shaped aggregates within a soil profile is related to the relative intensity of shrink-swell processes (Mermut et al., 1996; Vadivelu and Challa, 1985), but this relationship is complicated by other factors such as surface topography, and should not be used as a proxy for strength of rainfall seasonality (Tabor and Myers, 2015).

Exact measurements of ped size are of little use as paleoenvironmental proxies and should not be prioritized in descriptions. The Soil Survey Staff (1999) uses five classes to categorize ped size in modern soils, a simpler classification comprising three size classes (e.g., fine: <2 cm, medium: 2–5 cm, coarse: >5 cm) for paleosols does not sacrifice valuable genetic information.

Ped Coatings

Coatings of illuviated materials often accumulate on ped faces. These ped coatings, sometimes called cutans (Brewer, 1964), may form in single layers or accretions of multiple layers (Fig. 3). These coatings may consist of clay, silt, organic material, or a variety of minerals (e.g., calcite, hematite, goethite). The terminology and classification used for modern ped coatings (Soil Survey Staff, 1999) are appropriate for use with paleosols and require no additions or modifications. Ped coatings provide important evidence for illuviation and weathering processes that are linked to climatic, environmental, or drainage conditions (Chadwick et al., 1995), and therefore warrant description when they are encountered in the field. Detailed analysis of the fabric and mineralogical composition of ped coatings should be conducted later in a laboratory setting.

Soil Color

Soils may exhibit a broad palette of colors, including various hues of black, brown, gray, green, yellow, or red. Colors in both soils and paleosols are assessed using the Munsell color scale (Munsell Color, 1975), which characterizes colors according to their hue, value, and chroma. Black or dark brown colors are attributable to the presence of OM or manganese oxide minerals such as birnessite. Brown colors may also result from oxyhydroxides (e.g., ferrihydrite, lepidocrocite, goethite), and are consistent with well-aerated soils that undergo brief periods of saturation. Gray or green colors are typically the result of reduction of ferric minerals or removal of OM and oxides, processes associated with water saturation and anoxia. Red and yellow colors often indicate the presence of coarsely crystalline hematite or goethite, which form under well-aerated conditions.

Biochemical reactions that occur during burial may substantially alter original soil colors during the course of paleosol formation. For example, gray colors often develop in the upper horizons of profiles when paleosols are inundated by marine transgressions following burial (e.g., Rosenau et al., 2013a). Contrary to modern soils, paleosols rarely exhibit brown and yellow coloration. Instead, many paleosol profiles display vivid red colors, which typically result from progressive dehydration and oxidation of hydrated minerals and transformation to coarsely crystalline hematite within burial environments.

Despite the susceptibility of soil colors to alteration, the coloration of soil matrix and soil features such as mottles and root traces can provide important information about climate and/or drainage conditions during pedogenesis (e.g., Kraus and Hasiotis, 2006;

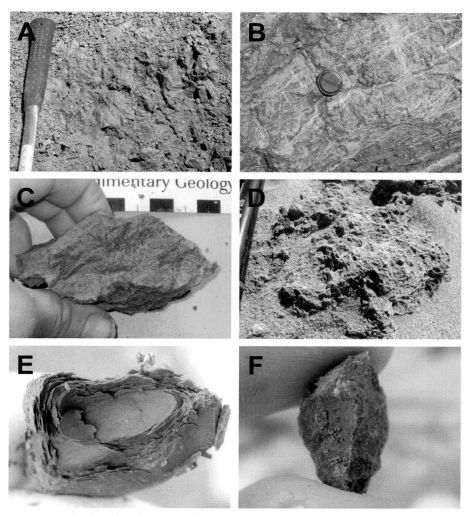

Fig. 3 Examples of paleosol ped coatings. (A) Clay-coated, wedge-shaped aggregates in an Argillisol in Oligocene strata, Chilga, western Ethiopia. (B) Reduced and calcified coatings around red-colored peds in a BC horizon in an Upper Carboniferous paleosol profile in the Lower Permian Supai Group, Arizona. (C) Wedge-shaped aggregate with pressure faces of oriented clay in a gleyed Vertisol in upper Oligocene strata, Chilga, western Ethiopia. (D) Strongly cemented calcareous coatings within subangular blocky matrix of a Calcisol in Upper Permian strata of the Moradi Formation, Niger. Vertical field of view (bottom to top of photo) is ~30 cm. (E) Subangular blocky ped composed almost entirely of translocated clay materials from an Argillisol in the Upper Permian Wutongguo Formation, Xinjiang, western China. This is an extraordinary example of clay coatings, but is recognizable because the friable exterior is made of clay-sized materials whereas the core matrix is composed of very fine sand to coarse silt material. (F) Subangular blocky ped with carbonate coatings from the Upper Permian Wutongguo Formation, Xinjiang, western China.

Vepraskas, 1992) and should always be noted in paleosol descriptions. Proper assignment of Munsell values depends on lighting conditions, the amount of moisture in the sample, and the ability of the researcher to accurately identify colors. From a practical perspective, Munsell values serve only to standardize color terminology and facilitate communication of color information, and assessment of Munsell colors in the field requires a substantial amount of time. Therefore, time constraints may dictate use of qualitative assessments of color, which still provide the same amount and quality of information necessary for interpretations of pedogenic processes or paleoenvironmental conditions. Munsell values should be recorded for horizons and structures within profiles that will be illustrated and described in detail for descriptive purposes (e.g., pedotypes; Retallack, 1994a,b; Tabor and Montañez, 2004; Tabor et al., 2006), but qualitative field assessments of color will be sufficient in most cases, especially when many paleosol profiles must be described. Munsell values of paleosol samples may be assessed in a laboratory setting, but these values may not be equivalent to those recorded in the field due to differences in lighting and sample moisture levels (e.g., Retallack, 1988).

Mottling

Mottles, areas within a soil horizon that differ from the dominant matrix color, typically manifest as patches of oxidized (red–yellow) or reduced (gray–green) colors (Fig. 4). These colors develop due to differential oxidation/depletion of matrix caused by reduction of iron or manganese, usually as the result of water saturation. The process of reduction and removal of iron or manganese, known as gleying, creates low chroma, grayish matrix colors (redox depletions), whereas concentration of these elements in other parts of a soil profile produces reddish or yellowish mottles (redox concentrations). Standard descriptions and definitions of mottles in modern soils include information about abundance, size, distinctness, and contrast (Soil Survey Staff, 1999). Of these characteristics, mottle abundance and distinctness are most directly related to climate and/or drainage conditions during pedogenesis. The strength of mottle development and the relative abundance of mottles within a paleosol profile are positively correlated with the intensity and duration of redoximorphic conditions, typically resulting from alternating periods of saturation/reduction and drying/oxidation (Vepraskas, 1992). Note, however, that mottle hue, value, chroma, and contrast are susceptible to the same postburial alteration that affects overall soil color.

Nodules and Concretions

The mineralogic, elemental, and isotopic composition of nodules and concretions, sometimes referred to collectively as glaebules (Fig. 5), often provide invaluable information about climatic and environmental conditions during pedogenesis. Nodules are internally homogeneous, whereas concretions are composed of concentric layers. In addition to

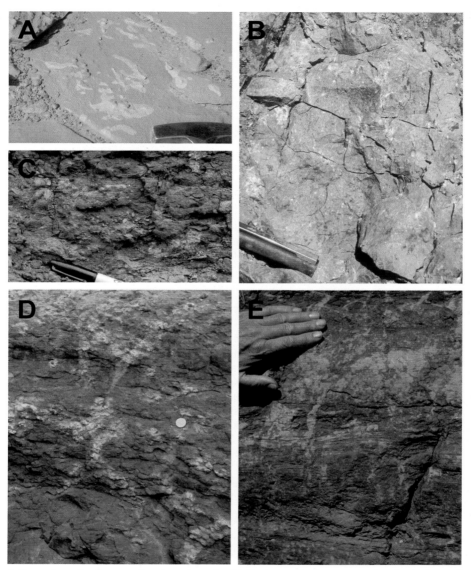

Fig. 4 Examples of color mottling in paleosols. (A) Drab vermicular to pseudospherical mottles in a calcic Protosol in the Upper Permian Moradi Formation, Niger. (B) Brown, yellow, and gray mottled subangular to angular blocky matrix of a Vertisol Bss horizon in the Lower Permian of Texas, United States. (C) Drab vermicular and irregular mottles in a calcic Protosol in the Upper Jurassic Lourinhã Formation, Portugal. (D) Coarse, prominent tubular and branching gray mottles in a calcic Protosol in the Upper Jurassic Lourinhã Formation, Portugal. (E) Drab irregular to vermicular mottles in the AC and C horizons of a Protosol in the Upper Permian Bundsandstein Formation, Italy.

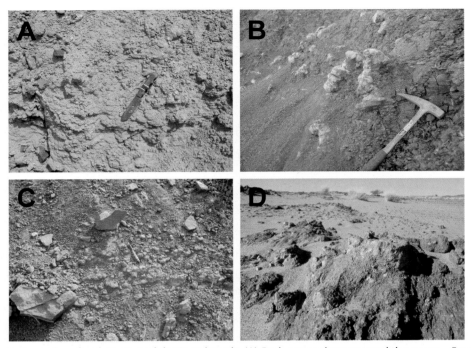

Fig. 5 Examples of carbonate nodules in paleosols. (A) Drab gray calcareous nodules up to ~5 cm in diameter and decimeter-long calcareous tubules about 1 cm in diameter in a Bk horizon in a Calcisol from Upper Oligocene strata of west Turkana, Kenya. (B) Stacked calcite nodules (rhizoliths *sensu* Blodgett, 1988) in a calcic Protosol in the Bursum Formation, central New Mexico, United States. (C) Calcite nodules in a Calcisol in Middle Triassic strata of the Argana Basin, western Morocco. (D) Calcite nodules within angular blocky matrix with calcite-cement coatings in a Calcisol in the Upper Permian Moradi Formation, Niger.

composition, which may be difficult to determine with certainty in the field, shape, size, and abundance are useful descriptors for nodules and concretions. Size, morphology, abundance, and location of glaebules within a paleosol profile can yield valuable information about pedogenic processes and environmental conditions during soil development (Gile et al., 1966; Machette, 1985). As with many other pedogenic features, spatial context within the soil profile is key, but quantitative measurements of glaebule size or spatial density are of little practical use in reconstructing paleoclimate. In order to assess the size and shape characteristics of a representative sample of glaebules, one should select meter-wide area within an exposed paleosol profile and conduct a thorough survey of the glaebules contained therein. Glaebules intended for elemental or isotopic analysis may be collected from anywhere within a paleosol profile, but it is imperative that calcium carbonate sampled for use with geochemical proxies of ancient atmospheric or soil CO_2 concentrations come from depths *at least* 50 cm below the original surface of the profile. Samples collected below this depth formed beneath the near-surface mixing zone in which the proportions of atmospheric and soil-respired CO_2 vary considerably (Ekart et al., 1999). Indurated layers of pedogenic carbonate are not appropriate for pCO_2 estimates, and may be difficult to differentiate from groundwater deposits.

Root Structures

Rooting structures (i.e., rhizoliths; Klappa, 1980) and burrows may provide paleoenvironmental information regarding water table depth, soil drainage characteristics, soil moisture regime, or relative amounts of precipitation (Cohen, 1982; Kahmann and Driese, 2008; Kraus and Hasiotis, 2006; Wright et al., 1988). Evidence of rooting in the geologic record (Fig. 6) has garnered much interest (e.g., Retallack, 1988; Sarjeant, 1975), but there is little consensus about how to evaluate the influence of root systems in paleosols. Pfefferkorn and Fuchs (1991) proposed a comprehensive classification of rooting structures that has been expanded subsequently (Kraus and Hasiotis, 2006). This classification—which evaluates the morphology, intensity, and preservation style of root structures—recognizes two primary types of subsurface root systems: (1) fibrous shallow roots that exploit near-surface soil water and (2) deep tap roots that provide stability and access to deep soil water (Pfefferkorn and Fuchs, 1991). Delicate fibrous roots are seldom preserved in the fossil record, but larger, more robust tap roots mineralize much more frequently. Root structures in paleosols preserve a record of vascular land plants extending back to *at least* the Devonian Period (e.g., Mora et al., 1996). Root traces may be differentiated from vermicular mottles using morphological attributes. In contrast to mottles, root traces typically exhibit downward tapering, ramifying, or bifurcating morphologies (Sarjeant, 1975).

Micromorphology

Micromorphology, the microscopic and/or spectroscopic study of soil materials, provides information complementary to macromorphological field observations of modern soils and paleosols. A number of different nomenclatural systems have been developed for pedologic micromorphology (Brewer, 1964; Bullock et al., 1985; FitzPatrick, 1984; Stoops, 2003; Stoops et al., 2010), creating a morass of terminology in which different terms are often used to describe the same features and processes. Among this profusion of micromorphological terminology, the system developed by FitzPatrick (1984, 1993) and modified by Bullock et al. (1985) and Stoops (2003) is the most intuitive and easiest to communicate to researchers with little prior experience in pedologic micromorphology. This system utilizes simple descriptive terms such as "coating," "nodule," or "concretion," and avoids creation of complex jargon that is unrelated to pedogenic processes.

Modern soils contain a variety of features that are most easily discerned in thin section, such as ped cross-sectional shape; birefringence fabric (b-fabric); clay, iron-manganese, or calcite coatings on and within peds (Fig. 7A); pedogenic (and postpedogenic) carbonate

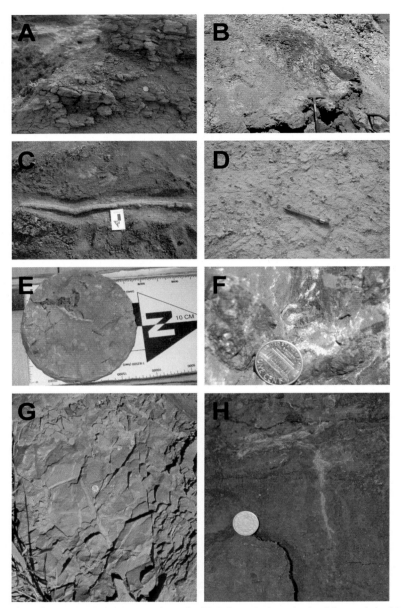

Fig. 6 Examples of rooting structures in paleosols. (A) Subvertical stacked calcite nodule rhizoliths in a Calcisol in the Upper Jurassic Lourinhã Formation, Portugal. (B) A silicified tree trunk and root system in the AC horizon of a Protosol developed in volcanic ash in upper Oligocene strata, Chilga, western Ethiopia. (C) Plan view of a laterally oriented rhizolith in the Bw horizon of a Protosol in the Miocene Hiwegi Formation, Rusinga Island, Kenya. (D) Plan view of calcite tubules that have replaced rooting structures in a Calcisol in Holocene lacustrine strata, Niger. (E) Carbon-compression root structures (black features) in the gleyed A horizon of a vertic Gleysol in a core cutting from the Upper Carboniferous Matoon Formation, Illinois Basin, United States. (F) Cross-section of a root system cemented by iron-oxide in a Protosol from Mississippian strata of south-central Colorado, United States. (G) Clay-lined tubules in the Bw horizon of an argillic Protosol in the Upper Carboniferous-Lower Permian Maroon Formation of Colorado, United States. (H) Drab-halo root structures in the AC horizon from a Protosol in the Upper Jurassic Lourinhã Formation, Portugal.

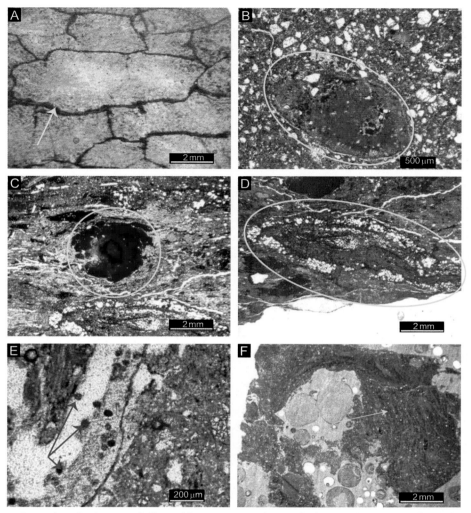

Fig. 7 Examples of micromorphological features in modern soils that can be preserved in paleosols. These include (A) clay-lining of peds *(yellow arrow)*, (B) hard carbonate nodules with discrete boundaries and shrinkage cracks *(yellow oval)*, (C) iron-manganese concretions *(yellow circle)*, (D) roots with cellular structure *(yellow oval)*, (E) orbited mite fecal material in roots *(red arrows)*, and (F) illuviated clay coatings *(yellow arrow)*.

nodules (Fig. 7B); and iron–manganese concretions (Fig. 7C) or nodules. Trace fossils such as back-filled burrows, roots (Fig. 7D), and fecal material from a variety of animals (Fig. 7E) preserve evidence of the biological component of soils, and the abundance of these features may indicate the relative intensity of biological activity. Iron–manganese, calcite, and clay coatings (Fig. 7F) along pore spaces and ped boundaries (Fig. 7A) may reflect drainage history (Kahmann and Driese, 2008) and offer information about pedogenic processes and environments not represented by macromorphological features. In some instances, clay coatings visible in thin section may provide the only evidence of illuviated clay in soil and paleosol profiles (Mack and James, 1992).

Micromorphology may also be used to differentiate the effects of chemical weathering from those of physical weathering. Paleosols formed in cold environments with minimal chemical weathering may still preserve evidence of physical weathering processes related to freeze-thaw cycles. Freeze-thaw processes create microstructure features such as "banded fabric," ice lenses, and features created by frost jacking, frost shattering, and silt caps (Kovda and Lebedeva, 2013) that would likely go unnoted in macromorphological observations.

Some macroscale features are also visible in thin section, so micromorphological analysis can reinforce or verify field observations. For example, the groundmass may have its own microstructure, which is the result of the coarse-to-fine ratio of fragments (c/f ratio) and the ped structure. Micromorphology may also be used to reconstruct a historical sequence of events that occurs during pedogenesis. For example, a soil may initially form on a well-drained landscape with substantial clay illuviation, but after clay completely fills the available pore space, the profile will become poorly drained. Studying the microscopic cross-cutting relationships within the profile allows this polygenetic pedogenic history to be deduced.

Micromorphological analysis may be used to determine whether paleosol carbonate accumulations are pedogenic or inherited from underlying parent material (Driese et al., 2005; Hsieh and Yapp, 1999; Kraimer and Monger, 2009; Michel et al., 2013). Stable isotopic values of paleosol carbonate may reflect inherited $\delta^{13}C$ and $\delta^{18}O$ values, as well as stable isotopic information from current weathering. Micromorphology may also help constrain diagenetic processes in paleosols and refine subsampling procedures for stable isotope analysis (Deutz et al., 2001). This application is tremendously important because all paleosols are affected to some degree by diagenesis.

CLASSIFICATION

Classification is the process of grouping and naming things according to their shared properties, thereby helping to organize, understand, and communicate information about those things. Inherent to the process of classification is assignment of names, providing verbal shorthand for communication.

A useful classification system is constructed using the following procedures:
(1) Development of uniform standards of comparison,
(2) Unbiased application of standards to the things being classified,
(3) Establishment of groups based on similarity, and
(4) Naming of established groups (development of taxonomy).

Classification systems proceed from more general, larger groups of things to progressively more specific, smaller groups. The number of different groups reflects the complexity of the system under consideration and the needs of the user. Classification systems may be tailored to fit the needs of different users, depending on the types of things classified and the properties used to classify them. Problems arise when a customized classification system developed for a specific purpose is applied under different circumstances.

The current USDA Soil Taxonomy was initially developed in 1965 in order to facilitate agronomic practices. Consequently, there is an inescapable disconnect between the modern USDA classification scheme and paleosol studies in terms of the standards of classification and overall objectives. The most recent version of the USDA classification (Soil Survey Staff, 1999), which comprises 12 Soil Orders (the largest divisions) and ~16,000 Soil Series (the smallest divisions), is based on detailed physical and chemical soil properties that are relevant to agronomic uses of soil, including recognition of anthropogenic influences such as irrigation and fertilization that have no bearing on prehistoric soils. Epipedons, the upper soil horizons that provide the diagnostic basis for much of the USDA classification system, are susceptible to erosion during the transition from soil to paleosol. Furthermore, the strict chemical parameters used to define these taxonomic groups are not preserved in paleosols that have undergone alteration. Because paleosols have undergone various physical and chemical modifications, they represent a pedogenic remnant visible only through the darkened lens of diagenetic alteration.

As an example, Mollisols are defined in the USDA taxonomy as soils with a dark-colored surface horizon, called a Mollic epipedon, and are base rich (i.e., a series of chemical criteria). Mollic epipedons are specifically defined by the USDA as layers with a color value and chroma <3.0 (moist) or <5.0 (dry), >0.6% organic carbon by weight, and >50% base saturation (as determined by ammonium oxalate extraction). Also, these are aggregate layers (not massive and potentially consisting of multiple horizons) that are usually >25 cm thick or 1/3 of the depth to restricting features such as argillic horizons or indurated materials. Mollic epipedons also have <250 ppm Bray solution-soluble (bioavailable) P_2O_5 and are not too dry for plant growth ½ of the time. Many Mollisols support grassland ecosystems, but the USDA definition of a Mollisol does not include presence of grass as a diagnostic criterion. The specific chemical criteria used to define the Mollisol Order may be measured in paleosol profiles, but OM loss and chemical changes that occur during burial and lithification would likely preclude accurate assessment of these diagnostic properties and render it impossible to identify Mollisols in the rock record.

There is currently no consensus regarding an appropriate classification scheme for paleosols. Some researchers use informal designations like calcrete or laterite (e.g., Alonso-Zarza, 2003; Bárdossy, 1995, 2013; McPherson, 1979), and others apply the modern USDA classification system (e.g., Nordt and Driese, 2013; Retallack, 1988, 1993; Tabor and Montañez, 2004) or group paleosols according to their most prominent genetic features (e.g., paleosols with Bt horizons; Thomas et al., 2011). Eschewing paleosol classification altogether hinders communication among researchers, and informal names should be avoided because they lack precision. Some informal categories also apply to deposits that are not pedogenic in origin (e.g., groundwater calcretes and laterites). Recent studies have attempted to use paleosol chemical compositions and transfer-functions to classify paleosols with the modern USDA Soil Taxonomy (Nordt et al., 2006, 2012, 2013; Nordt and Driese, 2010a,b). However, chemical analyses cannot be conducted practically in the field, so application of classification systems that rely on transfer functions is restricted to a laboratory setting. The paleosol classification system presented by Mack et al. (1993), which is designed to be applied in the field, is the best classification currently available to field scientists with interests in sedimentology and stratigraphy.

Paleosol Classification

The Mack classification system (Mack et al., 1993), developed specifically for use with paleosols, emphasizes pedogenic features that have a high preservation potential in the rock record because they are resistant to diagenetic alteration and destruction. This paleosol classification eschews characteristics—such as color, horizon thickness, and chemical concentration of cations—that form the basis of the modern USDA soil classification, but are especially susceptible to postburial alteration.

The Mack classification is based on assessment of the relative prominence in a paleosol profile of six features that record the effects of pedogenic processes: (1) OM accumulation, (2) horizonation, (3) redox-sensitive indicators, (4) evidence of in situ mineral alteration, (5) accumulations of soluble minerals, and (6) accumulations of illuviated insoluble minerals. The most prominent of these six features defines the paleosol Order, which always is capitalized. Determination of the relative prominence of diagnostic features in a paleosol profile is left to the individual investigator, so application of the Mack classification is a subjective exercise. However, this approach provides flexibility and avoids use of rigid rules that must apply to all paleosols. Paleosols with multiple, prominent pedogenic features are accommodated with subordinate modifiers that precede the Order name (Table 4). If more than one subordinate modifier is required, they are arranged in order of increasing prominence approaching the Order name. For example, a paleosol with a prominent calcic horizon that also contains vertic features and poorly developed gley features could be classified as a "gleyed vertic Calcisol." However, another investigator may decide that the vertic features are most conspicuous and identify the paleosol as a "gleyed calcic Vertisol." Despite the disjunction created by application of different taxonomic designations to the same paleosol profile, both identifications convey the same general information about the characteristic features of the paleosol. In the following section we discuss different groups of paleosol Orders within the Mack classification and the major processes associated with them.

Table 4 Subordinate paleosol modifiers, modified from the terminology presented by Mack et al. (1993)

Subordinate modifier	Description
Albic	E horizon with concentrations of recalcitrant chemicals created by weathering and illuviation; rare
Allophanic	Allophane, imogolite, or other amorphous Si and Al compounds; extremely rare (due to diagenesis)
Argillic	Accumulation of illuviated clay materials; frequently associated with a Bt horizon; common
Calcic	Pedogenic carbonate nodules, tubules, or ped-coatings; abundant
Carbonaceous	Dark-colored mineral matrix containing in situ or illuvial concentrations of OM; not peats or coals, but often associated with these deposits
Concretionary	Indurated (disorthic) glaebules with concentric internal structure; often associated with carbonate and iron-oxide accumulations
Dystric	Concentrations of base cation-poor materials due to intense chemical weathering, identified using petrographic analysis and X-ray diffraction; uncommon
Eutric	Concentrations of base cation-rich materials due to minimal chemical weathering, as deduced from petrographic and X-ray diffraction analysis; common to abundant
Ferric	Iron nodules, tubules, concretions or ped-coatings; common
Fragic	Semiindurated to indurated layer formed during pedogenesis, usually composed of tubular carbonate and/or ferric accumulations changing upward from vertically oriented to laterally oriented; may also include laminar cemented layer at top of horizon; rare
Gleyed	Evidence of periodic waterlogging, including Fe-oxides, sulfides, or strong gray/white/pale green or yellow/red/orange mottles; often associated with coals and peats; common
Gypsic	Pedogenic gypsum accumulations of diffuse cement, nodules, tubules, or ped-coatings; uncommon
Nodular	Indurated (disorthic), internally massive glaebules; abundant
Petrocalcic	Indurated, laterally extensive layers of pedogenic carbonate; uncommon
Petrogypsic	Indurated, laterally extensive layers of pedogenic gypsum; rare
Plinthic	Iron-cemented layers that may or may not be laterally continuous; B or C horizon modifier; equivalent to plinthite; uncommon
Ochric	Light-colored A horizon; rare
Salic	Cements or pseudomorphs after minerals more soluble than gypsum (e.g., cubic salt pseudomorphs); rare
Silicic	Silica accumulations, including diffuse cements, nodules, and tubules; often associated with volcanic ash deposits, otherwise rare
Vertic	Evidence of shrink-swell processes including desiccation cracks and clastic dikes, wedge-shaped aggregates, hummock-and-swale (gilgai) microtopography, mukkara subsurface structures, or slickensides; abundant
Vitric	Concentrations of glassy materials during pedogenesis, including relict glass shards, vesicular fragments, and pumice; rare

Paleosol Orders

OM accumulation is an essential part of soil development, but most modern soils and paleosols possess mineral-derived matrix comprising silicate (e.g., quartz, phyllosilicate) or chemical (e.g., carbonate, sulfate) components. Only occasionally are OM accumulations substantial enough to be the dominant matrix-supporting component of a soil profile (e.g., peats, coals). Paleosol Orders in the Mack classification are divisible into organic-dominated and mineral-dominated profiles.

Paleosols With OM-Derived Matrix

Histosol profiles, not to be confused with modern Histosols defined in the USDA Soil Taxonomy, contain a layer of concentrated OM that accumulated in situ (Fig. 8). Although some OM concentrations in continental environments are deposited by sedimentary transport processes (e.g., liptinites, sporonites), most laterally continuous beds of peat and coal are derived from in situ accumulations of vascular plant material (Taylor, 1998); therefore, all coals and peats fall into the Histosol Order (Mack et al., 1993). Coal layers often undergo significant compaction as they transition into the rock record, so no minimum thickness is necessary for a coal to qualify as a Histosol. However, discontinuous layers of coalified fallen tree trunks and other macerated materials, organic-rich shales, and coalified root systems in mineral-supported paleosol profiles are not Histosols because they either are allochthonous or occur within a predominantly mineral matrix (Fig. 8). It is critical to carefully inspect the mineral-supported sedimentary bed that underlies a coal. If this underlying mineral layer shows evidence of pedogenic modification, one must determine whether pedogenesis occurred prior to or contemporaneous with peat deposition and coal formation (e.g., Mack et al., 1993; Rosenau et al., 2013a,b; Staub and Cohen, 1978, 1979).

Based on the stratigraphic record of Histosols, deposition of peat in terrestrial environments appears to have been nearly continuous since the Silurian (Retallack, 1986), except for a brief episode in the Early Triassic (Retallack et al., 1996). Histosols are especially abundant in tropical paleolatitudes during the Late Carboniferous and Early Permian, and across a broad range of paleolatitudes in the Late Cretaceous (Boucot et al., 2013).

Paleosols With Mineral-Derived Matrix
Paleosols Characterized by Minimal Pedogenic Alteration and Horizon Development

Protosols are characterized by any evidence of occupation by terrestrial biota, typically in the form of rooting structures, with little to no development of pedogenic horizonation (Fig. 9). Sometimes weak reddening occurs toward the top of the profile, or there may be incipient development of pedogenic features such as poorly developed ped structure and weakly defined zones of mineral accumulation (e.g., carbonate, clay, Fe-oxides). Protosols are paleosols that lack obvious diagnostic characteristics of other paleosol Orders.

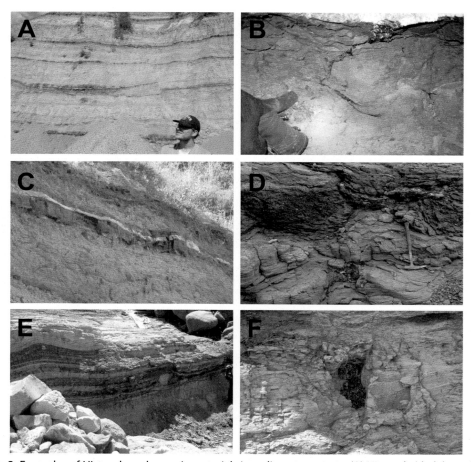

Fig. 8 Examples of Histosols and organic materials in sedimentary strata. (A) Histosols (dark layers) in Upper Oligocene strata, Chilga, western Ethiopia. (B) Carbon compressions of fossil roots in a gleyed Vertisol in the Upper Carboniferous Anna Shale cyclothem in the Eagle River Mine, Illinois, United States. The scale is a gloved finger (~8 cm long), holding a camera case. This paleosol profile contains abundant OM, but it is not a Histosol *sensu* Mack et al. (1993). (C) A Histosol (dark layer) with an interbedded volcanic ash layer in the Upper Oligocene strata near Chilga, western Ethiopia. (D) Allochthonous, carbonized tree trunks in fluvial strata of the Upper Jurassic Lourinhā Formation. This is not a Histosol because the trees do not constitute an in situ accumulation of OM on a soil surface. (E) Multiple Histosol profiles (i.e., coal layers) intercalated with volcanic ash deposits in the lower Miocene strata of Mush Valley, northeastern Ethiopia. The large orange clast is a volcanic bomb that was deposited and subsequently enveloped by in situ plant growth. The backpack at far right is ~60 cm tall. (F) Transported carbon compression of a tree trunk in the Abo tongue of the Hueco Formation (Lower Permian), south-central New Mexico, United States. This type of macerated, transported material does not qualify as a Histosol.

Fig. 9 Examples of Protosol profiles. (A) The dusky-red Bw horizon of a Protosol with angular blocky matrix, overlain by light grayish pink cross-bedded sandstones of the Arroyo member of the Lower Permian Clear Fork Formation, north-central Texas, United States. (B) Partially obliterated tree trunk (dark gray vertical structure at center) with root structures splayed laterally into an AC horizon of a Protosol developed on reworked volcanic ash in lower Miocene strata in the Mush Valley, northeastern Ethiopia. (C) Bedded muddy sandstones (above hammer head) with vertically oriented root halo structures intercalated with mudstones in a cumulate/compound AC horizon of a Protosol in the Abo tongue of the Hueco Formation (Lower Permian), south-central New Mexico, United States.

Protosols are more likely to form when the pedogenic evolution of a soil profile is truncated by sedimentary processes like episodic deposition and burial of the profile (Kraus, 1999; Marriott and Wright, 1993). Climatic conditions that inhibit or retard pedogenesis, such as extreme aridity or cold, also promote development of Protosols. Protosols are often the most abundant paleosol type in sedimentary successions (Mack et al., 2003; Myers et al., 2012; Nelson et al., 2001; Tabor et al., 2006; Tabor and Montañez, 2004), although other types of paleosols, such as Histosols, may be common in certain areas (Jacobs et al., 2005; Rosenau et al., 2013a,b).

Vertisols are characterized by evidence of cyclic shrinking and swelling that mix soil materials vertically and laterally within a profile, a process called pedoturbation. Pedogenic features formed by shrink-swell processes include desiccation features such as vertically oriented open channels that are often preserved as mud- or sand-rich dikes extending from the original paleosol surface into subsurface horizons (Fig. 10). Pedoturbation also affects soil structure, creating wedge-shaped ped aggregates that are often associated with small pressure faces and slickensides (Fig. 10), or sometimes larger, low-angle slickenplanes and slickensides. Pedogenic shrink-swell processes may also produce hummock and swale surface topography—referred to as gilgai microtopography—that is visible in outcrop, or unique subsurface features called mukkara (Fig. 10). Vertic features develop in response to soil expansion during wet intervals and contraction during dry periods. This type of shrink-swell behavior is often, but not always, attributed to the presence of expansible 2:1 phyllosilicate minerals such as smectite. The Mack classification does not use presence of expansible clays as a diagnostic characteristic of Vertisols because smectite may be transformed into illite, chlorite, or kaolinite during diagenesis (e.g., Nesbitt and Young, 1989). Vertisols are common throughout the Precambrian and Phanerozoic (Retallack, 1986), and appear to be especially abundant in upper Paleozoic and Mesozoic strata of Laurasia and Gondwana (Michel et al., 2015; Myers et al., 2011, 2012; Rosenau et al., 2013a,b; Smith et al., 2015; Tabor et al., 2006, 2011; Tabor and Montañez, 2004). Published reports of Vertisols in the Cenozoic (e.g., Driese et al., 2016) suggest lower abundance compared to Paleozoic and Mesozoic strata. Like Histosols, paleosols classified as Vertisols (*sensu* Mack et al., 1993) should not be conflated with Vertisols as defined in the USDA Soil Taxonomy.

Paleosols With Evidence for Strong Variations in Redox Conditions

Gleysols are characterized by subsurface horizons with prominent redoximorphic features formed by intermittent to long-term dysoxia or anoxia. Features indicative of redoximorphic conditions include low chroma (gray, green) soil matrix colors and reduced iron-bearing minerals (e.g., pyrite) and/or iron/manganese oxides occurring as pore linings, ped coatings, or cements in displacive nodules or concretions (Fig. 11). A mottled subsurface horizon is insufficient to identify a paleosol profile as a Gleysol, although intensely mottled horizons may occur in Gleysols with redoximorphic

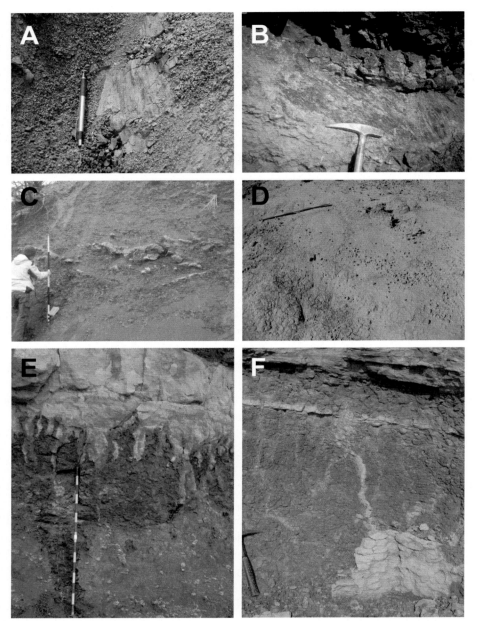

Fig. 10 Examples of Vertisols and their characteristic morphologies. (A) Large-scale slickenplane in a Vertisol in Lower Permian strata of the upper Nocona Formation, north-central Texas, United States. (B) Large slickenplane, wedge-shaped aggregates, and angular blocky structure in a calcic Vertisol in the Lower Permian Abo tongue of the Hueco Formation, south-central New Mexico, United States. (C) Subsurface mukkara structures (antithetic slickenplanes), highlighted by micritic calcareous cements in a calcic Vertisol in the Lower Permian Bursum Formation, central New Mexico, United States. (D) Plan view of gray sandy mudstones separated by red mudstone-filled clastic dikes in the upper Bw horizon of a Vertisol in Upper Permian strata of the Quartermaster Formation, north-central Texas, United States. (E) Gray sandstone-filled clastic dikes in a calcic Vertisol in the Upper Jurassic Lourinhã Formation, Portugal. (F) Sandstone-filled clastic dikes weathering out of a Vertisol in the Upper Jurassic Lourinhã Formation, Portugal.

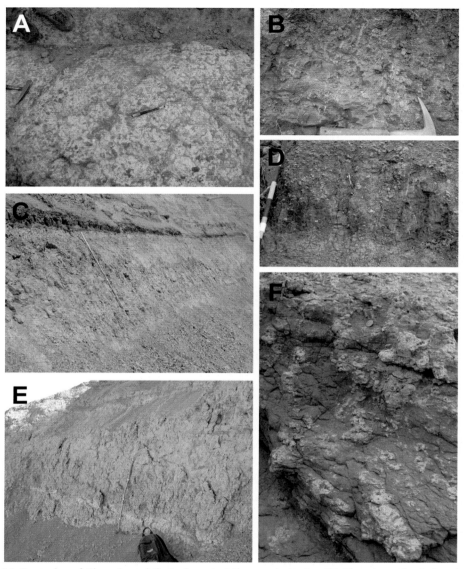

Fig. 11 Examples of Gleysols and their characteristic morphologies. (A) Plan view of a drab-colored Gleysol with angular blocky structure and hematite-cemented nodules and tubules in the Upper Triassic Ischigualasto Formation, Argentina. (B) Red and gray mottled vertic Gleysol with large slickenplanes in the Upper Jurassic Lourinhã Formation, Portugal. (C) Drab gray vertic Gleysol with siderite nodules, overlain by 30-cm-thick coal layer (Histosol) in upper Oligocene strata, Chilga, western Ethiopia. (D) Drab gray calcic Gleysol with calcium carbonate nodules and coarse red and yellow mottles in the Upper Jurassic Lourinhã Formation, Portugal. (E) Yellowish-brown mudstone with diffuse red coloration from partial oxidation of siderite cements in Upper Oligocene strata, Chilga, western Ethiopia. (F) Drab gray calcic Gleysol with abundant, amorphous calcium carbonate accumulations in the Upper Jurassic Lourinhã Formation, Portugal.

features. Gleyed horizons are usually attributable to influence of a shallow or fluctuating groundwater table that inhibits drainage and soil gas flux, creating low-oxygen conditions. Gleysols occur throughout the Phanerozoic and are a common component of fluvial floodplain and lake-plain environments (Tabor et al., 2006; Tabor and Montañez, 2004; Thomas et al., 2011). A problem inherent to identification of Gleysols (*sensu* Mack et al., 1993) is that gley features visible in outcrop may be coincident with pedogenesis or diagenesis. For example, a soil formed under well-drained conditions may undergo redoximorphic changes and gleying during burial (e.g., Rosenau et al., 2013a,b).

Paleosols Characterized by Accumulation of Soluble Minerals (Salts)

Calcisols are characterized by subsurface accumulations of calcium carbonate or dolomite (i.e., calcic horizons; Mack et al., 1993). However, dolomite in soils is exceedingly rare and as such its presence in paleosol profiles should be regarded as a diagenetic product typically (Michel et al., 2016). These paleosols are similar to sedimentary deposits that have been called cornstones, calcretes, and caliches and are usually, but not always, associated with dry climates. Quantification of the relationship between the presence and/or amount of carbonate in soil (and paleosol) profiles and various environmental factors like precipitation has been a topic of interest since the beginnings of soil science (Arkley, 1963; Jenny, 1941; Retallack, 1994b, 2005; Royer, 1999). When present, calcic horizons are often the most prominent feature of a paleosol profile. These horizons are typically composed of layers of nodular and/or tubular carbonate nodules and/or concretions in a mineral-supported matrix (Fig. 12). On occasion, entire subsurface horizons may be cemented by carbonate. Calcic paleosols are exceptionally abundant in the geological record, second only to Protosols among the mineral-based paleosol Orders. Calcisols and their equivalents have been described from every continent at various times throughout the Phanerozoic (e.g., Boucot et al., 2013; Retallack, 1986).

Gypsisols are characterized by subsurface accumulations of calcium sulfate minerals such as gypsum or anhydrite (i.e., gypsic horizons). Based on the distribution of modern soil analogs, gypsic paleosols are indicative of extremely dry environments where there is <25 cm annual precipitation (Watson, 1985) and leaching processes are incapable of removing sulfates from soil profiles. Pedogenic gypsum typically occurs as ped coatings, biscuit-shaped nodules, or semiindurated to indurated subsurface horizons cemented by sulfate minerals (Fig. 13). Because of the solubility of gypsum, these features are often preserved in paleosols as satin-spar calcite replacing the original pedogenic gypsum. Although gypsic paleosols are relatively rare, they have been recognized in upper Paleozoic and lower Mesozoic strata in North America (Tabor and Montañez, 2002; DiMichele et al., 2006; Retallack and Huang, 2010), Europe (Michel et al., 2015), and Africa (Tabor et al., 2011).

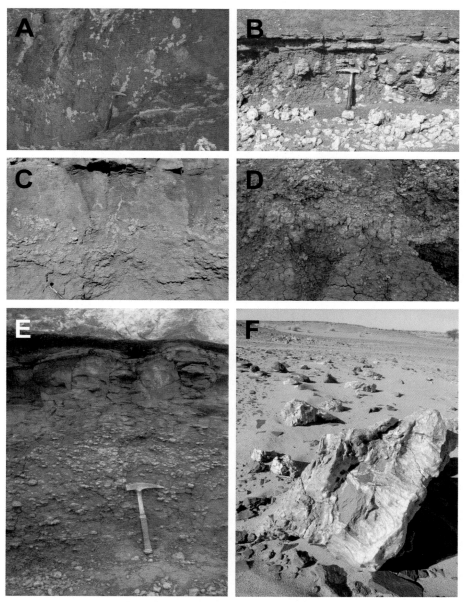

Fig. 12 Examples of Calcisols and their characteristic morphologies. (A) A Bk horizon with carbonate nodules and tubules, grading downward into mixed mudstones and stratiform calcite cements in a Calcisol in the Lower Permian Halgaito Formation, Utah, United States. (B) Calcisol with carbonate nodules and stacked carbonate nodule rhizoliths, overlain by thinly bedded sandstones, in the Lower Triassic Teloua Formation, central Niger. (C) Calcisol with red and gray mottling and a Bk horizon with densely packed calcium carbonate nodules in the Upper Jurassic Lourinhã Formation, Portugal. (D) Fully cemented Bkm horizon in a Calcisol in the Upper Jurassic Lourinhã Formation, Portugal. (E) A Bk horizon with abundant calcite nodules, grading upward into an ABw horizon in a Calcisol in Lower Permian strata of the Cedar Mesa Formation, Utah, United States. (F) Displaced blocks from a Calcisol with an indurated, calcite-cemented Bkm horizon (white layer) that occludes dusky-red mudstones in the Upper Permian Moradi Formation, Niger.

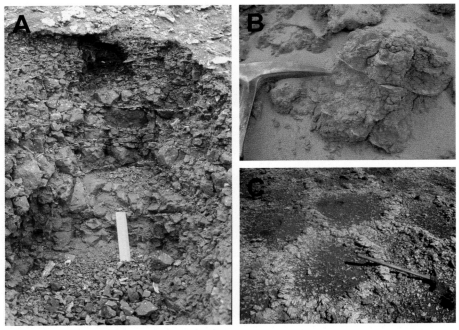

Fig. 13 Examples of Gypsisols and their characteristic morphologies. (A) Vertic Gypsisol with wedge-shaped aggregates, angular blocky structure, and spherical gray mottles in the Lower Permian Blaine Formation, north-central Texas, United States. Gypsum-lined slickenplanes and swallow-tail displacive gypsum crystals are visible in the profile. The yellow ruler at lower center is six inches long. (B) Gypsum-lined slickenplanes in a vertic Gypsisol in the Upper Permian Moradi Formation, Niger. (C) Plan view of gypsum-filled clastic dikes in a vertic Gypsisol in the Lower Permian San Angelo Formation, north-central Texas, United States.

Paleosols Characterized by Accumulation of Insoluble Materials

Argillisols include an argillic horizon, a concentration of clay minerals and associated oxy-hydroxides created by downprofile translocation. Identification of an argillic horizon requires no minimum thickness, but there must be evidence for an increase of clay-sized materials downward from the top of the paleosol profile to a subsurface zone of maximum accumulation (Fig. 14). Evidence for clay illuviation includes macroscopic or micro-scopic coatings of oriented clay (1) on ped surfaces, (2) within pores in the soil matrix, and (3) between detrital grains (grain bridges). Well-developed soil structure is not a pre-requisite for identification of an argillic horizon. Paleosol profiles that are eroded prior to burial, exposing a clay-enriched subsurface horizon, should be classified as argillic Pro-tosols (Mack et al., 1993) because they preserve evidence for argillic processes, but do not preserve evidence in such a state of subsurface enrichment. Identification of argillic hori-zons and Argillisols indicates leaching of base cations from the upper parts of a profile during pedogenesis, an indication of at least intermittent free drainage. Argillic paleosols

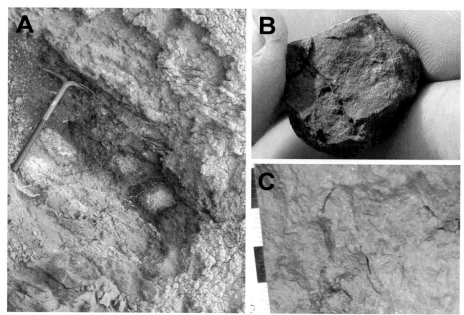

Fig. 14 Examples of Argillisols and their characteristic features. (A) A Bt/BCt horizon with spheroidally weathering basalts (light gray, circular features adjacent to pick) surrounded by highly oriented red clays in a Vertisol in the Upper Triassic Ischigualasto Formation, Argentina. (B) Clay-lined tubules in hand sample from the Bt horizon of an Argillisol in the Wutongguo Formation (Upper Permian), Xinjiang, western China. (C) Clay-coated surfaces (shiny surfaces) and organic-rich clay channels (dark gray) in a gleyed Argillisol in upper Oligocene strata, Chilga, western Ethiopia.

have a long history, extending back through the upper Paleozoic (Boucot et al., 2013; Retallack, 1986), but they appear to be relatively rare in upper Paleozoic and lower Mesozoic paleosols compared with reports of their occurrence in upper Mesozoic and especially Cenozoic strata (Mack and James, 1992; Sheldon, 2006). The causes underlying the unusual stratigraphic and temporal distribution of Argillisols are unknown.

Spodosols are any paleosols containing subsurface accumulations of OM and iron oxides (i.e., a spodic horizon; Fig. 15). This definition should not be conflated with the substantially different set of criteria used to identify spodic horizons in the USDA Soil Taxonomy, which is meant to identify substantial concentrations of illuvial organo–metallic substances (Mack et al., 1993; Soil Survey Staff, 1999). Assessment of the abundance of OM in paleosol profiles is possible, but OM is susceptible to oxidation during burial diagenesis, so quantitative analyses may produce equivocal results (e.g., Gastaldo et al., 2014). Changes in the concentrations of iron oxides within a paleosol profile may also be estimated using petrographic and geochemical analyses. However, it may be difficult to determine whether ferric materials are actually products of spodic processes or are derived from unrelated sources. In some instances reduced–iron sources

Fig. 15 A possible example of a Spodosol with prismatic and secondary angular blocky structure and abundant goethite and hematite nodules in the Upper Permian Wutongguo Formation, Xinjiang, western China. Field of view from left to right is ~1 m.

may accumulate in the subsurface during pedogenesis, through either aqueous transport to a soil-oxidized layer where ferric-oxide crystallization occurred or direct accumulation of ferrous iron minerals that later underwent postburial oxidation. These alternative scenarios would also lead to apparent downprofile increases in iron oxide minerals in a paleosol profile, achieved through processes entirely different from those associated with accumulation of sesquioxides and organometallic compounds in modern spodic horizons and Spodosols.

Spodosols have been reported from the Carboniferous through the recent (Retallack, 1986). Yet reports of Spodosols are relatively rare compared to most other paleosol Orders, possibly due to difficulties in conclusive identification. To put the complexities of this paleosol Order into perspective, the authors of this chapter have provided a single possible occurrence of a Spodosol paleosol profile (Fig. 15).

Paleosols With Evidence for Extreme Chemical Weathering

Oxisols, paleosols that have undergone extensive in situ chemical weathering, may be identified by the presence of a subsurface horizon composed primarily of base cation-poor minerals such as 1:1 phyllosilicates (e.g., kaolinite) and sesquioxides (Fig. 16). Petrographic microscopy and X-ray diffraction of the coarse-silt- and sand-size fractions should reveal no more than 10% coarse labile minerals such as volcanic fragments, feldspars, and dark igneous grains. Oxisols may or may not exhibit pedogenic structure, and there is no minimum requisite thickness for weathered subsurface horizons. Oxisol profiles do not contain argillic horizons, which are formed by translocation of clays rather than in situ weathering. The degree of in situ chemical weathering in an Oxisol may be assessed by comparison with stratigraphically associated, unweathered sedimentary and volcanic deposits. Oxisols, as defined by the USDA Soil Taxonomy, contain

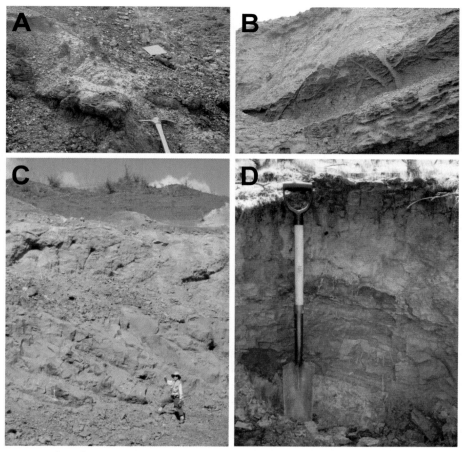

Fig. 16 Examples of Oxisols and their characteristic features. (A) Hematite- and goethite-cemented polygonal networks in a kaolinite-rich Oxisol from the Upper Carboniferous Markley Formation, north-central Texas, United States. (B) Hematite-cemented boxwork (sheetlike features in positive relief) separating kaolinite-rich claystone layers in an Oxisol derived from deeply weathered upper Oligocene basalts, Chilga, western Ethiopia. (C) Relict Upper Oligocene Oxisol derived from Upper Oligocene basalt flows, Chilga, western Ethiopia. (D) Relict Eocene Oxisol composed of kaolinite-dominated mudstones enriched with hematite and goethite, in the Ione Formation, northern California, United States.

some fraction of amorphous Fe- and Al-mineraloids, which are highly susceptible to postburial diagenetic transformation into crystalline minerals like kaolinite, goethite, hematite, or gibbsite.

Oxisols (*sensu* Mack et al., 1993) are relatively rare in the geologic record, but many examples have been reported in certain Phanerozoic stratigraphic successions (e.g., Abbott et al., 1976; Kahmann and Driese, 2008; Retallack, 1986, 2010; Singer and Nkedi-Kizza, 1980; Tabor and Yapp, 2005; Yapp, 2008). The intense chemical weathering conditions necessary for Oxisol development are currently limited to the humid tropical zone. In the past, particularly the late Paleozoic and Paleogene, environmental conditions conducive to Oxisol formation may have extended as far as 60° from the Equator (Retallack, 2010; Tabor and Yapp, 2005).

CONCLUSION

Although paleosols are a relatively minor component of the rock record in terms of volume, the paleoclimatic and paleoenvironmental information they contain makes them especially valuable to geoscientists. The rapid increase in the number of paleosol studies in peer-reviewed literature over the past 30 years ensures that paleopedology will remain a significant discipline for the foreseeable future. Given the close relationship of paleopedology with sedimentology and sedimentary petrology, it is imperative that sedimentologists and stratigraphers have a working knowledge of paleopedology and be capable of recognizing, describing, and communicating information about paleosols.

Paleosol research requires careful attention to observation, measurement, and recording of details in the field. Paleopedology studies should also incorporate laboratory-based petrographic, mineralogic, elemental, and isotopic measurements in order to provide additional data related to paleoecology, paleoenvironments, and diagenesis. The information and photos presented here offer a broad overview of many types of paleosols and paleosol features as they appear in the field. We have attempted to differentiate the rare and exotic examples of features from those that are frequently encountered. We have tried to indicate which soil features are resistant to diagenesis and are likely to be preserved in paleosols without significant alteration, and tried to contrast these with mutable characteristics that should not be relied on for paleoenvironmental reconstruction. We also hope that we have convinced readers that modern soils and paleosols are fundamentally different entities that require different methods of analysis and classification. While no single classification scheme has been developed to date that appears to adequately address the needs of field-based sedimentologists, stratigraphers, and paleopedologists, continued research and communication among these workers may arrive at a reasonable system to bridge the importance of modern soils and their fossil records, and help to inform our understanding of past environments. Most of all, we hope that this guide conveys the utility and versatility of paleosols as complex data storage media and will inspire readers to continue to explore the expanding discipline of paleopedology.

REFERENCES

Abbott, P.L., Minch, J.A., Peterson, G.L., 1976. Pre-eocene paleosol south of Tijuana, Baja California, Mexico. J. Sediment. Res. 46, 355–361.
Alonso-Zarza, A.M., 2003. Palaeoenvironmental significance of palustrine carbonates and calcretes in the geological record. Earth-Sci. Rev. 60, 261–298.

Arkley, R.J., 1963. Calculation of carbonate and water movement in soil from climatic data. Soil Sci. 96, 239–248.

Bárdossy, G., 1995. Carboniferous to Jurassic bauxite deposits as paleoclimatic and paleogeographic indicators. Can. Soc. Pet. Geol. Mem. 17, 283–293.

Bárdossy, G., 2013. Karst Bauxites. Elsevier, New York, NY.

Besly, B.M., Fielding, C.R., 1989. Palaeosols in Westphalian coal-bearing and red-bed sequences, central and northern England. Palaeogeogr. Palaeoclimatol. Palaeoecol. 70, 303–330. http://dx.doi.org/10.1016/0031-0182(89)90110-7.

Birkeland, P.W., 1999. Soils and Geomorphology, third ed. Oxford University Press, New York, NY.

Blodgett, R.H., 1988. Calcareous paleosols in the Triassic Dolores Formation, southwestern Colorado. In: Paleosols and Weathering Through Geologic Time. Geological Society of America Special Paper, vol. 216. Geological Society of America, Boulder, CO, pp. 103–121.

Blokhuis, W.A., 1982. Morphology and genesis of Vertisols. In: Presented at the 12th International Congress of Soil Science, Vertisols and Rice Soils of the Tropics, Managing Soil Resources to Meet the Challenges to Mankind, pp. 23–47.

Boucot, A.J., Xu, C., Scotese, C.R., 2013. Phanerozoic Paleoclimate: An Atlas of Lithologic Indicators of Climate. SEPM (Society for Sedimentary Geology), Tulsa, OK.

Bown, T.M., Kraus, M.J., 1987. Integration of channel and floodplain suites, I. Developmental sequence and lateral relations of alluvial paleosols. J. Sediment. Petrol. 57, 587–601.

Brewer, R., 1964. Fabric and Mineral Analysis of Soils. Wiley, New York, NY.

Bullock, P., Fedoroff, N., Jongerius, A., Stoops, G., Tursina, T., 1985. Handbook for Soil Thin Section Description. WAINE Research Publications, Wolverhampton.

Buol, S.W., Southard, R.J., Graham, R.C., McDaniel, P.A., 2003. Soil Genesis and Classification, fifth ed. Iowa State Press, Ames, IA.

Capo, R.C., 1993. Micromorphology of a Cambrian paleosol developed on granite: Llano Uplift region, Central Texas, U.S.A. In: Ringrose-Voase, A.J., Humphreys, G.S. (Eds.), Developments in Soil Science, Soil Micromorpohlogy: Studies in Management and Genesis. Elsevier, Amsterdam, pp. 257–264.

Chadwick, O.A., Nettleton, W.D., Staidl, G.J., 1995. Soil polygenesis as a function of Quaternary climate change, northern Great Basin, USA. Geoderma 68, 1–26.

Cohen, A.S., 1982. Paleoenvironments of root casts from the Koobi Fora Formation, Kenya. J. Sediment. Petrol. 52, 401–414.

Deutz, P., Montañez, I.P., Monger, H.C., Morrison, J., 2001. Morphology and isotope heterogeneity of Late Quaternary pedogenic carbonates: implications for paleosol carbonates as paleoenvironmental proxies. Palaeogeogr. Palaeoclimatol. Palaeoecol. 166, 293–317.

DiMichele, W.A., Tabor, N.J., Chaney, D.S., Nelson, W.J., 2006. From wetlands to wet spots: environmental tracking and the fate of Carboniferous elements in Early Permian tropical floras. In: Greb, S.F., DiMichele, W.A. (Eds.), Wetlands Through Time. Geological Society of America Special Paper, vol. 399. Geological Society of America, Boulder, CO, pp. 223–248.

Driese, S.G., Nordt, L.C., Lynn, W.C., Stiles, C.A., Mora, C.I., Wilding, L.P., 2005. Distinguishing climate in the soil record using chemical trends in a Vertisol climosequence from the Texas coast prarie, and application to interpreting Paleozoic paleosols in the Appalachian Basin, U.S.A. J. Sediment. Res. 75, 339–349.

Driese, S.G., Peppe, D.J., Beverly, E.J., DiPietro, L.M., Arellano, L.N., Lehmann, T., 2016. Paleosols and paleoenvironments of the early miocene deposits near Karungu, Lake Victoria, Kenya. Palaeogeogr. Palaeoclimatol. Palaeoecol. 443, 167–182. http://dx.doi.org/10.1016/j.palaeo.2015.11.030.

Ekart, D.D., Cerling, T.E., Montañez, I.P., Tabor, N.J., 1999. A 400 million year carbon isotope record of pedogenic carbonate: implications for paleoatmospheric carbon dioxide. Am. J. Sci. 299, 805–827.

FitzPatrick, E.A., 1984. Micromorphology of Soils. Chapman and Hall, London.

FitzPatrick, E.A., 1993. Soil Microscopy and Micromorphology. Wiley, New York, NY.

Gastaldo, R.A., Knight, C.L., Neveling, J., Tabor, N.J., 2014. Latest Permian paleosols from Wapadsberg Pass, South Africa: implications for Changhsingian climate. Geol. Soc. Am. Bull. 126, 665–679. http://dx.doi.org/10.1130/B30887.1.

Gerard, J., 1987. Alluvial Soils. Hutchinson Ross, New York, NY.

Gile, L.H., Peterson, F.F., Grossman, R.B., 1966. Morphological and genetic sequences of carbonate accumulation in desert soils. Soil Sci. 101, 347–360.

Gile, L.H., Hawley, J.W., Grossman, R.B., 1981. Soils and geomorphology in the basin and range area of southern New Mexico. In: Guidebook to the Desert Project, Memoir 39. New Mexico Bureau of Mines and Mineral Resources, Socorro, NM, p. 222.

Holliday, V.T., 1989. Paleopedology in archaeology. Paleopedology 16, 187–206.

Hsieh, J.C.C., Yapp, C.J., 1999. Stable carbon isotope budget of CO_2 in a wet, modern soil as inferred from $Fe(CO_3)OH$ in pedogenic goethite: possible role of calcite dissolution. Geochim. Cosmochim. Acta 63, 767–783. http://dx.doi.org/10.1016/S0016-7037(99)00062-9.

Jacobs, B., Tabor, N., Feseha, M., Pan, A., Kappelman, J., Rasmussen, T., Sanders, W., Wiemann, M., Crabaugh, J., García Massini, J.L., 2005. Oligocene terrestrial strata of northwestern Ethiopia: a preliminary report on paleoenvironments and paleontology. Palaeontol. Electron 8, 1–19.

Jenny, H., 1941. Factors of Soil Formation: A System of Quantitative Pedology. Dover Publications, Inc., New York, NY

Kahmann, J.A., Driese, S.G., 2008. Paleopedology and geochemistry of late Mississippian (Chesterian) Pennington formation paleosols at Pound Gap, Kentucky, USA: implications for high-frequency climate variations. Palaeogeogr. Palaeoclimatol. Palaeoecol 259, 357–381.

Klappa, C.F., 1980. Rhizoliths in terrestrial carbonates: classification, recognition, genesis and significance. Sedimentology 27, 613–629.

Kovda, I., Lebedeva, M., 2013. Modern and relict features in clayey cryogenic soils: morphological and micromorphological identification. Span. J. Soil Sci. 3, 1–18.

Kraimer, R.A., Monger, H.C., 2009. Carbon isotopic subsets of soil carbonate—a particle size comparison of limestone and igneous parent materials. Geoderma 150, 1–9. http://dx.doi.org/10.1016/j.geoderma.2008.11.042.

Kraus, M.J., 1999. Paleosols in clastic sedimentary rocks: their geologic applications. Earth Sci. Rev. 47, 41–70.

Kraus, M.J., Hasiotis, S.T., 2006. Significance of different modes of rhizolith preservation to interpreting paleoenvironmental and paleohydrologic settings: examples from Paleogene paleosols, Bighorn Basin, Wyoming, U.S.A. J. Sediment. Res. 76, 633–646.

Machette, M.N., 1985. Calcic soils of the southwestern United States. In: Weide, D.L. (Ed.), Soils and Quaternary Geology of the Southwestern United States. Geological Society of America Special Paper, vol. 203, Geological Society of America, Boulder, CO, pp. 1–21.

Mack, G.H., James, W.C., 1992. Calcic paleosols of the plio-pleistocene camp rice and palomas formations, southern Rio Grande rift, USA. Sediment. Geol. 77, 89–109. http://dx.doi.org/10.1016/0037-0738(92)90105-Z.

Mack, G.H., James, W.C., Monger, H.C., 1993. Classification of paleosols. Geol. Soc. Am. Bull. 105, 129–136.

Mack, G.H., Leeder, M., Perez-Arlucea, M., Bailey, B.D.J., 2003. Early Permian silt-bed fluvial sedimentation in the Orogrande basin of the Ancestral Rocky Mountains, New Mexico, USA. Sediment. Geol. 160, 159–178. http://dx.doi.org/10.1016/S0037-0738(02)00375-5.

Marriott, S.B., Wright, V.P., 1993. Palaeosols as indicators of geomorphic stability in two Old Red Sandstone alluvial suites, South Wales. J. Geol. Soc. Lond. 150, 1109–1120.

Maynard, J.B., Sutton, S.J., Robb, L.J., Ferraz, M.F., Meyer, F.M., 1995. A paleosol developed on hydrothermally altered granite from the hinterland of the Witwatersrand Basin: characteristics of a source of basin fill. J. Geol. 103, 357–377.

McPherson, J.G., 1979. Calcrete (caliche) palaeosols in fluvial redbeds of the Aztec siltstone (upper Devonian), Southern Victoria Land, Antarctica. Sediment. Geol. 22, 267–285. http://dx.doi.org/10.1016/0037-0738(79)90056-3.

Mermut, A.R., Dasog, G.S., Dowuona, G.N., 1996. Soil morphology. In: Ahmad, N., Mermut, A. (Eds.), Vertisols and Technologies for Their Management. In: Developments in Soil Science, vol. 24. Elsevier, New York, NY, pp. 89–114.

Michel, L.A., Driese, S.G., Nordt, L.C., Breecker, D.O., Labotka, D.M., Dworkin, S.I., 2013. Stable-isotope geochemistry of Vertisols formed on marine limestone and implications for deep-time paleoenvironmental reconstructions. J. Sediment. Res. 83, 300–308.

Michel, L.A., Tabor, N.J., Montañez, I.P., 2016. Paleosol diagenesis and its deep-time paleoenvironmental implications, Pennsylvanian-Permian Lodève Basin, France. J. Sediment. Res. 86, 813–829.

Michel, L.A., Tabor, N.J., Montañez, I.P., Schmitz, M.D., Davydov, V.I., 2015. Chronostratigraphy and paleoclimatology of the Lodève Basin, France: evidence for a pan-tropical aridification event across the Carboniferous–Permian boundary. Palaeogeogr. Palaeoclimatol. Palaeoecol. 430, 118–131.

Miller, K.B., McCahon, T.J., West, R.R., 1996. Lower Permian (Wolfcampian) paleosol-bearing cycles of the U.S. midcontinent: evidence of climatic cyclicity. J. Sediment. Res. 66, 71–84.

Mora, C.I., Driese, S.G., Colarusso, L.A., 1996. Middle to late Paleozoic atmospheric CO_2 levels from soil carbonate and organic matter. Science 271, 1105–1107.

Munsell Color, 1975. Munsell Soil Color Charts. Munsell Color, Baltimore, MD.

Myers, T.S., Tabor, N.J., Jacobs, L.L., 2011. Late Jurassic paleoclimate of Central Africa. Palaeogeogr. Palaeoclimatol. Palaeoecol. 311, 111–125.

Myers, T.S., Tabor, N.J., Jacobs, L.L., Mateus, O., 2012. Palaeoclimate of the late Jurassic of Portugal: comparison with the Western United States. Sedimentology 59, 1695–1717.

Nelson, W.J., Hook, R.W., Tabor, N.J., 2001. Clear fork group (Leonardian, Lower Permian) of north-central Texas. In: Johnson, K.S. (Ed.), Pennsylvanian and Permian Geology and Petroleum in the Southern Midcontinent, 1998 Symposium: Oklahoma Geological Survey Circular, vol. 104, pp. 167–169.

Nesbitt, H.W., Young, G.M., 1989. Formation and diagenesis of weathering profiles. J. Geol. 97, 129–147.

Nordt, L.C., Driese, S.G., 2010a. A modern soil characterization approach to reconstructing physical and chemical properties of paleo-Vertisols. Am. J. Sci. 310, 37–64.

Nordt, L.C., Driese, S.G., 2010b. New weathering index improves paleorainfall estimates from Vertisols. Geology 38, 407–410.

Nordt, L.C., Driese, S.G., 2013. Application of the critical zone concept to the deep-time sedimentary record. Sediment. Rec. 11, 4–9.

Nordt, L., Orosz, M., Driese, S., Tubbs, J., 2006. Vertisol carbonate properties in relation to mean annual precipitation: implications for paleoprecipitation estimates. J. Geol. 114, 501–510.

Nordt, L.C., Hallmark, C.T., Driese, S.G., Dworkin, S.I., Atchley, S.C., 2012. Biogeochemistry of an ancient critical zone. Geochim. Cosmochim. Acta 87, 267–282.

Nordt, L.C., Hallmark, C.T., Driese, S.G., Dworkin, S.I., 2013. Multi-analytical pedosystem approach to characterizing and interpreting the fossil record of soils. In: Driese, S.G., Nordt, L.C. (Eds.), New Frontiers in Paleopedology and Terrestrial Paleoclimatology. SEPM Special Publication, vol. 104, Society of Economic Geology, Tulsa, OK, pp. 89–107.

Pettijohn, F.J., 1975. Sedimentary Rocks. Harper and Row, New York, NY.

Pfefferkorn, H.W., Fuchs, K., 1991. A field classification of fossil plant substrate interactions. Neues Jahrb. Für Geol. Paläontol. Abh. 183, 17–36.

Retallack, G.J., 1986. The fossil record of soils. In: Wright, V.P. (Ed.), Paleosols: Their Recognition and Interpretation. Blackwell Scientific Publications, Oxford, pp. 1–57.

Retallack, G.J., 1988. Field recognition of paleosols. In: Reinhardt, J., Sigleo, W.R. (Eds.), Paleosols and Weathering Through Geologic Time. Geological Society of America Special Paper, vol. 216. Geological Society of America, Boulder, CO, pp. 1–20.

Retallack, G.J., 1993. Classification of paleosols: discussion and reply. GSA Bull. 105, 1635–1637.

Retallack, G.J., 1994a. A pedotype approach to latest Cretaceous and earliest Tertiary paleosols in eastern Montana. GSA Bull. 106, 1377–1397.

Retallack, G.J., 1994b. The environmental factor approach to the interpretation of paleosols. Factors of Soil Formation: A Fiftieth Anniversary Retrospective. SSSA Special Publication, vol. 33. Soil Science Society of America, Madison, WI, pp. 31–63.

Retallack, G.J., 2005. Pedogenic carbonate proxies for amount and seasonality of precipitation in paleosols. Geology 33, 333–336.

Retallack, G.J., 2010. Lateritization and bauxitization events. Econ. Geol. 105, 655–667.

Retallack, G.J., Huang, C., 2010. Depth to gypsic horizon as a proxy for paleoprecipitation in paleosols of sedimentary environments. Geology 38, 403–406.

Retallack, G.J., Veevers, J.J., Morante, R., 1996. Global coal gap between Permian–Triassic extinction and middle Triassic recovery of peat-forming plants. Geol. Soc. Am. Bull. 108, 195–207. http://dx.doi.org/10.1130/0016-7606(1996)108<0195:GCGBPT>2.3.CO;2.

Rosenau, N.A., Tabor, N.J., Elrick, S.D., Nelson, W.J., 2013a. Polygenetic history of paleosols in middle-upper Pennsylvanian cyclothems of the Illinois Basin, U.S.A.: Part I. Characterization of paleosol types and interpretation of pedogenic processes. J. Sediment. Res. 83, 606–636.

Rosenau, N.A., Tabor, N.J., Elrick, S.D., Nelson, W.J., 2013b. Polygenetic history of paleosols in Middle-Upper Pennsylvanian cyclothems of the Illinois Basin, U.S.A.: Part II. Integrating geomorphology, climate, and glacioestasy. J. Sediment. Res. 83, 637–668.

Royer, D.L., 1999. Depth to pedogenic carbonate horizon as a paleoprecipitation indicator? Geology 27, 1123–1126.

Ruhe, R.V., 1956. Geomorphic surfaces and the nature of soils. Soil Sci. 82, 441–456.

Ruhe, R.V., 1969. Quaternary Landscapes in Iowa. Iowa State University Press, Ames, IA.

Ruhe, R.V., Olson, C.G., 1980. Soil welding. Soil Sci. 130, 132–139.

Sarjeant, W.A.S., 1975. Plant trace fossils. In: The Study of Trace Fossils. Springer-Verlag, New York, NY, pp. 163–179.

Schaetzl, R.J., Anderson, S., 2005. Soils: Genesis and Geomorphology. Cambridge University Press, New York, NY.

Sheldon, N.D., 2006. Using paleosols of the picture Gorge Basalt to reconstruct the middle Miocene climatic optimum. PaleoBios 26, 27–36.

Sheldon, N.D., Tabor, N.J., 2009. Quantitative paleoenvironmental and paleoclimatic reconstruction using paleosols. Earth-Sci. Rev. 95, 1–52.

Singer, M.J., Nkedi-Kizza, P., 1980. Properties and history of an exhumed Tertiary Oxisol in California. Soil Sci. Soc. Am. J. 44, 587–590. http://dx.doi.org/10.2136/sssaj1980.03615995004400030031x.

Smith, S.Y., Manchester, S.R., Samant, B., Mohabey, D.M., Wheeler, E., Baas, P., Kapgate, D., Srivastava, R., Sheldon, N.D., 2015. Integrating paleobotanical, paleosol, and stratigraphic data to study critical transitions: a case study from the late Cretaceous-Paleocene of India. In: Earth-Life Transitions: Paleobiology in the Context of Earth Systems Evolution. The Paleontological Society, Cambridge, UK, pp. 137–166. Paleontological Society Short Course.

Soil Survey Staff, 1999. Soil Taxonomy: A Basic System of Soil Classification for Making and Interpreting Soil Surveys, second ed. Natural Resources Conservation Service, U.S. Department of Agriculture, Washington, DC. Agriculture Handbook 436.

Staub, J.R., Cohen, A.D., 1978. Kaolinite-enrichment beneath coals; a modern analog, Snuggedy Swamp, South Carolina. J. Sediment. Res. 48, 203–210.

Staub, J.R., Cohen, A.D., 1979. The Snuggedy Swamp of South Carolina: a back-barrier estuarine coal-forming environment. J. Sediment. Res. 49, 133–143.

Stoops, G., 2003. Guidelines for Analysis and Description of Soil and Regolith Thin Sections. Soil Science Society of America, Madison, WI.

Stoops, G., Marcelino, V., Mees, F. (Eds.), 2010. Interpretation of Micromorphological Features of Soils and Regoliths. Elsevier, Amsterdam.

Tabor, N.J., Montañez, I.P., 2002. Shifts in late paleozoic atmospheric circulation over western equatorial Pangea: insights from pedogenic mineral $\delta^{18}O$ compositions. Geology 30, 1127–1130.

Tabor, N.J., Montañez, I.P., 2004. Morphology and distribution of fossil soils in the Permo-Pennsylvanian Wichita and Bowie Groups, north-central Texas, USA: implications for western equatorial Pangean palaeoclimate during icehouse-greenhouse transition. Sedimentology 51, 851–884.

Tabor, N.J., Myers, T.S., 2015. Paleosols as indicators of paleoenvironment and paleoclimate. Annu. Rev. Earth Planet. Sci. 43, 333–361.

Tabor, N.J., Yapp, C.J., 2005. Coexisting goethite and gibbsite from a high-paleolatitude (55°N) Late Paleocene laterite: concentration and 13C/12C ratios of occluded CO_2 and associated organic matter. Geochim. Cosmochim. Acta 69, 5495–5510.

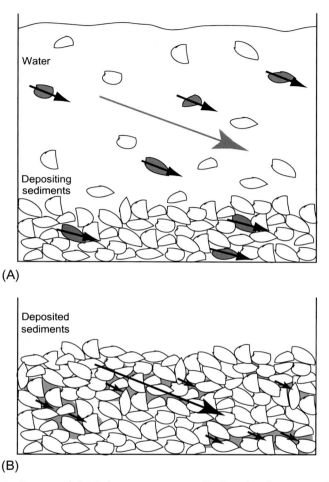

Fig. 1 (A) Schematic diagram of detrital remanent magnetization showing magnetic mineral grains settling out of a water column with some compaction (and thus inclination shallowing) in the lower sediments. (B) Schematic diagram of chemical remanent magnetization showing development of a magnetic mineral cement, such as hematite, in interstitial spaces. In both, *large arrow* represents the ambient magnetic field and *smaller arrows* represent the magnetization of the grain or cement.

to the geomagnetic field. As subsequent material is deposited on top of it, the dip the grain initially had from the horizontal may be flattened significantly, leading to shallowing of the inclination of the paleomagnetic vector (Fig. 1). Several authors have explored the issues related to inclination shallowing and developed statistical techniques to calculate the degree of shallowing expected for various sediments (e.g., King, 1955; Anson and Kodama, 1987; Deamer and Kodama, 1990; Jackson et al., 1991; Sun and Kodama; 1992; Hodych and Bijaksana, 1993; Kodama and Davi, 1995; Tan et al., 2002, 2007; Kent and Tauxe, 2005; Bilardello and Kodama, 2010; Kent and Irving, 2010).

CRM occurs after sediments have been deposited and is due to the formation of secondary magnetic minerals that acquire a magnetization oriented parallel to the geomagnetic field as the crystals grow (Fig. 1). In some cases, the secondary minerals are intergranular cements (Walker, 1967, 1974; Collinson, 1974; Walker et al., 1978, 1981; Turner, 1980; Larson et al., 1982; Butler, 1992; Roberts et al., 2005; Tauxe, 2010). Hematite is the most common cementing agent that can carry a remanent magnetization (e.g., Larson et al., 1982; Molina-Garza et al., 2003; Zeigler and Geissman, 2011) and is the mineral that causes the red to purple colors observed in many terrestrial depositional sequences (e.g., Walker, 1967). Whereas the issue of inclination shallowing is significantly less in the case of CRM, the problem that arises in interpreting a CRM is estimating the amount of time that has passed between deposition of the rock and the growth of the secondary magnetic minerals. In older sequences of rocks, a few thousand years between deposition and growth of the CRM is not significant compared to the age of the rock, but for researchers working on very young strata (e.g., Plio–Pleistocene), this time lag must be taken into consideration because it is a significant proportion of the rock's age.

It should come as no surprise that the complexities of depositional systems result in most sedimentary rocks having both DRM and CRM magnetizations. For example, Zeigler and Geissman (2011) demonstrated that red beds deposited in the Late Triassic in northern New Mexico carried detrital hematite and magnetite, providing a DRM, as well as coarse-grained pigmentary hematite as a secondary cement, yielding a chemical remanent magnetization. For rocks over 200 million years old, the difference in timing between the DRM and CRM is, for the most part, effectively insignificant. However, it is useful to use a variety of techniques to tease apart the different phases of magnetization recorded in the strata being studied.

SAMPLING SCHEMES AND CHOICE OF LITHOLOGY

Butler (1992) discusses variables associated with collecting an appropriate number and type of samples for paleomagnetic data acquisition, and Kodama and Hinnov (2015) provide an excellent discussion of sampling techniques specifically targeted toward crafting rock magnetic cyclostratigraphic frameworks for depositional systems. Here we blend the experiences of many paleomagnetists into a summary of the choices to be considered for sampling for magnetostratigraphic data sets. When considering how to sample a given outcrop of terrestrial strata, trade-offs must be made in order to determine the most efficient means of accurately detecting the magnetic polarity history of the rock sequence. Terrestrial depositional sequences are frequently lithologically heterogeneous, and different rock types require different means of obtaining samples. Well-indurated materials, such as siltstone or well-cemented sandstone, can be drilled with a water-cooled drill bit, creating cylindrical samples ∼5–8 cm long. These samples can be cut on a rock saw into 2-cm-long individual specimens.

Mudstone and claystone, as well as poorly cemented sandstone, can present a challenge for sampling. Johnson's et al. (1975) block-sampling method is a preferred method of obtaining samples from fine-grained or poorly cemented strata (Fig. 2). This methodology involves extracting individual blocks of rock from beneath the weathered mantle

Fig. 2 Photographs of the steps in block sampling for fine-grained sedimentary rocks. (A) A block sample still in situ with strike and dip marked on the most accessible face *(white circle)*. (B) Sample to be cut, showing strike and dip markings extended and repeated for ease of cutting. Repeated labeling during cutting is essential to prevent loss of the orientation markings should part of the sample crumble during cutting. (C) Individual oriented specimens with the strike and dip and specimen identification clearly labeled. Also included: the leftover piece of the sample, clearly labeled for future use.

and orienting them via a marked strike and dip on a flat surface. These samples are then cut into cubes on a rock saw with no water back in the laboratory. In the case of poorly cemented sandstone or siltstone, the addition of diluted sodium silicate in the field can help stabilize the material enough to collect reasonably sized samples and can be used during the cutting process to produce cube-shaped specimens that will hold together relatively well. If these samples are heated during thermal demagnetization, coating the specimens with nonmagnetic alumina cement can often hold the specimens together long enough to achieve reasonably robust data (e.g., Connell et al., 2013). If the samples will be analyzed via alternating field (AF) demagnetization, specimens can be shaved out of the outcrop in a cubic shape and inserted into nonmagnetic plastic boxes, or the nonmagnetic plastic boxes can be hammered into the outcrop if it is unconsolidated enough to not break the box. A strike and dip is then measured on the box itself for orientation of the sample.

It is important to note that paleomagnetists have developed a specific vocabulary for sampling for paleomagnetic data (Butler, 1992, Fig. 3). A specific outcrop targeted for sampling is a locality. Within a locality, the individual sedimentary bed sampled is a site. Paleomagnetists refer to a site as a "geological instant in time." In sampling igneous rocks, a lava flow would yield one paleomagnetic site. It is preferable to try to take samples horizontally along a single stratigraphic horizon as opposed to vertically across several horizons, to make sure one's site is restricted in time and hence samples only one geomagnetic field polarity. The block samples or individually drilled cores are oriented with respect to north, usually with a magnetic compass, and with respect to the horizontal with a bubble level. The individually oriented block sample or core taken from the site is referred to as a sample and the cubes or 2-cm-long cylinders cut from that individually oriented sample are called specimens. Usually multiple specimens are cut from individually oriented samples.

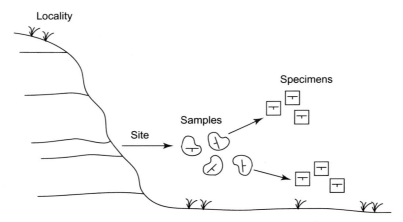

Fig. 3 Terms related to the sampling hierarchy for paleomagnetic samples.

An important choice must be made when heading into the field: developing a reasonable sampling scheme that considers the number of sites to be sampled and the number of samples to be collected. The decision regarding spacing of sample sites for a given outcrop and the number of samples collected boils down to a question of balancing the amount of time that can be devoted to analyzing all the samples from the outcrop in question and the precision of the paleomagnetic vector averaged from a site. The more samples collected and measured at a site, the smaller the uncertainty of the mean site direction. If there are multiple localities to be sampled, this also needs to be taken into account. Closely spaced sampling sites, coupled with large numbers of samples per horizon, will lead to lengthy sample preparation and analytical time, but frequently produces statistically robust data sets. A lower resolution sampling effort saves on sample preparation time and analytical efforts, but incurs a loss of information and short-duration polarity events may be missed entirely. As sampling expeditions are planned, one means of assessing the most efficient site spacing is to review the geomagnetic polarity time scale (GPTS) for the age of the rocks that are to be studied (Walker and Geissman, 2009; Gradstein et al., 2012). A very active geomagnetic field will result in a large number of short-duration polarity events that would be missed by a widely spaced sampling scheme. Consider the sampling scheme that would need to be developed to capture the majority of the rapid polarity changes in the Early Cretaceous, Jurassic, or the Triassic when the field reversed approximately five times per million years. In contrast, sampling strata of early Tertiary age where field reversals take place approximately every million years could be done at a more broadly spaced sampling scheme (Fig. 4).

In developing a local polarity stratigraphy, it is critical to sample at a fairly close stratigraphic spacing to detect all the polarity intervals (i.e., reversals) recorded in the rocks. Therefore, samples from many sites are usually collected. Because only geomagnetic field polarity, rather than a precise paleomagnetic direction, is required at a given site, researchers often favor taking only three to four individually oriented samples from each horizon to determine the horizon's polarity. This is entirely reasonable from the perspective of a quick means of providing basic polarity data and is the minimum number of specimens needed for proper statistical analysis (Fisher, 1953; Butler, 1992; Tauxe, 2010). But this choice may pose a problem if two of the samples from one horizon indicate opposite polarities, and the third is indeterminate in polarity. Taking a larger number of samples per site leads to a more robust determination of the polarity of the horizon and usually provides enough leftover material to conduct rock magnetic or chemical experiments or to create thin sections, but adds time to the entire process through both sample preparation and paleomagnetic analyses.

In addition, if the plan is to sample multiple outcrops, especially for correlation purposes, the number of samples to be processed for the entire endeavor can quickly balloon. Zeigler et al. (Chapter 6) chose to sample a locality at high stratigraphic

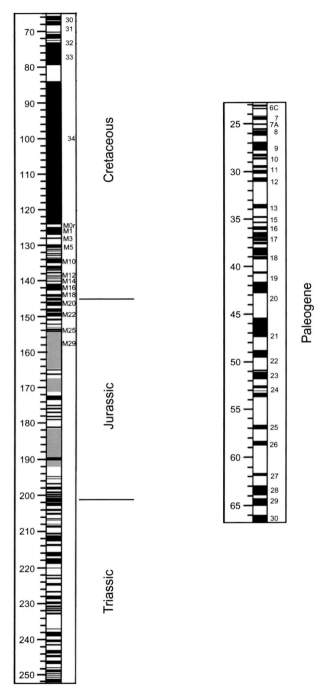

Fig. 4 The Paleogene and Mesozoic geomagnetic polarity time scale (GPTS) from Walker and Geissman (2009). Note the difference in polarity behavior between the Late Cretaceous, dominated by the Cretaceous Long Normal and the Early Cretaceous or Triassic where there are numerous rapid polarity changes.

resolution (1-m stratigraphic spacing between sites) with six to eight individual samples collected per site. This resulted in a huge number of samples, but only a subset of samples per site was prepared for thermal demagnetization. Once the initial set of samples had been analyzed, further material was prepared to examine sites that had poorly defined magnetic polarity data. Extra sample material was also prepared for rock magnetism experiments. Thus there was a large amount of material brought in from the field and available for study, even though not every sample was prepared and analyzed. This saved the authors repeated trips into the field to collect more samples for sites that needed more analysis.

Important in-field observations to make while collecting samples include any structural complications, diagenetic features, and high topography within the sampling locality that may be prone to lightning strikes (Butler, 1992). If the strata intended for sampling are not flat-lying, obtain a representative measurement of the average orientation of the bedding planes to correct the attitude of the resulting paleomagnetic vector data. If one is working in a structurally complex area with fault offsets or changes in bedding orientation, carefully record all of this information either on a geologic map and/or in a field notebook. Also, examine samples closely for signs of alteration that might affect the quality of the paleomagnetic data. Brecciated materials, in-filled fractures, color mottling, and/or growth of diagenetic phases like limonite or goethite are signs that there may be significant overprinting and/or remagnetization of the material. If possible, consider shifting sampling up or down section enough to avoid deeply weathered and/or significantly altered horizons. Keep track of any changes in sampling scheme. In the field, there are frequently places where obtaining perfect sampling spacing is simply not safe or even possible. Thick benches of cliff-forming strata, steep terrain, distance to and difficulties of access to the outcrop should all be considered prior to sampling.

One last note to be made in terms of field sampling methods involves labeling. It is, of course, important to label each sample as it is removed from the outcrop. Mark the orientation of the core or block sample and its identification on the sample directly. Also, write down the orientation of the sample in a field notebook. For block samples, which are frequently wrapped protectively for transport back to the laboratory, consider writing the identification and orientation on the protective wrapping as well. The more redundant labeling applied to the suite of samples collected, the better. It is frustrating to return home and unpack a sample and prepare to process it only to find it has no information associated with it or the markings have rubbed off (or broken off). If a sample fractures too badly to be prepared as an oriented specimen for analysis in a magnetometer, the sample is still useful. Unoriented material can still be used for rock magnetism and bulk susceptibility analyses, as well as chemical experiments and thin sections.

SAMPLE PROCESSING: TO AF OR TO THERMAL DEMAGNETIZE?

Once samples have been collected and cut to the correct size for measurement in the rock magnetometer, one must plan a demagnetization strategy. If the specimens contain magnetite or titanomagnetite, AF demagnetization will be the preferred method. However, if the samples are dominantly detrital or pigmentary hematite or goethite, AF demagnetization will not be as helpful because the coercivities of these minerals are too high to be affected by AF demagnetization with standard laboratory equipment, and for these samples thermal demagnetization would be the preferred analysis technique. In examining a hand sample, it is not possible to determine what magnetic minerals might be present. A few techniques for determining the correct demagnetization method include separation of the magnetic particles from crushed samples and subsequent scanning electron microscope examination, isothermal remanent magnetization (IRM) acquisition experiments and their modeling, or pilot AF and thermal demagnetization experiments on a few representative specimens. It helps to have multiple specimens per sample for these rock magnetic experiments. Butler (1992) and Tauxe (2010) both provide a comprehensive explanation of the rock magnetic theory behind AF and thermal demagnetization techniques.

Crushing and separating minerals using a strong (rare Earth) magnet can be very informative for identifying the magnetic phases present, but can be time consuming and destroys specimens. IRM acquisition experiments (discussed further in the "Rock Magnetism Experiments" section following) can be modeled to determine the coercivities of the magnetic minerals in the samples (Kruiver et al., 2001). A sample that achieves magnetic saturation at lower magnetic fields (<0.1 T) is fairly rich in low coercivity phases, usually magnetite. A sample that takes much higher magnetic fields to achieve magnetic saturation (>2–3 T) is dominated by high coercivity phases, like hematite or goethite. Frequently specimens will show a mixture of both behaviors by nearly saturating at relatively low fields, but requiring final steps at much higher fields to saturate (see Rock Magnetism Experiments below).

A pilot AF demagnetization consists of demagnetizing the specimens at several low field AF steps in the demagnetizer. A typical demagnetization sequence consists of AF demagnetization steps of 5, 10, 15, and 20 mT after measurement of the natural remanent magnetization (NRM). If the specimen responds to AF demagnetization, and the intensity of magnetization begins to decrease, even at these lower fields, there is probably enough magnetite or titanomagnetite present to continue a complete AF demagnetization sequence. If the magnetic intensity of the specimen does not appreciably decrease, then there is probably not a volumetrically significant quantity of low coercivity minerals present. This approach is a nice way to check specimens because it physically preserves the specimen as well as keeps the original magnetization of the specimen intact. IRM experiments will totally overprint the NRM, so they should only be conducted after the ancient paleomagnetic information has been extracted from the sample.

If AF demagnetization is the appropriate technique to use, run a few specimens from different lithologies and/or different sites through a closely spaced set of demagnetization steps. If there are low coercivity phases present, these will be demagnetized at relatively low fields and interesting overprint data may be lost if the initial steps are not closely enough spaced. Additionally, if a low coercivity phase is the only carrier of magnetization, you do not want to use bigger steps and potentially lose data. Once you have a sense of the field at which the specimens lose most of their magnetization, you can potentially go to fewer steps and focus on the appropriate fields to document the characteristic remanent magnetization (ChRM) of the sample.

Similarly, for a thermal demagnetization pilot study, run a small suite of specimens from different sites and of different lithologies at closely spaced temperature steps to determine the range of demagnetization behavior. Thermal demagnetization is a powerful demagnetization technique because it will demagnetize both high and low coercivity magnetic minerals at temperatures easily achievable in the laboratory. The only issue is that the time required to heat a group of specimens, analyze, heat, analyze, can be significant; however, it is an extremely effective technique for identifying the ChRM of ancient sedimentary rocks, so it is widely employed in magnetostratigraphy studies (e.g., Schmidt, 1993; Peppe et al., 2009).

ROCK MAGNETISM EXPERIMENTS

A wide array of rock magnetism experiments can help ascertain the proportions of and types of magnetic minerals present in a rock. This information is useful in assessing not only the magnetic minerals carrying the magnetic signature of a given lithology, but also has implications for diagenesis and depositional environments that can be tied to paleoclimate (Kodama and Hinnov, 2015). The data resulting from these experiments can be used to construct rock magnetic cyclostratigraphies, which can be used to correlate different outcrops and assign high-resolution chronostratigraphies to sedimentary sequences (e.g., Latta et al., 2006; Gunderson et al., 2012; Hinnov et al., 2013; Kodama and Hinnov, 2015). In rock magnetic cyclostratigraphy, rock magnetic parameters detect magnetic mineral concentration variations that beat to astronomically forced (Milankovitch) global climate cycles. Experiments and measurements frequently used on sedimentary rocks include bulk magnetic susceptibility, anhysteretic remanent magnetization (ARM), IRM, and thermal demagnetization of a multiple component IRM (Lowrie, 1990). As is the case in planning any paleomagnetic or rock magnetic study, it is important to consider the balance of the number of analytical steps used versus the time available to process samples.

Bulk susceptibility measurements are easy and quick to perform and give a general sense of the concentration of magnetic minerals in a sample. A stratigraphic sequence of susceptibility measurements can provide a first pass look at susceptibility variations to assess climate-driven processes, for example, and is frequently used for an initial attempt to correlate closely spaced outcrops. However, bulk susceptibility measurements will include contributions from ferromagnetic minerals as well as paramagnetic (e.g., iron-rich silicate minerals) and diamagnetic (e.g., quartz or carbonate) minerals, so more detailed study would be needed to understand the source of the susceptibility signal.

For full characterization of the magnetic mineralogy of the samples collected, more than just bulk susceptibility measurements will be needed to properly quantify the concentrations and grain sizes of different ferromagnetic minerals in the rock (Fig. 5). IRM acquisition experiments are helpful for determining if the dominant magnetic mineralogy of a sample is a high or low coercivity mineral (Kruiver et al., 2001; Fig. 5A–D), but they yield only relative concentration information for the different coercivity components. In the IRM acquisition experiment, a sample is exposed to sequentially higher DC magnetic fields using an impulse magnetizer or an induction coil. The sample, oriented the same way for each step, is "zapped" in the impulse magnetizer and then measured in the magnetometer repeatedly until the specimen reaches saturation (note the plateau of the curve in Fig. 5B–D). Since different magnetic minerals have different coercivities, they will acquire magnetic saturation at different peak magnetic fields. Magnetite, for example, saturates smaller magnetic fields than does hematite. However, a problem in magnetic mineral identification can arise because different magnetic minerals can have similar coercivities. Distinguishing hematite from goethite can be difficult based on IRM acquisition data alone. The addition of thermal demagnetization can help tell minerals apart by using both their coercivity and unblocking temperature.

Multiple component IRM-thermal demagnetization experiments are very useful for magnetic mineral identification because they yield information about both coercivities and unblocking temperatures and can be used to assess the proportions of different magnetic minerals in a specimen (Lowrie, 1990; Fig. 5E and F). This particular experiment is effectively a combination of an IRM acquisition and thermal demagnetization. A specimen is placed in an impulse magnetizer or coil and is given IRMs along two or three orthogonal axes using different strength magnetic fields along each axis. This method magnetizes different coercivity fractions, so the magnetic fields used depend on the coercivities of the expected magnetic minerals in the sample. Lowrie (1990) used 5, 0.4, and 0.12 T in his experiments. Subjecting the specimen to subsequent thermal demagnetization gives information about the Curie points of the different mineral fractions. This experiment produces two or three data sets for a given specimen: one for each coercivity fraction. For example, sample CK1-1b (Fig. 5F) shows a low concentration of low coercivity minerals (shown by the 0.03 T curve) and a high concentration

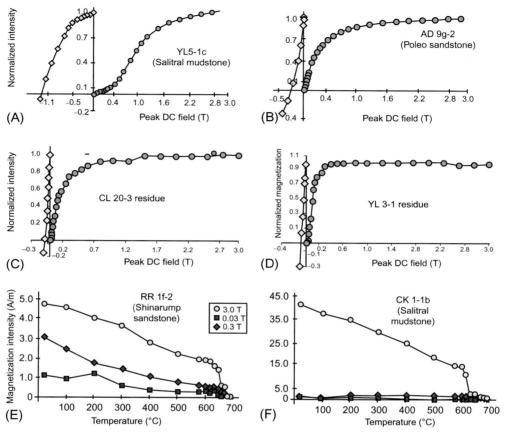

Fig. 5 Examples of multiple component IRM-thermal demagnetization and IRM acquisition results for specimens from Triassic red beds. (A, B) Multiple component IRM-thermal demagnetization responses for a sandstone and a mudstone specimen. Note the contributions of different coercivity phases in (A) versus the near-single phase contribution in (B). (C, D) IRM acquisition and backfield demagnetization curves for mudstone and sandstone specimens. Note the inflection point in (C), suggesting two distinct coercivity phases. (E, F) IRM acquisition and backfield demagnetization curves for residues left over after leaching experiments on red bed mudstone specimens, showing saturation being reached below 0.8 T, indicating a fine-grained, low coercivity phase, probably magnetite. *(From Zeigler, K.E., Geissman, J.W., 2011. Magnetostratigraphy of the Upper Triassic Chinle Group of New Mexico: implications for regional and global correlations among Upper Triassic sequences. Geosphere 7, 802–829.)*

of high coercivity minerals that have an unblocking temperature near 660°C (shown by the 3.0 T curve).

ARM is another method that is useful for quantifying ferromagnetic minerals in a sample. In this experiment, samples are subjected to an alternating field with an initial peak value of around 100 mT that then decays to zero. During this process, a small and constant biasing magnetic field ($\sim$50–100 µT) is applied to the sample, giving an ARM to the sample. The only drawback to this particular method is that it is limited to lower coercivity ferromagnetic minerals (e.g., magnetite, titanomagnetite) and has no effect on higher coercivity minerals like hematite or goethite based on the limitations of typical laboratory equipment.

It takes a combination of different experiments to truly characterize the type, grain size, and concentration of magnetic minerals present in any given sample. Kodama and Hinnov (2015) offer an excellent review of different combinations of experiments that can provide further information about the magnetic minerals in the sedimentary sequence being studied. Once this rock magnetic information is gathered, vector data can be examined for potential remagnetization, and inferences can be drawn about the source areas for the magnetic minerals.

CHEMICAL DEMAGNETIZATION

Another means of assessing the characteristic magnetic mineralogy of a given set of specimens is to apply chemical demagnetization methods (Henry, 1979; Zeigler and Geissman, 2011). One simple experiment consists of soaking specimens in hydrochloric acid for sequentially longer time steps and measuring the magnetization after each application of acid (Henry, 1979). Well-cemented samples that have been drilled should have a series of shallow grooves cut along the sides of the core with a rock saw or small holes drilled to increase surface area and allow access of the acid to the interior of the sample. If possible, small cuts can be made in the edges of cubic samples as well, depending on the competency of the material. The objective of this experiment is to remove pigmentary hematite and isolate the detrital component, assuming it is carried by large detrital hematite grains. Just before the sample loses its competency entirely, the acid baths can be stopped and the specimens run through the same thermal demagnetization steps other specimens are being analyzed with.

Another means of using hydrochloric acid baths to understand magnetic mineralogy was developed by Geissman (Zeigler and Geissman, 2011) and consists of soaking individual specimens in hydrochloric acid until the specimen has completely broken down and has been bleached of all color. The acid is decanted and refreshed frequently until there is nothing left in the beaker but a fully bleached slurry. The resulting material is then dried, powdered further if needed, mixed with alumina cement, and packed into nonmagnetic plastic cubes or even plastic vial caps to create a faux specimen. These specimens are then run through the standard IRM steps applied to other specimens in the assemblage. This experiment can reveal more information about the detrital component once all pigmentary contributing minerals are removed by the acid (Fig. 5C and D).

QUALITY OF THE DATA: STATISTICAL ANALYSIS

A suite of statistics is applied to the paleomagnetic vectors determined in a particular study. The first round of calculations is the maximum angular deviation (MAD) of the vector resolved during demagnetization (Kirschvink, 1980). This is an assessment of how linear the demagnetization steps are during the demagnetization process, be it AF or thermal. Fig. 6 includes some examples of linear, low MAD behavior, compared to nonlinear, high MAD demagnetization behavior, shown as vector endpoint diagrams (Zijderveld, 1967). Generally, MAD values $<10°$ are considered excellent, $10°–20°$ are acceptable, and anything above $20°$ should only be used with caution. After MAD values have been calculated for individual magnetization components (recalling that more than one vector component may be present), the dispersion of the demagnetized directions within the sampling site is examined using Fisher statistics (Fisher, 1953), which assumes that the directions at the site are unit vectors that are sampled from a two-dimensional normal, or Gaussian, distribution on the surface of a sphere (the stereonet). Two values result from this analysis: α_{95} and k. The first, α_{95}, effectively describes how tightly clustered the vectors are on a sphere. Smaller α_{95} values are preferred, and as with MAD values, $<10°$ is excellent, $10°–20°$ is acceptable, etc. The α_{95} is the 95% confidence interval around the observed mean direction calculated from the demagnetization data. There is a 95% probability that the true mean of the theoretical Fisher distribution sampled by the data lies within the α_{95} around the observed mean direction calculated from the data. The second statistic, k, is also a measure of the clustering of the vectors and is an estimate of the Fisher precision parameter, κ. Contrary to the α_{95} values, higher k values are preferred because they indicate more tightly clustered directions.

Once the site mean direction and the α_{95} and k values for a site have been calculated, the grand mean direction for a stratigraphic unit (member or formation-rank strata) can then be determined. Here, all the site mean directions of the nondominant polarity are inverted into the other hemisphere, and all vectors together are then averaged to determine a mean magnetic polarity direction for the entire unit, complete with α_{95} and k. For example, if a stratigraphic unit is predominantly reversed polarity, one would invert the normal polarity directions, calculate the mean, and present the results as predominantly reverse polarity. Sometimes the quality of data for a given set of specimens is poor enough that one might hesitate to discuss the results of the study. However, if the vectors generally show normal and/or reversed polarity behavior, even if it is poorly defined, there is still some relevant information to be shared.

ASSESSING THE POSSIBILITY OF REMAGNETIZATION

When processing a suite of samples from a sedimentary sequence, it is important to bear in mind the possibility that the outcrop sampled may have been remagnetized at a significantly younger time than original deposition and lithification, either by chemical

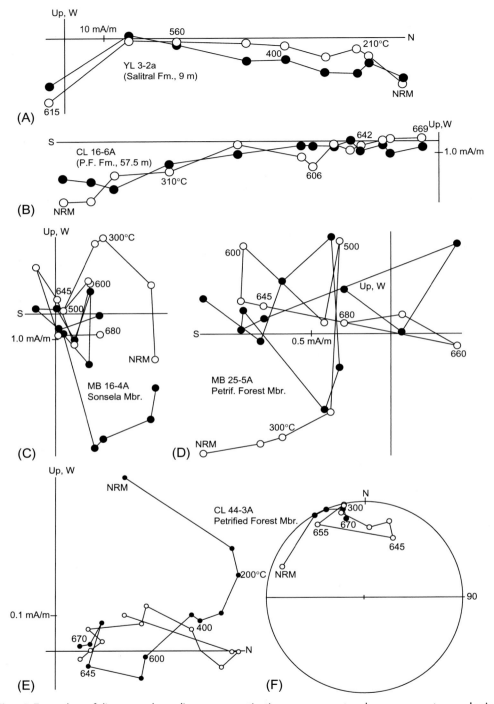

Fig. 6 Examples of linear and nonlinear magnetization components, shown as vector endpoint demagnetization diagrams, showing different responses to progressive thermal demagnetization. (A, B) Two examples of nearly linear vectors dominated by a single component from Triassic strata in New Mexico. (C, D) Examples of uninterpretable vectors. (D) Could be very tentatively identified as representing a reverse polarity vector because it is more or less south-seeking in its orientation, but this is a very poor-quality response and should not be relied on to determine polarity for a specimen. (E, F) Example of a specimen exhibiting multiple vector components and its attendant equal area net.

processes (e.g., Jackson et al., 1992; Wang et al., 2005; Elmore et al., 2012; Van der Voo and Torsvik, 2012) and/or the application of heat via nearby igneous intrusions or extrusive rocks (for an interesting example of magnetite destruction during contact metamorphism, see Gillett, 2003). Being aware of local and regional tectonic events is essential for knowing whether or not a given locality is at risk of having been remagnetized. The first sign that a locality may have been remagnetized is specimens from the entire locality showing the same polarity. Unless the rocks were deposited during one of the superchrons (e.g., the Cretaceous Long Normal, Kiaman reversed polarity superchron), in recent sediments capturing the Brunhes polarity epoch (late Pleistocene to Holocene), or the sequence was deposited unusually quickly, there should be at least a few polarity intervals captured by even a fairly coarse-resolution sampling scheme.

Deliberately sampling tilted and/or folded rocks is a means of providing a test for constraining the age of magnetization and assessing the possibility of remagnetization. A simple fold test can be used to determine if the paleomagnetic directions observed in a sequence are original or remagnetized at some time after or even during folding (Graham, 1949; McFadden and Jones, 1981; Butler, 1992; Tauxe and Watson, 1994; Enkin, 2003). If strata sampled are on different limbs of a fold, or from variably tilted strata from an area, and they all have the same paleomagnetic direction once the bedding has been rotated back into the horizontal, the paleomagnetic data has passed the fold test and the magnetization is older than the structural tilting event. Thus it is advantageous to try to sample localities that are tilted or folded so that a fold test can be applied to constrain the age of the rocks' magnetization.

Another test used for constraining the age of magnetization is the conglomerate test. In this test conglomerate clasts are sampled. If a conglomerate was remagnetized at some time after deposition, all of the individual clasts within the conglomerate will have the same paleomagnetic direction. If no remagnetization has occurred, the clasts should have random directions of magnetization. Comparing the mean paleomagnetic direction determined from the conglomerate's matrix to the individual clast results can be instructive about timing of remagnetization. As always, careful observations of the rock itself can also aid in determining potential postdepositional influences. It is usually well worth crafting a handful of thin sections to observe grain size, grain composition, cement type, and cement composition.

DEVELOPING A MAGNETIC POLARITY CHRONOLOGY AND WRESTLING WITH CORRELATIONS
Developing a Magnetostratigraphy

Once the specimens have been completely demagnetized, one can begin to construct a local magnetic polarity stratigraphy. The easiest way to determine a horizon's polarity is to calculate the corresponding virtual geomagnetic pole (VGP) for each sample's characteristic remanence and the paleomagnetic pole for the formation studied. Because plates move through geologic time, the paleomagnetic pole of ancient rocks will not necessarily be close to the current geographic north or south poles. Therefore, the great circle distance between the paleomagnetic pole for the formation, assuming it was the north paleomagnetic pole, and the horizon's or sample's VGP will indicate whether the horizon or sample was normal (usually within 45° of the north paleomagnetic pole) or reversed (within 45° of the south paleomagnetic pole).

If the paleomagnetic data is reported as local geomagnetic vectors (declination and inclination), then depending on when the rocks were deposited and where the place of deposition was geographically, the paleomagnetic direction of any given sample will be different. Assuming the geomagnetic field is an axial, geocentric dipole (Butler, 1992; Tauxe, 2010), the local inclination depends on the latitude of deposition. For sites collected near the ancient equator, where the paleomagnetic vector is nearly flat, the declination of the paleomagnetic direction can be used to indicate polarity. For samples that were deposited at high latitudes where the paleomagnetic direction is very steep, inclination can be used to indicate polarity.

For example, paleomagnetic directions from Upper Triassic strata in the American Southwest are north- or south-directed with very shallow inclinations (Fig. 7) indicating that the Upper Triassic paleomagnetic pole was about 90° away from the American Southwest. In this case the north-directed declinations indicate normal polarity and south-directed reversed polarity. In samples obtained from Cretaceous strata in western North America (Fig. 6), the inclinations are significantly steeper; hence western North America had moved northward into higher latitudes during the middle Mesozoic and was closer to the north geographic pole. Typically, the latitude of the individual site's VGP is plotted as a function of stratigraphic position in the sequence studied. A positive 90° latitude indicates a normal polarity sample and a negative 90° latitude indicates a reverse polarity sample. However, incomplete demagnetization or ancient variations in the local geomagnetic vector causes VGP latitudes to vary from +90° or −90° (Fig. 8). Depending on the amount of scatter in the data, VGP latitudes as low as +45° or −45° can be used to indicate normal or reversed polarity samples, respectively. Samples that can be assigned neither a reversed or normal polarity are designated to be indeterminate and cannot contribute to the local magnetostratigraphy.

When there are magnetic overprints, there may be more than one magnetization component present in a sample or in samples from a given horizon, each representing a magnetization event of a different age (e.g., Fig. 6E and F). Sometimes, small temperature or AF demagnetization steps can help define these additional vectors enough to be able to determine their direction and thus a possible age of magnetization (e.g., Molina-Garza et al., 1995). Sometimes a weighting scheme is involved in interpreting these results. Because the field is currently normal polarity, a present-day viscous

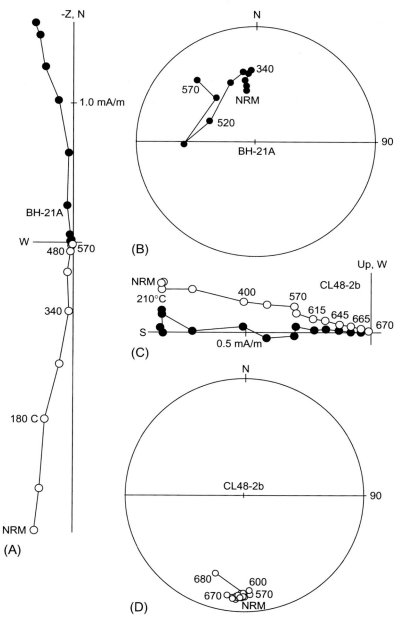

Fig. 7 Vector endpoint demagnetization diagrams and equal area nets for Cretaceous and Triassic specimens, highlighting the differences in inclination values. (A, B) Vector endpoint demagnetization diagram and equal area net for Cretaceous example. (C, D) Vector endpoint demagnetization diagram and equal area net for a Triassic example.

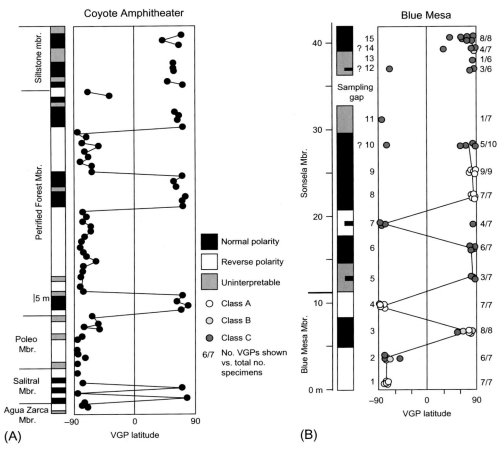

Fig. 8 Examples of both site-level and specimen-level VGPs calculated for two Late Triassic sequence in the southwestern United States. (A) Site-level VGPs for Triassic strata from northern New Mexico (Zeigler and Geissman, 2011) showing a single VGP calculated for each site mean direction. (B) Specimen-level VGPs showing the scatter of individual specimen VGPs and interpreted magnetic polarity (Zeigler et al., this volume).

overprint could affect reversed polarity samples more than normal polarity samples; hence when interpreting the polarity of horizons with mixed polarities, more weight could be given to reversed polarity samples than normal polarity samples (e.g., Spahn et al., 2013). The reversal test (McFadden and McElhinny, 1990; Butler, 1992; Tauxe, 2010) is a means of assessing the possibility of secondary NRM components by using the concept that the time–averaged geomagnetic field during normal and reverse polarity intervals should be 180° from one another. If the normal and reversed mean directions for a sequence do not coincide with one another, this indicates that either secondary NRM components are present or the sampling scheme used does not adequately time-average geomagnetic secular variation during one of both of the polarity intervals compared.

Once the local magnetic stratigraphy of the study area has been determined, it should be compared to the GPTS (Gradstein et al., 2012). Additional geochronological information is needed (e.g., biostratigraphy, radiometric dating) to determine the rough age of the stratigraphic sequence and only then can the local polarity stratigraphy be tied to the GPTS. The GPTS is based largely on the magnetic anomaly pattern (so-called "zebra stripes") created by seafloor spreading at the midocean ridges. The oldest seafloor preserved on the planet is Jurassic in age, so the GPTS is well established back to the Jurassic only. For stratigraphic sequences older than the Jurassic, the GPTS is based on the correlation of multiple magnetostratigraphies of terrestrial sequences that have been compiled using additional geochronological data to ensure correlation between sequences is as accurate as possible.

Correlation of Magnetic Polarity Chronologies

Most magnetostratigraphic studies are aimed at developing better correlations between individual outcrops or outcrop belts, some of which may be separated by tens to hundreds of kilometers or may even be on opposite sides of the globe. It is important to recognize that magnetic polarity chronologies cannot be compared to one another without either correlation to the GPTS or the use of other geochronologic data. For example, consider the correlation between upper Triassic strata in northern New Mexico and the "gold standard" Late Triassic magnetic polarity chronology of the Newark Supergroup in eastern North America. With no other information, it is nearly impossible to determine which chrons in the New Mexico section are correlative to those in the Newark Supergroup record (Fig. 9). Normal and reversed polarity diagrams are best utilized when there is robust biostratigraphic or, even better, absolute age determinations, coupled with the polarity chronology (see Zeigler et al., this volume).

It is in the comparison of polarity chronologies that the trade-off between number of horizons sampled and time spent analyzing samples comes into play. High-resolution sampling efforts make for a much more robust correlation, but the time required to run the requisite number of samples is that much longer. Consider the following polarity chronology from the Chinle Formation of Petrified Forest National Park in Arizona (Fig. 10). The Martha's Butte section was sampled at 1-m resolution with six to eight independently oriented block samples recovered from each horizon. Had we chosen to only sample every 3 m, note the difference in the structure of the polarity chronology!

A criticism that has been leveled at magnetostratigraphic correlations is that the researchers involved are simply guessing at how well their local magnetostratigraphy best matches with the GPTS, and thus one could conceivably match any magnetic polarity chronology to any part of the GPTS if the pattern of chrons looks reasonable. However,

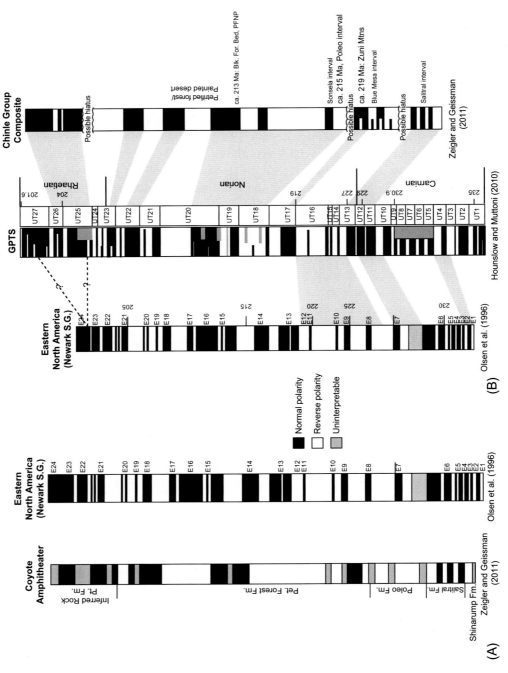

Fig. 9 (A) A magnetic polarity chronology for the Upper Triassic Chinle Formation from northern New Mexico compared to the Newark Supergroup polarity record in eastern North America with no other stratigraphic or geochronologic controls for correlation. (B) Magnetic polarity chronologies for the Chinle Formation and the Newark Supergroup with stratigraphic and geochronologic data added to develop more robust correlations. *(From Zeigler, K.E., Geissman, J.W., 2011. Magnetostratigraphy of the Upper Triassic Chinle Group of New Mexico: implications for regional and global correlations among Upper Triassic sequences. Geosphere 7, 802–829, but see Zeigler et al. (this volume).)*

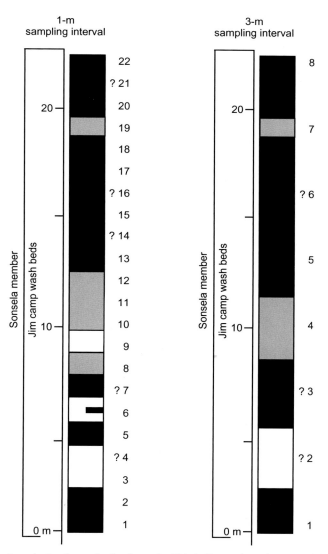

Fig. 10 Two magnetic polarity chronologies from the Chinle Formation of eastern Arizona showing the difference in the architecture of the magnetic polarity chronology that results from 1-m and 3-m resolution sampling schemes. Note the loss of information in the lower half of the section that results from the coarser sampling resolution.

the paleomagnetic directions help to constrain which parts of the GPTS a magnetic polarity chronology provides the best correlation. Paleomagnetic directions from samples of Triassic age in the southwestern United States cannot be compared to a Jurassic GPTS because the paleomagnetic directions obtained from Triassic samples will have significantly shallower inclinations than Jurassic samples (barring remagnetization, of course).

This criticism also serves to show that paleomagnetic data cannot stand entirely alone, but is best matched with other forms of stratigraphic and geochronologic data, including biostratigraphic information, for the most robust assessment of the age and correlations of any given strata.

FINAL THOUGHTS ON UTILIZING PALEOMAGNETIC AND ROCK MAGNETIC DATA

As noted by Kodama and Hinnov (2015), paleomagnetic data are a critical component of any stratigraphic study, given the long time frames over which paleomagnetic directions can remain stable in different rock types. The information obtained in any paleomagnetic study of sedimentary rocks includes the magnetizations, the stratigraphic ordering of those magnetizations (the magnetic polarity stratigraphy), and the suite of magnetic minerals that carry the paleomagnetic information. These three aspects of a data set can provide information about paleogeography, behavior of the magnetic field during the time of deposition, rates of sedimentation, depositional and diagenetic processes, and potentially climatic variation that may impact the concentration of detrital magnetic minerals. However, paleomagnetic and rock magnetic data cannot stand alone, but must be coupled with another form of stratigraphic data to improve correlations between outcrops.

With any methodology, there are trade-offs between the outcrop sampling resolution and the time required to process the samples collected. Hopefully the methodologies reviewed here provide a means of narrowing down the proper experiments and techniques to use for the most efficient exploration of a depositional sequence. In further chapters, we present two case studies focused on paleomagnetic data. In Zeigler et al. (this volume), Upper Triassic strata in eastern Arizona are examined, and absolute age determinations are coupled with high-resolution magnetic polarity data to refine correlation of these rocks with similar age strata in the American Southwest. In Maidment et al. (Chapter 7), biostratigraphic and geochronologic data for the Middle to Upper Jurassic Morrison Formation are coupled with a preliminary magnetic vector data set to begin deciphering the complex depositional systems of the Middle to Late Jurassic. In both these contributions, the authors explore the benefits and pitfalls of the paleomagnetic data sets.

ACKNOWLEDGMENTS

We are grateful to Dan Peppe for useful discussion and helpful comments on an earlier version of this manuscript.

REFERENCES

Anson, G.L., Kodama, K.P., 1987. Compaction-induced inclination shallowing of the post-depositional remanent magnetization in a synthetic sediment. Geophys. J. R. Astron. Soc. 88, 673–692.

Bilardello, D., Kodama, K.P., 2010. A new inclination shallowing correction of the Mauch Chunk Formation of Pennsylvania, based on high-field AIR results: implications for the Carboniferous North American APW path and Pangea reconstructions. Earth Planet. Sci. Lett. 229, 218–227. http://dx.doi.org/10.1016/j.epsl.2010.09.002.

Butler, R.F., 1992. Paleomagnetism: Magnetic Domains to Geologic Terranes. Blackwell, Boston, MA 319 p.

Collinson, D.W., 1965. Depositional remanent magnetization in sediments. J. Geophys. Res. 70, 4663–4668.

Collinson, D.W., 1974. The role of pigment and specularite in the remanent magnetism of red sandstones. Geophys. J. R. Astron. Soc. 38, 253–264.

Connell, S.D., Smith, G.A., Geissman, J.W., McIntosh, W.C., 2013. Climatic controls on nonmarine depositional sequences in the Albuquerque Basin, Rio Grande rift, north-central New Mexico. Geol. Soc. Am. Spec. Pap. 494, 1–44.

Cox, A., 1973. Tectonics and Geomagnetic Reversals. W.H. Freeman, San Francisco, CA, p. 702 p.

Cox, A., Doell, R.R., 1961. Palaeomagnetic evidence relevant to a change in the Earth's radius. Nature 189, 45–47.

Cox, A., Doell, R.R., Dalrymple, G.B., 1963. Geomagnetic polarity epochs and Pleistocene geochronometry. Nature 198, 1049–1051.

Creer, K.M., Tucholka, P., Barton, C.E., 1983. Geomagnetism of Baked Clays and Recent Sediments. Elsevier, Amsterdam 324 p.

Deamer, G.A., Kodama, K.P., 1990. Compaction-induced inclination shallowing in synthetic and natural clay-rich sediments. J. Geophys. Res. 95, 4511–4529.

Dunlop, D.J., Özdemir, O., 1997. Rock Magnetism: Fundamentals and Frontiers. Cambridge University Press, Cambridge 573 p.

Elmore, R.D., Muxworthy, A.R., Aldana, M., 2012. Remagnetization and chemical alteration of sedimentary rocks. Geol. Soc. Lond. Spec. Publ. 371, 1–21.

Enkin, R.J., 2003. The direction-correction tilt test: an all-purpose tilt/fold test for paleomagnetic studies. Earth Planet. Sci. Lett. 212, 151–166.

Evans, M.E., Heller, F., 2003. Environmental Magnetism: Principles and Applications of Enviromagnetics. Academic Press, Amsterdam, 299 p.

Fisher, R.A., 1953. Dispersion on a sphere. Proc. R. Soc. Lond. A 217, 295–305.

Gillett, S.L., 2003. Paleomagnetism of the Notch Peak contact metamorphic aureole, revisited: pyrrhotite from magnetite + pyrite under submetamorphic conditions. J. Geophys. Res. Solid Earth. 108. http://dx.doi.org/10.1029/2002JB002386.

Glen, W., 1982. The Road to Jaramillo: Critical Years of the Revolution in Earth Science. Stanford University Press, Stanford, CA. 459 p.

Gradstein, F.M., Ogg, J.G., Schmitz, M.D., Ogg, G.M., 2012. The Geologic Time Scale 2012. Elsevier, Oxford, 1144 p.

Graham, J.W., 1949. The stability and significance of magnetism in sedimentary rocks. J. Geophys. Res. 54, 131–167.

Gunderson, K.L., Kodama, K.P., Anastasio, D.J., Pazzaglia, F.J., 2012. Rock-magnetic cyclostratigraphy for the Late Pliocene-Early Pleistocene Stirone section, northern Apennine mountain front, Italy. Geol. Soc. Lond. Spec. Publ. 373, 26 p.

Henry, S., 1979. Chemical demagnetization—methods, procedures, and applications through vector analysis. Can. J. Earth Sci. 16, 1832–1841.

Hess, H.H., 1962. History of the ocean basins. In: Engel, A.E.J., James, H.L., Leonard, B.F. (Eds.), Petrologic Studies: A Volume in Honor of A.F. Buddington. Geological Society of America, New York, NY, pp. 599–620.

Hess, H.H., 1965. Mid-oceanic ridges and tectonics of the sea floor. In: Whittard, W.F., Bradshaw, R. (Eds.), Submarine Geology and Geophysics. Butterworth, London, pp. 317–332.

Hinnov, L.A., Kodama, K.P., Anastasio, D.J., Elrick, M., Latta, D.K., 2013. Global Milankovitch cycles recorded in rock magnetism of the shallow marine lower Cretaceous Cupido Formation, northeastern Mexico. In: Jovane, L., Herrero-Bervera, E., Hinnov, L.A., Housen, B.A. (Eds.), Magnetic Methods and the Timing of Geological Processes. Geological Society of London, London, pp. 1–12.

Hodych, J.P., Bijaksana, S., 1993. Can remanence anisotropy detect paleomagnetic inclination shallowing due to compaction? A case study using Cretaceous deep-sea sediments. J. Geophys. Res. 98, 22429–22441.

Hospers, J., 1955. Rock magnetism and polar wandering. J. Geol. 63, 59–74.

Hounslow, M.W., Muttoni, G., 2010. The geomagnetic polarity timescale for the Triassic: linkage to stage boundary definitions. Geol. Soc. London Spec. Publ. 334, 61–102.

Iriving, E., 1988. The paleomagnetic confirmation of continental drift. Eos Trans. AGU 69, 1001–1014.

Irving, E., 1958. Paleogeographic reconstruction from paleomagnetism. Geophys. J. R. Astron. Soc. 1, 224–237.

Jackson, M., Banerjee, S.K., Marvin, J.A., Lu, R., Gruber, W., 1991. Detrital remanence, inclination errors, and anhysteretic remanence anisotropy: quantitative model and experimental results. Geophys. J. Int. 2014, 95–103.

Jackson, M., Sun, W.-W., Craddock, J.P., 1992. The rock magnetic fingerprint of chemical remagnetization in midcontinental Paleozoic carbonates. Geophys. Res. Lett. 19, 781–784.

Johnson, N.M., Opdyke, N.D., Lindsay, E.H., 1975. Magnetic polarity stratigraphy of Pliocene-Pleistocene terrestrial deposits and vertebrate faunas, San Pedro Valley, Arizona. Geol. Soc. Am. Bull. 86, 5–12.

Kent, D.V., Irving, E., 2010. Influence of inclination error in sedimentary rocks on the Triassic and Jurassic apparent polar wander path for North America and implications for Cordilleran tectonics. J. Geophys. Res. 115, B10103. http://dx.doi.org/10.1029/2009JB007205.

Kent, D.V., Tauxe, L., 2005. Corrected Late Triassic latitudes for continents adjacent to the North Atlantic. Science 307, 240–244.

King, R.F., 1955. Remanent magnetism of artificially deposited sediments. Mon. Not. R. Astron. Soc. Geophys. Suppl. 7, 115–134.

Kirschvink, J.L., 1978. The Precambrian-Cambrian boundary problem: magnetostratigraphy of the Amadeus Basin, Central Australia. Geol. Mag. 115, 139–150.

Kirschvink, J.L., 1980. The least squares line and plane and the analysis of paleomagnetic data. Geophys. J. R. Astron. Soc. 62, 699–718.

Kodama, K.P., 2012. Paleomagnetism of Sedimentary Rocks: Process and Interpretation. Wiley-Blackwell, Oxford, 157 p.

Kodama, K.P., Davi, J.M., 1995. A compaction correction for the paleomagnetism of the Cretaceous Pigeon Point Formation of California. Tectonics 14, 1153–1164.

Kodama, K.P., Hinnov, L.A., 2015. Rock Magnetic Cyclostratigraphy. Wiley-Blackwell, Oxford, 165 p.

Kruiver, P.P., Dekkers, M.K., Heslop, D., 2001. Quantification of magnetic coercivity components by the analysis of acquisition curves of isothermal remanent magnetization. Earth Planet. Sci. Lett. 189, 269–276.

Larson, E.E., Walker, T.R., Patterson, P.E., Hoblitt, R.P., Rosenbaum, J.B., 1982. Paleomagnetism of the Moenkopi Formation, Colorado Plateau: basis for long-term model of acquisition of chemical remanent magnetism in red beds. J. Geophys. Res. 87, 1081–1106.

Latta, D.K., Anastasio, D.J., Hinnov, L.A., Elrick, M., Kodama, K.P., 2006. Magnetic record of Milankovitch rhythms in lithological noncyclic marine carbonates. Geology 34, 29–32.

Liu, Q., Roberts, A.P., Larrasoana, J.C., Banerjee, S.K., Guyodo, Y., Tauxe, L., Oldfield, F., 2012. Environmental magnetism: principles and applications. Rev. Geophys. 50RG4002http://dx.doi.org/10.1029/2012RG000393.

Lowrie, W., 1989. Magnetic polarity time scales and reversal frequency. In: Lowes, F.J., Collinson, D.W., Parry, J.H., Runcorn, S.K., Tozer, D.C., Soward, A. (Eds.), Geomagnetism and Paleomagnetism. Kluwer, Boston, MA, pp. 155–183.

Lowrie, W., 1990. Identification of ferromagnetic minerals in a rock by coercivity and unblocking temperature properties. Geophys. Res. Lett. 17, 159–162.

McFadden, P.L., Jones, D.L., 1981. The fold test in paleomagnetism. Geophys. J. R. Astron. Soc. 67, 53–58.

McFadden, P.L., McElhinny, M.W., 1990. Classification of the reversal test in paleomagnetism. Geophys. J. Int. 103, 725–729.

Molina-Garza, R.S., Geissman, J.W., Van der Voo, R., 1995. Paleomagnetism of the Dockum Group (Upper Triassic), northwest Texas: further evidence for the J-1 cusp in the North America apparent polar wander path and implications for rate of Triassic apparent polar wander and Colorado Plateau rotation. Tectonics 14, 979–993.

Molina-Garza, R.S., Geissman, J.W., Lucas, S.G., 2003. Paleomagnetism and magnetostratigraphy of the lower Glen Canyon and upper Chinle Groups, Jurassic-Triassic of northern Arizona and northeast Utah. J. Geophys. Res. 108http://dx.doi.org/10.1029/2002JB001909.

Nagata, T., 1961. Rock Magnetism. Maruzen, Tokyo, 350 p.

Olsen, P.E., Kent, D.V., Cornet, B., Witte, W.K., Schlische, R., 1996. High resolution stratigraphy of the Newark rift basin (Early Mesozoic, Eastern North America). Geol. Soc. Am. Bul. 108, 40–77.

Opdyke, N.D., 1969. The Jaramillo event as detected in oceanic cores. In: Runcorn, S.K. (Ed.), The Application of Modern Physics to Earth and Planetary Interiors. Wiley-Interscience, London, pp. 549–552.

Peppe, D.J., Evans, D.A.D., Smirnov, A.V., 2009. Magnetostratigraphy of the Ludlow Member of the Fort Union Formation (Lower Paleocene) in the Williston Basin, North Dakota. Geol. Soc. Am. Bull. 121, 65–79.

Roberts, A.P., Jiang, W.T., Florindo, F., Horng, C.-S., Laj, C., 2005. Assessing the timing of greigite formation and the reliability of the upper Olduvai polarity transition record from the Crostolo River, Italy. Geophys. Res. Lett. 32. http://dx.doi.org/10.1029/2004GL022137.

Schmidt, P.W., 1993. Paleomagnetic cleaning strategies. Phys. Earth Planet. Inter. 76, 169–178.

Spahn, Z.P., Kodama, K.P., Preto, N., 2013. High-resolution estimate for the depositional duration of the Triassic Latemar Platform: a new magnetostratigraphy and magnetic susceptibility cyclostratigraphy from basinal sediments at Rio Sacuz, Italy. Geochem. Geophys. Geosyst. 14, 1245–1257.

Sun, W.-W., Kodama, K.P., 1992. Magnetic anisotropy, scanning electron microscopy, and x-ray pole figure goniometry study of inclination shallowing in a compacting clay-rich sediment. J. Geophys. Res. 97, 19599–19615.

Tan, X., Kodama, K.P., Fang, D., 2002. Laboratory depositional and compaction-induced inclination errors carried by hematite and their implications in identifying inclination error of natural remanence in red beds. Geophys. J. Int. 151, 475–486.

Tan, X., Kodama, K.P., Gilder, S., Courtillot, V., 2007. Rock magnetic evidence for inclination shallowing in the Passaic Formation red beds from the Newark basin and a systematic bias of the Late Triassic apparent polar wander path for North America. Earth Planet. Sci. Lett. 354, 345–357.

Tauxe, L., 2010. Essentials of Paleomagnetism. University of California Press, Berkeley, CA, 489 p.

Tauxe, L., Kent, D.V., 1982. A time framework based on magnetostratigraphy for the Siwalik sediments of the Khaur area, northern Pakistan. Palaeogeogr. Palaeoclimatol. Palaeoecol. 37, 43–61.

Tauxe, L., Kent, D.V., 1984. Properties of a detrital remanence carried by haematite from study of modern river deposits and laboratory redeposition experiments. Geophys. J. R. Astron. Soc. 76, 543–561.

Tauxe, L., Watson, G.S., 1994. The fold test: an eigen analysis approach. Earth Planet. Sci. Lett. 122, 331–341.

Turner, P., 1980. Continental Red Beds. Elsevier, Amsterdam, 561 p.

Van der Voo, R., Torsvik, T.H., 2012. The history of remagnetization of sedimentary rocks: deceptions, developments and discoveries. Geol. Soc. Lond. Spec. Publ. 371, 23–53.

Vine, F.J., Matthews, D.H., 1963. Magnetic anomalies over oceanic ridges. Nature 199, 947–949.

Walker, T.R., 1967. Color of recent sediments in tropical Mexico: a contribution to the origin of red beds. Geol. Soc. Am. Bull. 78, 917–919.

Walker, T.R., 1974. Formation of red beds in moist tropical climates: a hypothesis. Geol. Soc. Am. Bull. 85, 633–638.

Walker, J.D., Geissman, J.W. (compilers), 2009. Geologic time scale: Geological Society of America. Doi:10.1130/2009.CTS004R2C.

Walker, T.R., Waugh, B., Grone, A.J., 1978. Diagenesis in first-cycle desert alluvium of Cenozoic age, southwestern United States and northwestern Mexico. Geol. Soc. Am. Bull. 89, 19–32.

Walker, T.R., Larson, E.E., Hoblitt, R.P., 1981. The nature and origin of hematite in the Moenkopi Formation (Triassic), Colorado Plateau: a contribution to the origin of magnetism in red beds. J. Geophys. Res. 86, 317–333.

Wang, X., Lovlie, R., Yang, Z., Pei, J., Zhao, Z., Sun, Z., 2005. Remagnetization of quaternary eolian deposits; a case study from SE Chinese Loess Plateau. Geochem. Geophys. Geosyst. 6, 14 p.

Zeigler, K.E., Geissman, J.W., 2011. Magnetostratigraphy of the Upper Triassic Chinle Group of New Mexico: implications for regional and global correlations among Upper Triassic sequences. Geosphere 7, 802–829.

Zijderveld, J.D.A., 1967. A.C. demagnetization of rocks: analysis of results. In: Collinson, D.W., Creer, K.M., Runcorn, S.K. (Eds.), Methods of Paleomagnetism. Elsevier, New York, NY, pp. 254–286.

The Lower Chinle Formation (Late Triassic) at Petrified Forest National Park, Southwestern USA: A Case Study in Magnetostratigraphic Correlations

K.E. Zeigler*, W.G. Parker[†], J.W. Martz[‡]
*Zeigler Geologic Consulting, Albuquerque, NM, United States
[†]Petrified Forest National Park, Petrified Forest, AZ, United States
[‡]University of Houston–Downtown, Houston, TX, United States

Contents

INTRODUCTION

The ability to correlate terrestrial deposits with any degree of certainty usually requires the use of multiple stratigraphic data sets. One means of providing a chronologic yardstick

Terrestrial Depositional Systems
http://dx.doi.org/10.1016/B978-0-12-803243-5.00006-6

for a depositional package is magnetic polarity stratigraphy, which is a history of the behavior of the Earth's magnetic field as it is recorded by various magnetic minerals in sedimentary rocks. Magnetic polarity stratigraphy at its most basic provides polarity data for a stratigraphic interval and the stratigraphic position of boundaries between alternating polarities. Additional information can be gleaned from characterizing the magnetic minerals in different facies via rock magnetic experiments that may shed light on changes in deposition and diagenesis. The alternations in polarity recorded in a vertical sequence create a unique signature for that particular locality. Matching polarity records between outcrops or across a larger area requires additional independent chronologic data in order to make correct lateral correlations. Magnetic polarity data implies no absolute age determination; it is simply a record of the Earth's magnetic field at or shortly after deposition. Here we present new magnetic polarity chronology data from the Late Triassic Chinle Formation of Petrified Forest National Park (PEFO) and suggest correlations between the PEFO Chinle Formation and sections in New Mexico, as well as tentative correlations to the magnetic polarity stratigraphy of the Newark Supergroup. This new data set also can be used to demonstrate some of the advantages and disadvantages of correlating using magnetostratigraphic data and provide a testable framework for additional larger-scale correlations.

THE LATE TRIASSIC CHINLE FORMATION

The Late Triassic age Chinle Formation is a prominent Mesozoic stratigraphic unit throughout the American Southwest. These strata are continental in origin and reflect deposition in a complex environment including fluvial, lacustrine, and aggradational fan systems (Blakey, 1989; Dubiel, 1987, 1989a,b, 1994; Weissmann et al., 2007; Trendell et al., 2013). Vertebrate biostratigraphy and palynostratigraphy were initially used to show that deposition of the Chinle Formation probably spanned nearly all of Late Triassic time (Stewart et al., 1972; Litwin, 1986; Lucas and Hunt, 1992; Hunt and Lucas, 1993a,b; Lucas et al., 2003, 2005). However, a growing body of radioisotopic age determinations and correlations via magnetostratigraphic data sets demonstrates that the Chinle Formation most likely represents deposition during mostly the Norian and Rhaetian stages of the Late Triassic (Riggs et al., 2003; Dickinson et al., 2010; Irmis and Mundil, 2008; Heckert et al., 2009; Ramezani et al., 2011; Irmis et al., 2011; Zeigler and Geissman, 2011; Atchley et al., 2013), while the oldest age of the Late Triassic, the Carnian, may be completely unrepresented. The complexity of the depositional environments, coupled with its predominantly fluvial environment and relatively low accumulation rate, means the presence of hiatuses and disconformities within the Chinle Formation is likely (e.g., Dickinson et al., 2010; Ramezani et al., 2011; Atchley et al., 2013), potentially leading to complications in efforts to correlate these strata locally, regionally, or globally.

The first efforts to construct a stratigraphically continuous magnetic polarity chronology for the Chinle Formation focused primarily on sampling sandstones and siltstones from independent sections of Chinle strata from different subbasins and outcrop belts. These data were then correlated and built into composite magnetic polarity records, based on lithologic and to a lesser extent biostratigraphic correlations (e.g., Reeve, 1975; Reeve and Helsley, 1972; Bazard and Butler, 1989, 1991; Molina-Garza et al., 1991, 1993, 1996, 1998a,b, 2003; Steiner and Lucas, 2000). Consequently, the polarity record of the mudstones, which is one of the principal rock types in the Chinle Formation, has been less well understood. Zeigler and Geissman (2011) published a magnetic polarity stratigraphy for the Chinle Formation in north-central New Mexico that focused on a complete section of Late Triassic strata, and they sampled not only the well-indurated lithologies, but also the mudstone horizons. This effort thus was among the first to develop a more robust, "continuous" magnetic polarity chronology for the Chinle Formation, although it was limited by a relatively coarse sampling interval (~3 m between sampled stratigraphic horizons). This study also presented magnetic polarity stratigraphies from individual stratigraphic units in the lower Chinle Formation from western New Mexico and the upper Chinle Formation from eastern New Mexico.

Our present study adds to and revises the magnetic polarity chronology for a portion of the Chinle Formation in Petrified Forest National Park (PEFO) that was initially developed by Steiner and Lucas (2000). We focus on outcrops in the southern half of the park, where there is strong lithostratigraphic, biostratigraphic, and geochronologic control among localities (Fig. 1). PEFO has long been a focus of significant research on Chinle Formation stratigraphy, palynology, and vertebrate paleontology (Martz and Parker, 2010 and references therein). The park contains outcrops that represent a nearly complete stratigraphic section of Chinle strata (Fig. 2), with lower Chinle strata exposed in southern PEFO and upper Chinle Formation cropping out in northern PEFO. Recent careful geologic mapping and lithostratigraphic description by Martz and Parker (2010) present a well-defined framework for a magnetic polarity stratigraphy for the lower Chinle Formation section in southern PEFO. In addition, recent work by Ramezani et al. (2011) and Atchley et al. (2013) has provided absolute age constraints for much of the lower part of the section as exposed in the southern half of the park.

The complete lithostratigraphic model for the Chinle Formation utilized here is summarized by Martz et al. (2012). The current study is confined to the lower part of the Chinle Formation in the southern part of PEFO, specifically the interval encompassing the upper Mesa Redondo Member, Blue Mesa Member, and the entire Sonsela Member (*sensu* Martz et al., 2012). For the following discussion, it is important to note that the name "Sonsela Member" or "Sonsela sandstone bed" was initially applied within southern PEFO only to the ledge-forming sandstone capping Blue

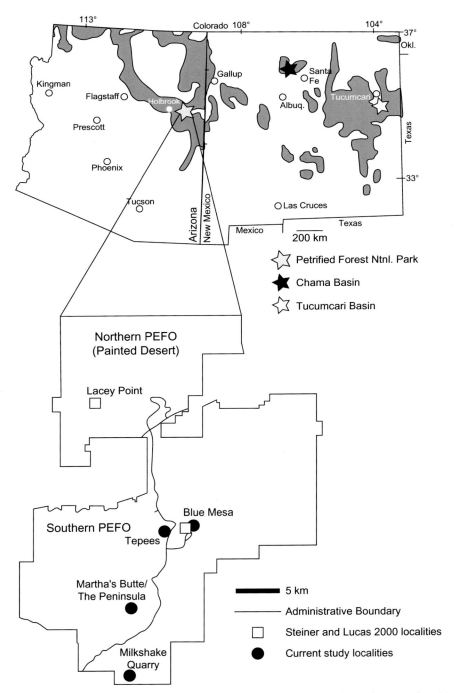

Fig. 1 Distribution of Triassic strata in Arizona and New Mexico with PEFO boundaries and positions of sampled localities for this study and for Steiner and Lucas (2000).

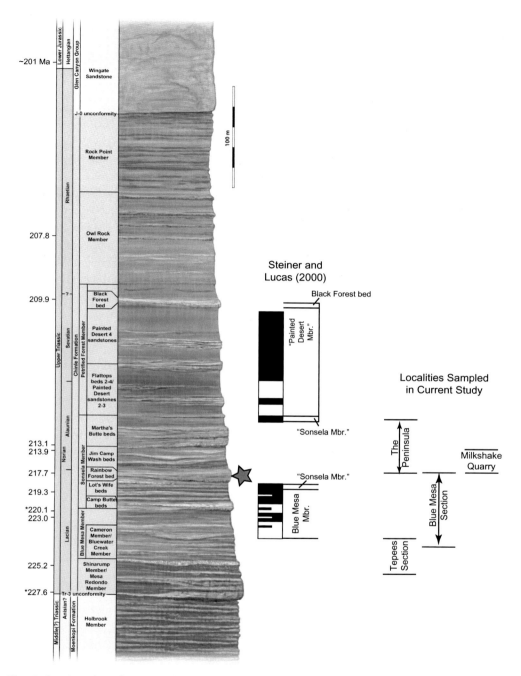

Fig. 2 Stratigraphy of PEFO (after Martz and Parker, 2010), showing the stratigraphic interval represented by Steiner and Lucas's (2000) study redrawn to the same scale, and the stratigraphic ranges for sampled localities in the current study. Note that in the original figure in Steiner and Lucas (2000), the two beds labeled "Sonsela Member" were considered stratigraphically equivalent. Radioisotopic dates along left margin are from Ramezani et al. (2011) and Atchley et al. (2013, denoted with *), rounded to the nearest 100 ky. Star indicates position of the Jasper Forest bed (see text for discussion of correlation to north PEFO).

Mesa (e.g., Cooley, 1957; Roadifer, 1966; Billingsley, 1985a,b; Ash, 1987; Lucas, 1993). Steiner and Lucas (2000) used it in this sense to correlate the two localities sampled in their study.

Later, the name "Sonsela Member" was expanded to encompass a much thicker sandstone-dominated sequence (Heckert and Lucas, 2002; Woody, 2006) that includes the upper part of the Blue Mesa Member-type section as originally formulated by Lucas (1993). The traditional "Sonsela sandstone bed" in PEFO capping the Blue Mesa locality lies roughly in the middle of what is now called Sonsela Member (Martz and Parker, 2010). In this new concept of the Sonsela Member, the traditional "Sonsela sandstone bed" has been referred to as the "Agate Bridge Bed" (Heckert and Lucas, 2002), "Rainbow Forest Bed" (Fig. 2; Heckert and Lucas, 2002), or "Jasper Forest bed" (Martz and Parker, 2010). For the sake of simplicity, we will rely primarily on the latter name, Jasper Forest bed, in the following discussion (see Martz and Parker, 2010 and Martz et al., 2012 for a much more detailed discussion on the complex stratigraphic and nomenclatural issues).

Correlation of the Jasper Forest bed to the northern part of PEFO has been highly contentious (e.g., Billingsley, 1985a,b; Ash, 1987; Murry, 1990; Lucas, 1993; Heckert and Lucas, 2002; Martz et al., 2012). However, this correlation is of considerable importance to developing a complete magnetostratigraphic model for the entire park, as correlations of the Jasper Forest bed have been critical to unifying the northern and southern regions of PEFO into a single lithostratigraphic model. Here we rely on published and ongoing work (Parker and Martz, 2011; Martz et al., 2012; Martz, J.W. and Parker, W.G., unpublished data) that correlates the Jasper Forest bed to white sandstones (the "Kellogg Butte beds" *sensu* Parker and Martz, 2011; Martz et al., 2012) in the Devil's Playground region of northern PEFO, occurring just below the Brown sandstone (Billingsley, 1985a,b). This is close to the placement of some authors (e.g., Murry, 1990) who considered the Jasper Forest bed correlative with the Brown sandstone, but well below beds in northern PEFO that many other authors (e.g., Billingsley, 1985a,b; Ash, 1987; Lucas, 1993) have considered correlative to the Jasper Forest bed, but which we consider to lie in the uppermost Sonsela Member (Parker and Martz, 2011; Martz et al., 2012). It is worth noting that there is no *a priori* reason a fluvial sandstone should be correlative over long distances, and one could make the case that in general they should not.

In addition to revising the preliminary magnetostratigraphic model of Steiner and Lucas (2000), sampling in this effort was also focused on the stratigraphic interval encompassing the level of the Adamanian-Revueltian biotic transition (Parker and Martz, 2011), the largest biotic turnover event evident in the PEFO section. This transition occurs immediately above a persistent red silcrete low in the Jim Camp Wash beds (middle Sonsela Member), with Adamanian taxa below the stratigraphic horizon of the silcrete and Revueltian taxa above. Parker and Martz (2011) lay out careful lithostratigraphic and geographic placement of numerous vertebrate fossil localities within the revised lithostratigraphy of Martz and Parker (2010) to locate the stratigraphic position of the faunal transition as precisely as possible. In addition, trends in the sedimentological record of the Sonsela Member suggest at least local aridification at the time of the Adamanian-Revueltian transition (Martz and Parker, 2010; Parker and Martz, 2011; Atchley et al., 2013; Nordt et al., 2015), as well as a possible link to the Manicouagan impact in Quebec (Parker and Martz, 2011; Olsen et al., 2011). Therefore, constraining the timing of these events as precisely as possible, using multiple stratigraphic methods, is imperative. The localities chosen for this current study are locations that have yielded significant biostratigraphic information, have detailed lithostratigraphic sections (Martz and Parker, 2010), and were included in Ramezani et al.'s (2011) and Atchley et al.'s (2013) sampling efforts for obtaining detrital zircon grains. These localities are thus part of the integrated framework needed to develop a robust and useful magnetostratigraphy for the lower Chinle Formation.

The Impact of Recent Lithostratigraphic Revisions of the Chinle Formation in PEFO on the Magnetostratigraphy of Steiner and Lucas (2000)

The work documented here revises the Chinle Formation magnetic polarity stratigraphy for Petrified Forest National Park (PEFO) that was initially developed by Steiner and Lucas (2000). This revision is necessary because recent lithostratigraphic revisions (Martz and Parker, 2010; Parker and Martz, 2011; Martz et al., 2012) reveal that there is a considerable stratigraphic gap in Steiner and Lucas's (2000) composite section caused by miscorrelation of the southern and northern park areas. Steiner and Lucas (2000: p. 25, 972-25, 974) collected magnetostratigraphic samples at two sections in Petrified Forest National Park: Blue Mesa in the southern part of PEFO, and near Lacey Point in the northern part of PEFO. These were the type sections of Lucas's (1993) Blue Mesa and Painted Desert Members of his Petrified Forest Formation of the Chinle Group. In southern PEFO, Steiner and Lucas (2000) identified the sandstone capping Blue Mesa as the "Sonsela Member" (Sonsela Sandstone bed) following previous workers (e.g., Cooley, 1957; Billingsley, 1985a,b), which would be the Jasper Forest bed of current usage (Martz and Parker, 2010; Martz et al., 2012).

The lower part of the type section of the Painted Desert Member (units 1–15 of Lucas, 1993) occurs in the northern part of PEFO at SE ¼, NE ¼, sec. 11, T19N, R23E (Lucas, 1993), at the southwestern flank of a large mesa called the Citadel (Martz et al., 2012). A prominent ledge-forming sandstone was identified here by Lucas (1993) and subsequently by Steiner and Lucas (2000) as the Sonsela Sandstone bed following Billingsley (1985b). This correlation enabled them to hypothetically link the two sections into a single composite section (Steiner and Lucas, 2000, Fig. 8). However, the bed in northern PEFO that Billingsley (1985b) and Steiner and Lucas (2000) identified as the Sonsela Sandstone bed (=Jasper Forest bed of current use) was mapped by Martz et al. (2012) as Painted Desert Sandstone #1, which is in the upper Martha's

Butte beds of Martz and Parker (2010). This correlation was made by carefully walking out strata from the Devil's Playground area of the park through a monocline to the base of the Citadel. The Citadel is capped by a prominent sandstone that fortunately all workers agree is the Painted Desert Sandstone #3, also referred to as the Lithodendron Wash bed (Billingsley, 1985b; Lucas, 1993; Steiner and Lucas, 2000; Martz et al., 2012). This correlation makes the interpretation of the Steiner and Lucas (2000) section easier and allows us to identify their error.

The sandstones below the Lithodendron Wash bed shown by Steiner and Lucas (2000, Fig. 8), including the one they identified as "PDS 1," are sandstones lying between the top of the Martha's Butte beds and the Lithodendron Wash bed (the "Painted Desert two" sandstones of Johns, 1988; Parker and Martz, unpublished data). Because Steiner and Lucas (2000) used the bed they misidentified as the Sonsela Sandstone bed (Jasper Forest bed of current use) to link the northern and southern sections, their composite section (Steiner and Lucas, 2000, Fig. 8) contains a significant stratigraphic gap of ~50 m (or more) that excludes the upper half of the Sonsela Member, including all of the Jim Camp Wash beds, and most or all of the Martha's Butte beds (Fig. 2), amounting to the addition of a third of their original total section thickness (50 m/160 m = 0.31). An additional issue requiring comment is that if the scale of the section in Steiner and Lucas (2000) is correct, and identification of the Black Forest bed is correct (as seems very likely), then the upper PEFO section can extend downward only to the Lithodendron Wash bed as identified by Martz and Parker (2010). Conversely, if they identified the Lithodendron Wash bed correctly in their section, the scale must be off by a factor of two (~70 m in Steiner and Lucas (2000) versus ~150 m in Martz and Parker, 2010), and if linearly scaled from the Black Forest bed, their section would extend nearly to the correlative position of the Jasper Forest bed.

These discrepancies cannot be reconciled within this paper, but we tentatively correlate the upper PEFO section of Steiner and Lucas (2000) with the section from the top of the Martha Butte beds (top Sonsela) of Martz and Parker (2010) to the base of the Black Forest bed, recognizing the need for additional work. The current study not only resamples the lower part of the Chinle Formation, but also describes the magnetostratigaphy of the upper Sonsela Member for the first time.

SAMPLING AND ANALYTICAL METHODS

The lower part of the Chinle Formation was sampled at four different localities in southern PEFO. Most of these sections have been at least partially described by Martz and Parker (2010). The localities (from south to north) with the approximately equivalent sections in Martz and Parker (2010) are: the Milkshake Quarry (lower section 6), Peninsula/Martha's Butte (sections 15 and 16), Blue Mesa (section 25), and Tepees localities (Fig. 3; Table 1). The Tepees section sampled here includes units that are

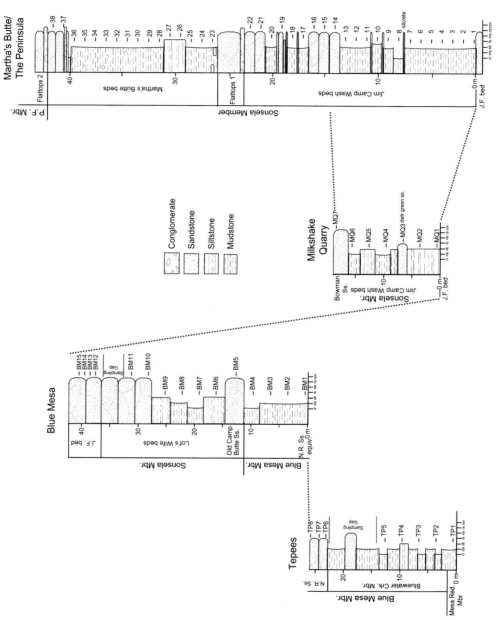

Fig. 3 Stratigraphy at each locality with position of sampled horizons and correlation between sections based on Martz and Parker (2010).

Table 1 Decimal degree latitude and longitude coordinates for the bases and tops of each sampled section

Locality	Location of base of sampled section	Location of top of sampled section	Comments
Milkshake Quarry	34.795°N, 109.796°W	34.796°N, 109.849°W	
Peninsula/ Martha's Butte	34.857°N, 109.810°W	34.850°N, 109.820°W	Between sampling site 10 and 11, sampling is moved to 34.854°N, 109.817°W due to Quaternary cover
Blue Mesa	34.944°N, 109.747°W	34.827°N, 109.856°W	
Tepees	34.938°N, 109.796°W	34.938°N, 109.796°W	

stratigraphically lower than any section presented by Martz and Parker (2010). Together, these localities encompass most of the lower part of the Chinle Formation, from the upper part of the Mesa Redondo Member (exposed at Tepees) through the uppermost Sonsela Member (Peninsula/Martha's Butte). These areas were chosen because they are highly fossiliferous (Parker and Martz, 2011; Chapter 2) and allow direct correlation of lithostratigraphic, biostratigraphic, and paleomagnetic data. The Milkshake Quarry, Blue Mesa, and Tepees localities were sampled at 3 m intervals, and the Peninsula was sampled at 1 m intervals. The Peninsula section was sampled at a higher sampling frequency to capture magnetic polarity information through the portion of the Sonsela Member that includes transition between the Adamanian and Revueltian biozones, marked by the persistent red silcrete horizon (Parker and Martz, 2011; Chapter 2).

In all cases, samples were collected at each site (discrete stratigraphic horizon) from a stratigraphic interval no more than 0.5–0.75 m thick and no wider than 2.0 m in order to avoid inadvertently sampling a different age polarity zone or missing any short-lived polarity intervals. Samples were taken at 1 m or 3 m intervals depending on the locality. At the same time that samples were collected, lithologic characteristics of the sampled site were recorded to tie these sections to the work of Martz and Parker (2010) (Fig. 3). In the case of conglomeratic units, such as the Jasper Forest bed, every effort was made to obtain samples from the finest grained part of the horizon sampled. Samples from well-indurated sandstone beds were obtained by drilling with a water-cooled diamond drill bit, and eight to ten independently oriented core samples were taken per horizon. These core samples were then cut into 2.2-cm-high cylinders, frequently yielding multiple specimens per sample. Mudstone and poorly consolidated siltstone units were sampled using the block sampling method of Johnson et al. (1975). Eight to ten independently oriented blocks were taken from each horizon, and each block was then dry sawed with a diamond blade into 1.0–2.0 cm^3 specimens.

Between four and ten specimens for each sampled horizon were subjected to progressive thermal demagnetization at the University of New Mexico and Baylor University paleomagnetism laboratories. At the University of New Mexico, measurements were conducted using a 2G Enterprises Model 760R superconducting rock magnetometer, equipped with DC SQuIDs. Thermal treatment steps were applied to at least two specimens per sample using a Shaw Magnetic Measurement Thermal Demagnetizer. At Baylor University, specimens were measured using an automated 2G 3-axis, DC-SQuID cryogenic magnetometer inside a two-layer magnetostatic shield with a background field typically less than 300 nT. Samples were demagnetized using thermal demagnetization performed in a controlled nitrogen atmosphere to reduce potential oxidation reactions (Peppe et al., 2009). Cubic samples analyzed at Baylor University were packed in plastic cubes for each measurement step and were unpacked for thermal demagnetization.

The resulting thermal demagnetization data were then analyzed using the principal component analysis methods of Kirschvink (1980) with individual vector segments accepted if the maximum angular deviation (MAD) values were less than or equal to 15°. Class A specimen data include vectors with MAD values $\leq 10°$, class B data $11° \leq MAD \leq 15°$, and class C are vector segments with MAD $>15°$. Vector segments with MAD values between 15° and 25° that showed Late Triassic affinities (shallow inclination and north- or south-directed declinations) were considered when calculating virtual geomagnetic pole (VGP) positions and developing the magnetic polarity stratigraphies for each locality. For specimens where no vector segments could be identified, these are considered "incoherent" (or uninterpretable) and were not used in VGP calculations. All data were anchored to the origin, but not forced. Estimated site mean directions were calculated using Fisher (1953) statistics and were termed excellent if the 95% confidence ellipse (α_{95}) was $\leq 10°$, good if α_{95} was 10°–20°, or poor quality if α_{95} was 20°–30° with clear Late Triassic affinities. If the α_{95} was greater than 30°, the results were considered uninterpretable.

Acquisition of isothermal remanent magnetization (IRM) was conducted using a home-built pulse magnetizer providing a maximum field of 2.97 T, which is capable of saturating most hematite grains, at the University of New Mexico Paleomagnetism Laboratory. Both purple to red mudstones and fine-grained sandstone samples were subjected to IRM. Multicomponent IRM-thermal demagnetization (Lowrie, 1990) in DC fields of 1.2 T, 0.4 T, and 0.12 T was conducted at the Imperial College London on representative lithologies, including dark purple mudstone, pale purple to blue mudstone, and red mottled muddy siltstone.

PALEOMAGNETIC RESULTS

Mesa Redondo Member

Only one site was sampled in the uppermost portion of the Mesa Redondo Member (TP-01), and these specimens yielded a well-defined demagnetization behavior that isolates reverse polarity characteristic remanent magnetization (ChRM) above 600°C (Fig. 4A). Average natural remanent magnetization (NRM) intensities are of 1.8–0.7 mA/m, and ChRM intensities range between 1.1 and 0.3 mA/m. The ChRM magnetization is distributed over 500–600°C with the specimens completely demagnetized by 630°C. The site mean direction is $D = 182.5°$, $I = 1.5°$, $\alpha_{95} = 8.2°$, $k = 55.76$, $N/N_0 = 7/8$ specimens (Table 2).

Blue Mesa Member

The Blue Mesa Member was sampled at the Tepees and Blue Mesa localities to create a composite magnetic polarity stratigraphy with a total of 11 sites. Changes in bedding orientation between localities prevent calculation of a geographic grand mean for the entire unit, so only a bedding-corrected or stratigraphic grand mean is described. The Blue Mesa Member is predominantly of reverse polarity with two stratigraphically thin intervals of normal polarity. Specimens from the Blue Mesa Member yield moderately well-defined demagnetization behavior that isolates both the reverse and normal polarity ChRM (Fig. 4B). NRM intensities range between 1.2 and 0.2 mA/m and ChRM intensities between 0.5 and 0.2 mA/m. A low-stability component is present to ~300°C in all specimens, representing a viscous magnetization with steep inclinations.

The ChRM component generally has a moderately distributed unblocking temperature ranging between 500°C and 615°C. In approximately a third of the specimens, the demagnetization behavior becomes effectively uninterpretable above 580–600°C, such that the component above these temperatures cannot always be successfully resolved. The bedding-corrected grand mean for the Blue Mesa Member is $D = 007.6°$, $I = 2.2°$, $\alpha_{95} = 13.9°$, $k = 19.71$, $N/N_0 = 7/11$ sites (Fig. 5A). The normal polarity corrected grand mean is $D = 009.8°$, $I = 24.1°$, $\alpha_{95} = 21.6°$, $k = 135.99$, $N = 2/11$ sites. The reverse polarity corrected grand mean is $D = 186.8°$, $I = 6.6°$, $\alpha_{95} = 11.9°$, $k = 42.33$, $N = 5/11$ sites (Table 2). These means do not pass the reversal test, with an observed γ of 30.84°.

Sonsela Member

The Sonsela Member was sampled at the Milkshake Quarry, Blue Mesa, and Peninsula localities for a total of 55 sampled horizons. The Sonsela Member is dominantly of normal polarity with some stratigraphically thin reverse polarity intervals in the Lot's Wife beds and Jim Camp Wash beds. As with the Blue Mesa Member, changes in bedding

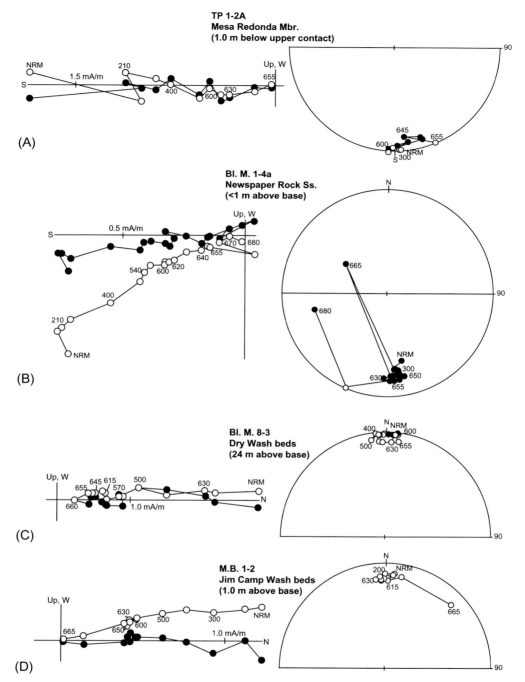

Fig. 4 Orthogonal demagnetization diagrams (Zijderveld, 1967) and equal area nets showing examples of response to progressive thermal demagnetization of specimens for the Mesa Redonda Member (A), Blue Mesa Member (B, C) and Sonsela Member (D–H). *NRM*, natural remanent magnetization. Diagrams show simultaneous projection of horizontal (north-south versus east-west) component of the magnetization *(filled symbols)* and the vertical (north-south versus up-down) component of the magnetization *(open symbols)*. For thermal demagnetization, peak demagnetizing temperatures are shown beside the vertical projections. Meters are position of the sampled horizon above base of the locality's section.

(Continued)

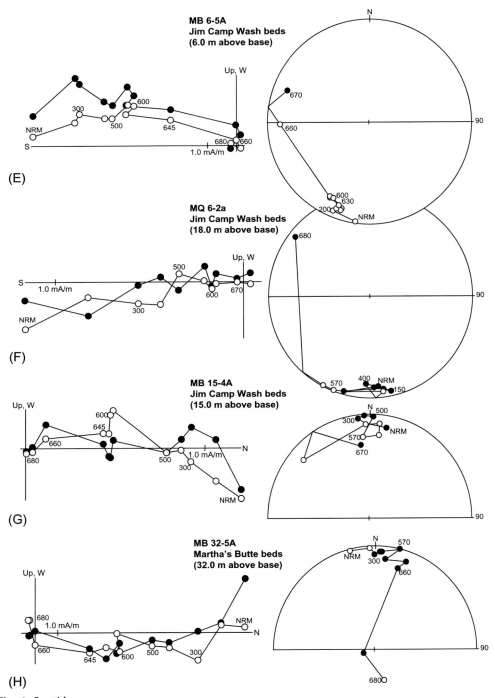

MB 6-5A
Jim Camp Wash beds
(6.0 m above base)

(E)

MQ 6-2a
Jim Camp Wash beds
(18.0 m above base)

(F)

MB 15-4A
Jim Camp Wash beds
(15.0 m above base)

(G)

MB 32-5A
Martha's Butte beds
(32.0 m above base)

(H)

Fig. 4, Cont'd

Table 2 Grand mean directions for each member of the lower Chinle Formation sampled in the southern Petrified Forest National Park

Unit	Geog. Decl.	Geog. Inc.	Strat. Decl.	Strat. Inc.	α_{95}	k	N/N$_0$
Sonsela Member							
Normal mean	–	–	005.2	3.7	2.5	119.41	29/54
Reverse mean	–	–	178.6	3.5	11.0	38.37	6/54
Grand mean	–	–	004.1	2.5	2.8	77.97	35/54
Blue Mesa Member							
Normal mean	–	–	009.8	24.1	21.6	135.99	2/11
Reverse mean	–	–	186.8	6.6	11.9	42.33	5/11
Grand mean	–	–	007.6	2.2	13.9	19.71	12/13
Mesa Redondo Member[a]							
	182.5	1.5	–	–	8.2	55.76	7/8

Abbreviations: α_{95}—cone of confidence, k—measure of data dispersion, N/N$_0$—number of sites accepted out of total number of sites sampled for that stratigraphic unit. For the Blue Mesa and Sonsela members, the orientation of the beds changed partway through the sampled locality and so only corrected stratigraphic directions are shown. The Mesa Redondo Member was sampled in a flat-lying area and has no stratigraphic correction.
[a]Mesa Redondo Member was sampled at only one site. The mean direction is for the site and is not representative of the grand mean direction of the entire unit.

orientations between and within the sampled localities necessitate using a stratigraphic grand mean direction for the whole unit. NRM intensities range from 2.7 to 1.4 mA/m and ChRM intensities from 1.2 to 0.4 mA/m. Unblocking temperatures for the ChRM are distributed across 630–680°C, but become more broadly distributed, from 570°C to 680°C, toward the top of the section (Fig. 4D–H). A low-stability component occurs in 90% of the specimens with a steep inclination. The bedding-corrected grand mean for the Sonsela Member is D=004.1°, I=2.5°, α_{95}=2.8°, k=77.97, N/N$_0$=35/54 sites (Fig. 5B). The normal polarity corrected grand mean is D=005.2°, I=3.7°, α_{95}=2.5°, k=119.41, N=29/54 sites, and the reverse polarity corrected grand mean is D=178.6°, I=3.5°, α_{95}=11.0°, k=38.37, N=6/54 sites (Table 2). These means yield a positive, class B reversal test with an observed angle of 9.76°.

Rock Magnetism

Multiple component IRM-thermal demagnetization (Lowrie, 1990) was applied to representative specimens of various lithologies, including red and dark purple mudstone,

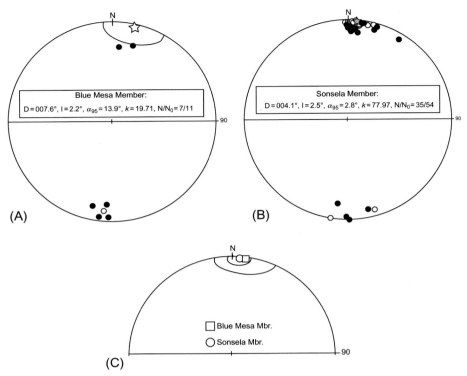

Fig. 5 Equal area projections showing estimated site mean directions for accepted sites. Includes site mean directions not accepted for grand mean calculations to show dispersion of all calculated vectors for each member. (A) Blue Mesa Member for both Tepees and Blue Mesa localities. (B) Sonsela Member at the Peninsula/Martha's Butte locality. (C) Equal area projection showing site grand mean directions for accepted sites from both members at all localities. Closed symbols are lower hemisphere projections and open symbols are upper hemisphere projections.

very pale blue mottled mudstone, and red mottled muddy siltstone (Fig. 6A–D). For the red and dark purple mudstone specimens, the high coercivity component (1.2 T) is of the highest intensity and highest laboratory unblocking temperature (650°C), indicating hematite as a significant contributor to the magnetic signal in the lithologies with red and dark purple hues (Fig. 6C and D). For the red mottled muddy siltstone, the intermediate coercivity component (0.4 T) is the primary contributor, and laboratory unblocking temperatures are widely distributed from ∼300°C to ∼600°C, indicating the presence of magnetite and potentially maghemite (Fig. 6A). The low coercivity component has a significant contribution in several of the samples (Fig. 6B and D). A marked decline in intensity around 200°–300°C in the intermediate and low coercivity components of some of the specimens suggests the presence of other minerals such as maghemite or titanomagnetite.

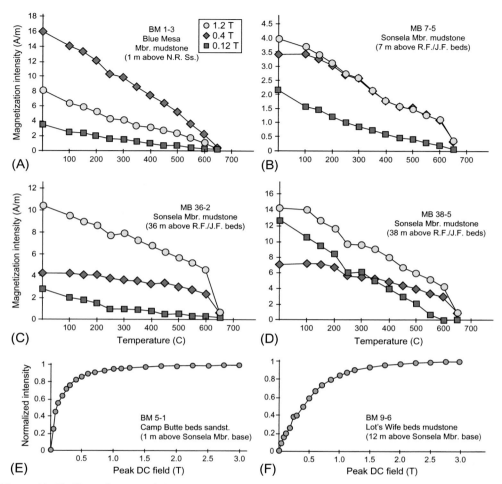

Fig. 6 (A–D) Plots showing three component progressive thermal demagnetization of selected samples representing different lithologies for the Blue Mesa Member (A) and Sonsela Member (B–D). Specimens of isothermal remanent magnetization (IRM) acquired along three orthogonal axes following Lowrie (1990). (E–F) Plots showing IRM acquisition to saturation or near-saturation for specimens from the Sonsela Member.

IRM acquisition response for both sandstone and mudstone specimens shows a concave downward response (Fig. 6E and F). The sandstone specimen shows a rapid acquisition of isothermal remanence, acquiring 80% of maximum IRM by 0.5 T (Fig. 6E), whereas the mudstone specimen shows a more distributed acquisition, gaining 80% of maximum IRM by 1.0 T (Fig. 6F). Both sandstone and mudstone specimens reach near saturation by ~3.0 T. The distributed response of the mudstone samples suggests at least two phases with distinctly different coercivities.

DISCUSSION
Interpretation of Paleomagnetic Data

Many of the samples analyzed from southern PEFO yielded fairly consistent paleomagnetic behavior with ChRMs with north- or south-directed declinations and inclinations that are generally shallower than would be anticipated for the Early Jurassic or a younger age for North America (Van der Voo, 1993; Besse and Courtillot, 2002; Molina-Garza et al., 2003; Kent and Irving, 2010; Zeigler and Geissman, 2011) (Tables 2 and 3). This supports a Triassic age for acquisition of ChRM in these strata. Paleolatitudes calculated from the grand mean directions place eastern Arizona at or just north of the equator during the Norian, which is consistent with paleogeographic reconstructions of the Southwest (e.g., Ziegler et al., 1983; Blakey and Gubitosa, 1983; Golonka, 2007) (Table 3; Appendix I). The average paleolatitude for the Blue Mesa Member is 1.1°S and for the Sonsela Member, 1.3°N. These results place the American Southwest further south than suggested by others (Ziegler et al., 1983; Blakey and Gubitosa, 1983; Scotese, 1994; Molina-Garza et al., 1995; Golonka, 2007; Zeigler and Geissman, 2011).

Variation in individual paleolatitudes (see Appendix II) is most likely due to a relatively high proportion of poor-quality data for both the Blue Mesa Member and the Sonsela Member and probably also reflects inclination flattening effects. Numerous studies have documented the impact of compaction-related shallowing in sedimentary rocks (e.g., Anson and Kodama, 1987; Deamer and Kodama, 1990; Tan et al., 2002, 2007; Kent and Tauxe, 2005, among others) that will result in erroneous paleolatitude interpretations. For the two units sampled here, we did not have a sufficient number of independent specimens to apply Kent and Tauxe's (2005) statistical correction approach, but the paleolatitudes calculated for these data are southerly enough to suggest a slight but significant flattening factor for these strata.

Samples from the Sonsela Member pass a reversal test with a class B positive reversal test (observed $\gamma = 9.76°$), but samples from the Blue Mesa Member do not, with an observed angular difference between the normal and reverse polarity mean directions of 30.84°, well beyond the classification for a positive reversal test of McFadden and

Table 3 Paleomagnetic pole positions for each stratigraphic unit

Unit	Latitude	Longitude	dp	dm	Paleolat.	Plat (+)	Plat (−)	N/N$_0$
Sonsela Mbr.	56.2°N	062.8°E	1.4	2.8	1.3	2.7	−0.2	35/54
Blue Mesa Mbr.	55.6°N	056.7°E	7.0	13.9	−1.1	5.9	−8.2	7/11
Mesa Redondo Mbr.[a]	54.3°N	065.9°E	4.1	8.2	0.8	4.9	−3.4	1/1

[a]Mesa Redondo Member was sampled at only one site, and the inferred pole position is calculated from a single site mean and is therefore not representative of the pole position for the entire unit.

McElhinny (1990). The reversal test is useful because the time-averaged geomagnetic field during normal and reverse polarity intervals should be 180° from one another. Therefore, given that the specimens from the Blue Mesa Member do not pass the reversal test, this indicates secondary NRM components or a younger overprint may be present. All specimens from the Blue Mesa Member and 90% of the specimens from the Sonsela Member include a low-stability vector component that is unblocked by 300–400°C and is interpreted as a younger secondary magnetization. There may be intermediate-age vector components present in some of the specimens analyzed, but these are not clearly distinguishable from the steep inclination low-stability component or the shallow inclination high-stability component. In addition, nonlinear aspects of the NRM and ChRM vectors could be a result of small errors introduced during the measurement process since it is highly unlikely that the cube-shaped samples were oriented precisely the same way for every measurement.

VGP positions were calculated for individual specimens (Fig. 7, Appendix II), including class A, B, and C vector segment data. The Tepees locality yielded mostly class A (33.3%) and class C (33.3%) data with a high proportion of incoherent results (20.4%) and minor class B (13.0%) for 54 specimens analyzed. At Blue Mesa, 16.4% of the 110 specimens analyzed are class A, 14.5% class B, 40.9% class C, and 28.1% incoherent. Milkshake Quarry specimens were primarily poor quality, with 3.8% class A, 26.9% class B, 38.5% class C, and 30.8% incoherent data for 26 specimens analyzed. At the Peninsula, the 274 specimens analyzed yielded mixed quality with 31.8% class A, 13.5% class B, 20.1% class C, and 34.7% incoherent.

In terms of site or sampled horizon data for each stratigraphic unit, almost a third of the sampled horizons yielded interpretable data (47/67 sites). For the Blue Mesa Member, 36.4% of all 11 sites sampled yielded excellent to good α_{95} values, 27.3% yielded poor-quality data, and 36.4% were uninterpretable. For the Sonsela Member, 58.2% of the 55 sampled horizons have excellent to good α_{95} values, 7.3% are poor quality, and 34.5% are uninterpretable. Horizons with uninterpretable vector data include conglomeratic and coarse-grained sandstone units (Jasper Forest bed, Camp Butte sandstone, Bowman sandstone) and, in some cases, mudstone or siltstone units with significant color mottling and generally lighter hues.

Magnetic Minerals, Color Variations, and Sediment Sources

The combination of IRM acquisition data and multiple component IRM-thermal demagnetization analyses indicate a variety of magnetic minerals carrying the ChRM in these strata. Compared to similar strata sampled in northern New Mexico that are dominated by one or two magnetic minerals, the southern PEFO Chinle Formation includes a range of low to high coercivity and low to high unlocking temperature magnetic minerals that could include hematite, magnetite, maghemite,

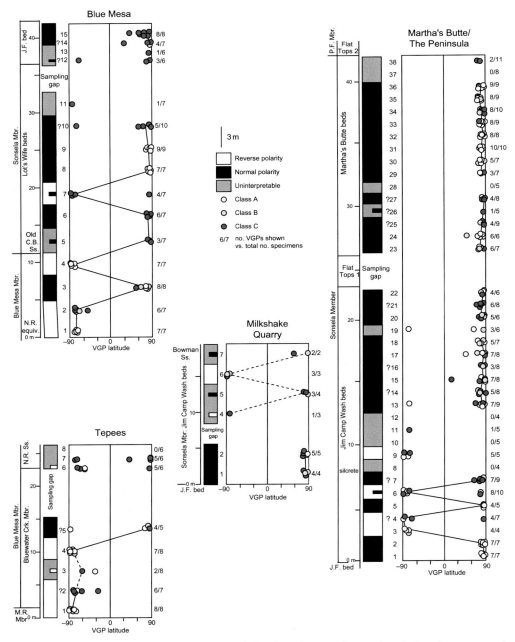

Fig. 7 Magnetic polarity stratigraphies for each locality along with associated virtual geomagnetic pole (VGP) latitudes calculated from site mean directions. Open circles are vectors with MAD values ≤10°, medium gray circles are class B data with 11°≤MAD≤°15, and dark gray circles are class C with MAD >15°. N/N₀ beside each sampled horizon represents the number of VGPs shown versus the number of specimens from that horizon (e.g., 3/4 indicates three VGPs are shown and the fourth was uninterpretable and is not shown). Horizons with no VGPs had completely uninterpretable results.

and titanomagnetite. Samples subjected to rock magnetic experiments from northern New Mexico were dominated by high coercivity and high unblocking temperature responses, and chemical experiments demonstrated a mixture of pigmentary and detrital hematite as the dominant magnetic mineral in those strata (Zeigler and Geissman, 2011).

The outcrops-sampled Tepees (uppermost Mesa Redonda Member and lower Blue Mesa Member) are pale blue to reddish purple mudstone to siltstone that changes up-section to sandy mudstone or unconsolidated sand that is very pale green to buff in color. At Blue Mesa, strata are reddish brown to dark purple in the lower third of the section in the upper Blue Mesa Member, and become increasingly more mottled up-section into the Sonsela Member. The Jasper Forest bed is primarily very coarse-grained to conglomeratic sandstone that yielded poor-quality data. Milkshake Quarry strata (Jim Camp Wash beds) are primarily purplish-brown mudstone and siltstone locally interbedded with pale gray to buff coarse-grained sandstone. The Sonsela Member strata sampled at the Peninsula/Martha's Butte section become increasingly darker hued up-section and the Martha's Butte beds are dark reddish purple to dark purple in color.

The quality of paleomagnetic data in the Chinle Formation at PEFO does bear a strong relationship to hue of the strata, as thoroughly documented by Steiner and Lucas (2000), and our rock magnetism and vector quality data support their observations. In addition, according to Atchley et al.'s (2013) interpretation of paleosol development in the lower Chinle Formation in southern PEFO, the highest potential for nonreduced iron is in the middle Blue Mesa Member and the upper Sonsela Member. The higher-quality data for this study for the Blue Mesa Member came from the lower and upper Blue Mesa Member, but we note that the middle part of the member was not well sampled. High-quality data is also documented in the upper Sonsela Member, reinforcing the interpretation of Atchley et al. (2013).

The Chinle Formation strata sampled in northern New Mexico are predominantly dark red to dark purple in color with much of the section being within the range of hues that suggests pigmentary hematite. Multiple component IRM-thermal demagnetization and IRM acquisition data for these strata indicate the dominant magnetic mineral is hematite. In eastern Arizona, Chinle Formation strata are a wider variety of hues, including red, purple, pale blue, pale green, tan or buff in color, suggesting less pigmentary hematite overall than is observed in the New Mexican section. As noted by Molina-Garza et al. (1998b), specimens from the Chinle Formation in the Zuni Mountains of western New Mexico that lacked pigmentary hematite tended to exhibit more complex, multivector magnetizations, and we observe similar behavior in many of the specimens in this study.

The variety of magnetic minerals observed in the results of the multiple component IRM-thermal analyses for the PEFO specimens may reflect a difference in source areas for sediments compared to the strata deposited in northern New Mexico. The section preserved at PEFO is within a few hundred kilometers of the Cordilleran uplift to the west and southwest, which exposed intrusive and extrusive igneous rocks as well as older sedimentary strata (Howell and Blakey, 2013; Atchley et al., 2013; Riggs et al., 2013, 2016). The Chama Basin section probably had primary source areas to the north, in remanent Ancestral Rocky Mountain uplifts (the Ute and Arapahoe uplifts of Dickinson et al., 2010) that were presumably significantly weathered. If the Chama Basin section represents a more distal area in the depositional system transporting material southwest and west from these ancient uplifts, many of the magnetic minerals observed in the PEFO section were presumably chemically weathered away or were deposited in more proximal parts of the system in southern Colorado and are thus not represented in the Chama Basin section. The difference in diversity of magnetic minerals in the PEFO section as compared to the Chama Basin section bears further investigation.

Magnetostratigraphy: Local and Regional Correlations

Within southern PEFO, lithostratigraphic and biostratigraphic correlations among the overlapping segments of strata at each locality are well studied (Martz and Parker, 2010; Parker and Martz, 2011), permitting the development of a composite magnetostratigraphic section for lower Chinle Formation strata from the top of the Mesa Redonda Member to the uppermost Sonsela Member (Fig. 8). The Newspaper Rock Sandstone in the upper part of the Blue Mesa Member, sampled at the top of the Tepees section, yields poor-quality, but possibly reverse magnetic polarity near its base, and laterally equivalent red mudstones (Parker, 2006) at the base of the Blue Mesa locality also are of reverse polarity. Strata sampled at the Milkshake Quarry, from just above the Jasper Forest bed to a sandstone correlated to the lower sandstone at Bowman locality site 1 (informally referred to as the Bowman sandstone (Martz and Parker, 2010), Appendix, Bowman 1-3 sections) encompasses the Jim Camp Wash beds of the lower Sonsela Member. The magnetic polarity stratigraphy for this locality compares moderately well to the more densely sampled Jim Camp Wash beds at the Peninsula locality, despite the overall poor quality of data from specimens from Milkshake Quarry. The Adamanian-Revueltian faunal transition identified by Parker and Martz (2011) as occurring immediately above the persistent red silcrete that occurs low in the Jim Camp Wash beds is in a zone of incoherent magnetic polarity data that is stratigraphically immediately above a normal polarity zone.

The current study compares well to Steiner and Lucas's (2000) magnetic polarity stratigraphy for the lower Chinle Formation. Their Blue Mesa Member and basal Sonsela Member sequence includes two reverse and two normal polarity chrons with two more

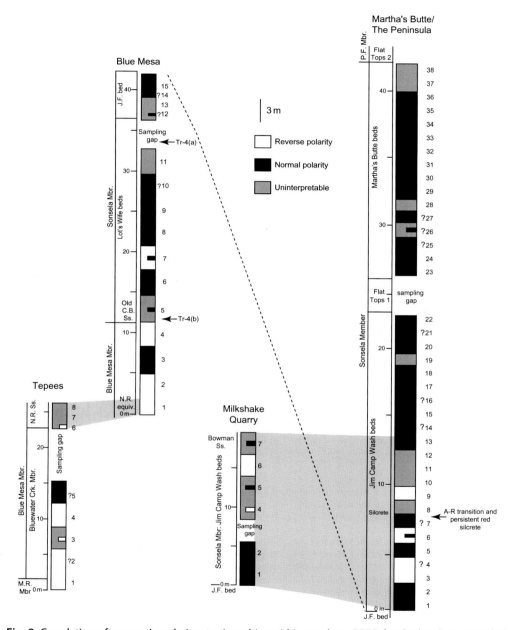

Fig. 8 Correlation of magnetic polarity stratigraphies within southern PEFO for the localities sampled for this study. *N.R.Ss.*, Newspaper Rock Sandstone. Tr-4(a) is the position of the Tr-4 unconformity of Lucas (1991, 1993) and Heckert and Lucas (1996) and Tr-4(b) the revised position of the unconformity of Heckert and Lucas (2002). Position of the Adamanian-Revueltian faunal transition ("A-R transition") from Parker and Martz (2011).

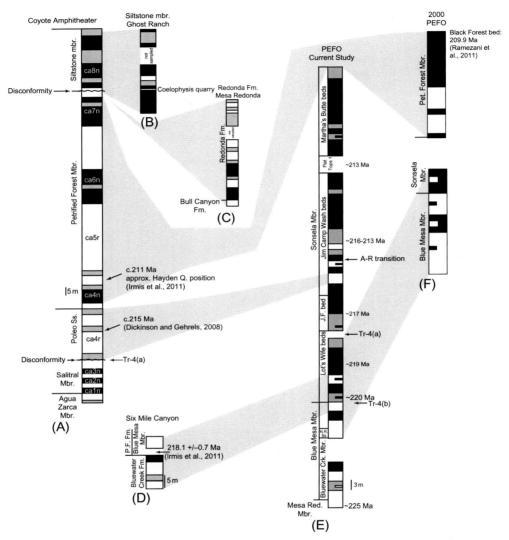

Fig. 9 Comparison of magnetic polarity stratigraphies from New Mexico (A–D; Zeigler and Geissman, 2011) with the current PEFO (E) and Steiner and Lucas (2000), corrected per Martz and Parker (2010) (F) magnetic polarity stratigraphies. *N.R.Ss.*, Newspaper Rock Sandstone. Tr-4(a) is the position of the Tr-4 unconformity of Lucas (1991, 1993) and Heckert and Lucas (1996) and Tr-4(b) the revised position of the unconformity of Heckert and Lucas (2002). Position of the Adamanian-Revueltian faunal transition ("A-R transition") from Parker and Martz (2011).

normal and reverse polarity chrons tentatively identified (Fig. 9). This matches the polarity zonations identified in the Blue Mesa Member and Lot's Wife beds of the Sonsela Member in this effort. Our sampling efforts ended at the base of their upper section, which is identified as primarily Petrified Forest Member, so the magnetic polarity stratigraphy they identify for that unit still remains to be examined. The magnetic polarity stratigraphic developed here for the Jasper Forest bed, Jim Camp Wash beds, and Martha's Butte beds fills in the sampling gap in the Steiner and Lucas (2000) study for a more complete magnetostratigraphy for PEFO.

In comparison to other magnetic polarity chronologies developed for Chinle strata in New Mexico, correlations on a regional scale become problematic (Fig. 9). For example, the complete Chinle Formation section examined in the Chama Basin of north-central New Mexico (Zeigler, 2008; Zeigler et al., 2008; Zeigler and Geissman, 2011) has very little age control aside from a detrital zircon maximum depositional age estimate for the Poleo Sandstone (~215 Ma, Dickinson et al., 2010) and a single U-Pb detrital zircon date of 211.9 ± 0.7 Ma from the Hayden Quarry in the lower part of the Petrified Forest Member at Ghost Ranch (Irmis et al., 2011). There is abundant vertebrate biostratigraphic data from restricted stratigraphic zones in the lower Petrified Forest Member, including vertebrate fossil material from the Hayden, Snyder, and Canjilon quarries, as well as for the uppermost siltstone member (formerly Rock Point Member) (Zeigler, 2008; Rinehart et al., 2009; Irmis et al., 2007, 2011). For the lower Chinle Formation in the Chama Basin, there is very little biostratigraphic data to help with correlation to strata outside of the Chama Basin aside from fragmentary remains (e.g., Zeigler et al., 2005). Zeigler and Geissman (2011) provided tentative correlations of this section to magnetic polarity stratigraphies from eastern New Mexico, the Newark Supergroup in eastern North America and sections in Europe, but these are to be considered tenuous at best. A coarse sampling resolution coupled with a lack of biostratigraphic data for much of the section hampers the potential for strong correlation of the northern New Mexico section to others.

The Poleo Sandstone in north-central New Mexico, with an estimated maximum depositional age of ~215 Ma (Dickinson et al., 2010), had been tentatively correlated to the traditional Sonsela Bed (Jasper Forest bed of current use) because both units were assumed to occupy the same stratigraphic position (e.g., Lucas, 1993). Previous sampling at PEFO suggested that the Sonsela bed was of normal polarity (Steiner and Lucas, 2000) due to their interpretation of the stratigraphy, whereas the Poleo Sandstone is entirely of reverse polarity (Zeigler and Geissman, 2011). However, in light of revised lithostratigraphic correlations, the Poleo Sandstone can be now tentatively correlated to the lower Jim Camp Wash beds in PEFO, which has maximum depositional age ranging from ~217 to ~213 Ma (Ramezani et al., 2011). The Petrified Forest Member in the Chama Basin is potentially correlative to the Petrified Forest Member at PEFO of Steiner and Lucas (2000), but agreement between the two sections is again hampered by a lack of geochronologic data for either section. However, the magnetic polarity zonation is similar between the two sections, with three normal polarity chrons and two reverse polarity chrons.

The correlation of the two sections of Petrified Forest Member to one another coupled with correlation of the Poleo Sandstone to the lower Jim Camp Wash beds suggests that the Chama Basin area witnessed a depositional hiatus or extreme reduction in accumulation rate at or near the base of the Petrified Forest Member that would include the upper Jim Camp Wash beds and Martha's Butte beds at PEFO (Fig. 9). This hiatus has not previously been recognized: The contact between the Poleo Sandstone and overlying Petrified Forest Member has been described as a gradational contact with interbedded sandstone, siltstone, and mudstone at the top of the Poleo Sandstone grading upwards in mudstone and siltstone in the lower Petrified Forest Member (Lucas et al., 2003, 2005).

The lowest Chinle Formation strata in the Chama Basin, the Agua Zarca Member and Salitral Member, have very little biostratigraphic data and no geochronologic data. These units were tentatively assigned to the very lowest stratigraphic position in Zeigler and Geissman's (2011) Chinle Formation composite magnetic polarity stratigraphy and were not correlated to other lower Chinle sections in western New Mexico or Arizona due to the lack of biostratigraphic or geochronologic control. A short stratigraphic section was sampled at the Six Mile Canyon locality in west-central New Mexico to include a detrital zircon maximum depositional age of 218.1 ± 0.7 Ma (Irmis and Mundil, 2008; Zeigler and Geissman, 2011; Irmis et al., 2011) and includes the upper Bluewater Creek and Blue Mesa members of Heckert and Lucas (2002). This magnetic polarity stratigraphy can be correlated to the lower Lot's Wife beds, which have an estimated age of ~219 Ma from near the base of the unit (Ramezani et al., 2011).

With these new correlations, the composite Chinle Formation magnetic polarity stratigraphy of Zeigler and Geissman (2011) is still reasonable for New Mexico. However, development of a composite magnetic polarity stratigraphy for both Arizona and New Mexico is still problematic given the inability to correlate the lowest Chinle Formation units in the Chama Basin with sections in either west-central New Mexico or Arizona.

Magnetostratigraphy: PEFO and the Newark Supergroup

Correlations among Upper Triassic outcrops on a regional scale are clearly still problematic. However, it is worth investigating the potential correlations of the PEFO and Chama Basin magnetic polarity stratigraphies to that of the Newark Supergroup in eastern North America (Fig. 10). The Newark Supergroup is considered a benchmark for Late Triassic magnetic polarity stratigraphy and has an associated robust cyclostratigraphic framework (e.g., Olsen et al., 1996, 2002, 2011; Olsen and Kent, 1999; Kent and Olsen, 1999, 2000). Comparing the Chama Basin sections to the Newark Supergroup with the minimal age control available suggests that the Poleo Sandstone correlates to chron E14r, and the New Mexico section of Petrified Forest Member strata tentatively correlates to the top of E15 through E18r. As documented in Zeigler and Geissman (2011), the uppermost strata in the Chama Basin are Rhaetian in age and have been correlated to E23 through E24. The Redonda Formation of eastern New Mexico may correlate to E21 to E22, but a lack of strong biostratigraphic or geochronologic control again creates a

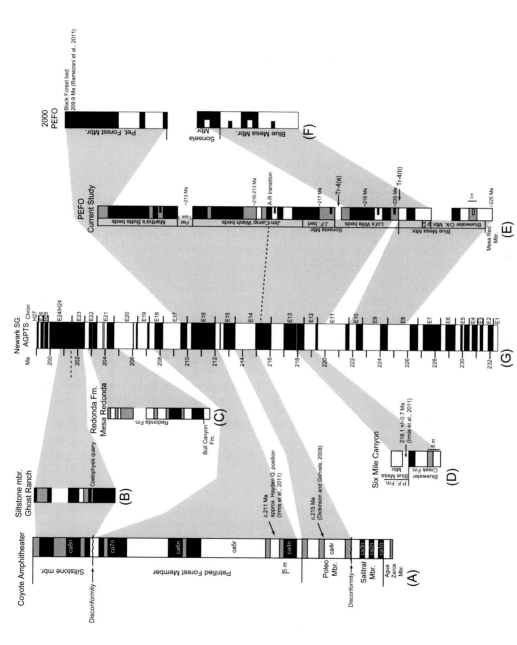

Fig. 10 Comparison of magnetic polarity stratigraphies from New Mexico (A–D), PEFO (E, F) and the Newark Supergroup (G) with cyclostratigraphically calculated age associations (from Olsen et al., 2011). Tr-4(a) is the position of the Tr-4 unconformity of Lucas (1991, 1993) and Heckert and Lucas (1996) and Tr-4(b) the revised position of the unconformity of Heckert and Lucas (2002). Position of the Adamanian-Revueltian faunal transition ("A-R transition") from Parker and Martz (2011).

tenuous correlation. The Six Mile Canyon section in west-central New Mexico is here correlated with E12r through E13r1. The lowest Chinle Formation strata in the Chama Basin are difficult to correlate even on the regional scale, so no attempt is made to correlate the chrons recorded in the Agua Zarca and Salitral members to those of the Newark Supergroup.

However, in contrast to the Chama Basin section, the current PEFO magnetic polarity stratigraphy can be correlated to the new astronomically tuned geomagnetic polarity time scale (AGPTS) of the Newark Supergroup (Olsen et al., 2011) with a somewhat higher degree of confidence, given the geochronologic data for the PEFO section from Ramezani et al. (2011) and Atchley et al. (2013). However, because the samples for this current study were not collected at the same time and from precisely the same lithostratigraphic unit as the detrital zircon samples used by Ramezani et al (2011) and Atchley et al. (2013), it is not possible to be certain that a particular polarity chron is the precise age yielded by their efforts. In Fig. 10 we denote this uncertainty by using tildes alongside the age determinations of Ramezani et al. (2011) and Atchley et al. (2013). In addition, these ages are maximum depositional ages and so the magnetic polarity data for those horizons could presumably be somewhat younger than the detrital zircon dates themselves.

The Blue Mesa Member is here correlated with E8r through E10r and the Camp Butte Sandstone (basal Sonsela Member) may correlate to the lower part of E12n. The Lot's Wife beds and Jasper Forest bed tentatively correlate to upper E12n through E13n1. The Jim Camp Wash beds and Martha's Butte beds are tentatively correlated to E13n1 through E15n1. The Petrified Forest Member sampled by Steiner and Lucas (2000) is correlated here to E15r1 through E17n. These correlations between the current PEFO magnetic polarity stratigraphy and the Newark Supergroup suggest that in spite of sedimentation rate variations, hiatuses and erosional events (see the following), the PEFO magnetic polarity stratigraphy developed here captures the majority of the magnetic polarity chrons in the middle Norian portion of the Newark Supergroup record. The PEFO Chinle Formation does not have a sedimentation pattern similar to the nearly continuous rift-fill sedimentation that occurred in the Newark rift basins that allows for the application of astronomic tuning (Olsen et al., 2011). With further revisions to the geochronology of the Chinle Formation, it may be possible to place these strata in an estimated time-based, rather than a stratigraphic thickness-based, framework that would allow for further investigation of sedimentation rates to corroborate the periodic nature of accommodation and deposition documented by Ramezani et al. (2011) and Atchley et al. (2013).

Zeigler and Geissman (2011) compared the New Mexico sections to other magnetic polarity stratigraphies from Europe, as well as the Newark Supergroup record. As they noted, the correlation of these magnetic polarity stratigraphies at the global level is even more tenuous, if not impossible. In these comparisons, terrestrial records with all their attendant problems must somehow be compared to marine records, with very different biostratigraphic frameworks. The marine geomagnetic polarity timescale (GPTS) for the Late Triassic proposed by Hounslow and Muttoni (2010) could be considered as a benchmark section, similar to the Newark Supergroup section, except that it is a composite magnetic polarity record built with data sets from many different localities with different depositional environments and highly variable age control. It has been suggested that the marine Late Triassic GPTS may not be an accurate compilation of the behavior of the geomagnetic field in the Late Triassic, and correlation of local stratigraphic sections to the GPTS are potentially problematic.

Tr-4 Unconformity

Previous authors have argued for the presence of a regional disconformity in medial Chinle strata, the so-called Tr-4 unconformity (Lucas, 1991, 1993), which has been postulated to lie at the base of the Jasper Forest bed (Lucas, 1991, 1993; Heckert and Lucas, 1996) and later reinterpreted to occur much lower in the section at the base of the Sonsela Member (Heckert and Lucas, 2002). Woody (2006) and Martz and Parker (2010) used detailed geologic mapping to refute the probability of a long-lived, large-scale erosional event and/or depositional hiatus at the base of the Sonsela Member in PEFO. In addition to their arguments, we note here that the presence of a well-developed unconformity would result in a lack of preservation of the associated magnetic polarity zones of that time interval. The reasonably robust correlation of the Blue Mesa Member and Sonsela Member to the Newark Supergroup record suggests that this sampling attempt captured most of the polarity behavior for the early to middle Norian. Missing magnetic polarity zones may be a function of small-scale depositional hiatuses, localized erosional events, or the broader sampling resolution employed for the Blue Mesa Member and lower Sonsela Member (3 m versus 1 m).

There are undoubtedly hiatuses within the section, such as within the Camp Butte sandstone to lower Lot's Wife beds and in the Jasper Forest bed interval, but they appear to have been fairly short-lived events (but see Ramezani et al., 2011). By contrast, the Chama Basin section includes several significant disconformities, including at the base of the Poleo Sandstone, and at the boundary between the Petrified Forest Member and uppermost "siltstone member." The Poleo Sandstone, which had previously been included as evidence of part of the Tr-4 regional erosional event (Lucas, 1993; Lucas et al., 2005), is here correlated to the lower Jim Camp Wash beds in PEFO, stratigraphically above either interval argued for by previous authors (Lucas, 1991, 1993; Heckert and Lucas, 1996, 2002). However, we agree with Ramezani et al. (2011) that the Sonsela Member probably includes multiple short-lived disconformities that may not be laterally continuous, as opposed to one single regional-scale disconformity.

Adamanian-Revueltian Transition

As thoroughly documented by Parker and Martz (2011), the strata sampled here include the Adamanian-Revueltian tetrapod faunal transition, estimated to occur in the lower Jim Camp Wash beds (lower Sonsela Member). Lucas (1991, 1993) suggested that the Adamanian and Revueltian faunal assemblages were separated by an extended period of erosion, the Tr-4 unconformity, that also coincided with the Carnian-Norian stage boundary. Per their argument, the seemingly abrupt reorganization of both fauna (Long and Ballew, 1985; Lucas, 1991, 1993; Hunt and Lucas, 1995) and flora (Litwin et al., 1991) was used as evidence that there was a region-wide disconformity. The careful documentation of both geographic and stratigraphic positions of numerous vertebrate fossil localities by Parker and Martz (2011) restricted the position of the faunal transition to a stratigraphic interval, at or near the persistent red silcrete, that has not been interpreted to represent a major hiatus in deposition.

Parker and Martz (2011) suggest a possible link between the Adamanian-Revueltian faunal transition and the Manicouagan impact in Quebec. The impact event has been dated to c.215.5 (Ramezani et al., 2005), and magnetic polarity data for the impact is of normal polarity (Robertson, 1967; Larochelle and Currie, 1967). At PEFO, the strata immediately below the silcrete are of normal polarity, and the overlying strata are uninterpretable. In addition, the silcrete horizon is bounded by maximum depositional ages of 213.9 and 217.7 Ma (Ramezani et al., 2011). Both the Adamanian-Revueltian faunal transition and the Manicouagan impact occur within (or immediately above) strata of normal polarity, and when coupled with the narrow age constraints of Ramezani et al. (2005, 2011), the possibility that the impact in Quebec contributed to the faunal turnover cannot be discounted. However, the sampling for both detrital zircons (Ramezani et al., 2011) and for magnetic polarity data (Steiner and Lucas, 2000; this study) is too coarse to be absolutely certain of the correlation between the impact and the faunal transition.

CONCLUSIONS

The current study provides new insight into the magnetic polarity stratigraphy of the Sonsela Member (*sensu* Martz and Parker, 2010) and helps to account for a sampling gap identified in the previous work of Steiner and Lucas (2000). Overall, the two magnetic polarity stratigraphies agree reasonably well. Comparison of the magnetic polarity data for overlapping stratigraphic units (e.g., Newspaper Rock Sandstone, Jim Camp Wash beds) show acceptable local correlations, in spite of a high proportion of poor-quality data in many of the sampled horizons. Data quality is probably a reflection of a combination of depositional environment (in terms of potential for nonreduced iron as described by Atchley et al., 2013), presence of younger low-stability vector components, younger overprints, and/or inclination shallowing effects. The variety of magnetic minerals present in specimens from southern PEFO is much greater than in specimens from northern New Mexico, presumably indicating significant differences in depositional environments and source areas for magnetic mineral grains.

At the regional level, a continued lack of biostratigraphic data in the lower section and geochronologic data for the entire section in New Mexico hampers correlation of the new PEFO magnetic polarity stratigraphy to sections documented by Zeigler and Geissman (2011). However, with the correlations suggested here, it becomes apparent that the northern New Mexico section, previously identified as a nearly "complete" section of the Chinle Formation, includes long-lived depositional hiatuses and/or erosional events. This further highlights the complications inherent in attempting long-distance lithostratigraphic correlations.

The New Mexico and Arizona magnetic polarity chronologies can also be compared to that of the Newark Supergroup, and geochronologic data from the sections at PEFO allows for a relatively robust correlation, indicating that deposition of the lower Chinle Formation at PEFO captured the magnetic polarity stratigraphy for part of the early Norian and much of the middle Norian. Correlations between the New Mexico sections and the Newark Supergroup are still tenuous at best. The comparison of the ability to correlate the New Mexico sections, which lack robust biostratigraphic or geochronologic control versus the PEFO section to the Newark Supergroup AGPTS, highlight one of the pitfalls of using magnetic polarity data: Without any other corroborating stratigraphic data sets, magnetic polarity stratigraphies cannot be easily used to reconcile regional or global correlations.

ACKNOWLEDGMENTS

This work was conducted under NPS permit #PEFO-2009-SCI-0012 and was funded by the Petrified Forest Museum Association. We are grateful to Radchagrit Supakulopas and Dr. Adrian Muxworthy at the Imperial College London for assistance processing samples for rock magnetic experiments. Dr. John Geissman and Dr. Daniel Peppe provided access to cryogenic magnetometers at both University of New Mexico and Baylor University. Mark Hounslow and Dan Peppe reviewed earlier versions of the manuscript, and Paul Olsen offered an abundance of informative and thoughtful comments. Amanda Huffman, Susie Davenport, Linda Donohoo-Hurley, Peter Reser, David Cleveland, and Vin Morgan assisted with sample collection and analysis. This is Petrified Forest Paleontological Contribution #48.

APPENDIX I

Site mean data for each locality. Bedding corrections are noted at the top of each table. Abbreviations: α_{95}—cone of confidence, k—measure of data dispersion, N/N_0—number of sites accepted out of total number of sites sampled for that stratigraphic unit. Geo. Dec. = geographic declination, Geo. Inc. = geographic inclination, Strat. Dec. = stratigraphic (bedding corrected) declination, Strat Inc. = stratigraphic (bedding corrected) inclination.

Tepees

Stratigraphy	Site ID	Geo. Dec.	Geo. Inc.	α_{95}	k	N/N$_0$
Blue Mesa Member						
Newspaper Rk. Ss	8	incoh.				
	7	incoh.				
	6	incoh.				
Lower Blue Mesa Mbr	5	14.8	22.3	20.6	36.84	3 of 5
	4	182.5	5.1	10.9	49.82	5 of 7
	3	incoh.				
	2	184.3	18.6	27.9	20.54	3 of 7
Mesa Redondo Member						
	1	182.5	1.5	8.2	55.76	7 of 8

Blue Mesa (bedding correction 345/04 through site 11, horizontal strata sites 12–15)

Stratigraphy	Site ID	Geo. Decl.	Geo. Inc.	Strat. Dec.	Strat. Inc.	α_{95}	k	N/N$_0$
Sonsela Member								
	15	incoh/N						
	14	incoh						
	13	incoh/N						
Jasper Forest Bed	12	incoh/N						
	11	incoh						
	10	incoh						
	9	359	2.3	359.1	6.2	5.8	109.99	7 of 9
	8	359	4.1	359.1	7.9	11.1	48.83	5 of 7
	7	164.6	−3.3	164.6	−7.3	33.3	14.79	3 of 7
	6	7.7	0.2	7.8	3.9	13.5	47.38	4 of 7
Camp Butte Ss	5	16.3	1.3	16.4	4.7	17.2	213.72	2 of 7
Blue Mesa Member								
	4	185.5	−7.9	185.7	−11.7	9.7	63.35	5 of 7
	3	4.1	21.9	4.7	25.7	5	178.99	6 of 8
	2	188.6	9	188.4	5.3	20.4	37.64	3 of 6
Newspaper Rk equiv.	1	193.8	18.7	193.2	15.2	17.4	12.97	7 of 7

Milkshake Quarry (bedding correction: 165/04)

Stratigraphy	Site ID	Geo. Dec.	Geo. Inc.	Strat. Dec.	Strat. Inc.	α_{95}	k	N/N$_0$
Sonsela Member								
Bowman Sandstone	7	incoh/N						
	6	181.8	−0.6	181.8	3.2	17.8	49.05	3
	5	incoh/N						
	4	incoh/R						
	3	Samples not analyzed						
	2	359.8	11.1	359.6	7.2	12.8	36.63	5
Jim Camp Wash beds	1	359.2	10.4	359.1	6.5	16.7	31.11	4

Peninsula/Martha's Butte (bedding correction: 165/05 for sites 1–22, 142/03 for sites 23–38)

Stratigraphy	Site ID	Geo. Dec.	Geo. Inc.	Strat. Dec.	Strat. Inc.	α_{95}	k	N/N$_0$
Sonsela Member								
Martha's Butte beds	38	incoh.						
	37	incoh.						
	36	4.9	12.7	4.5	10.5	6.5	73.12	8 of 9
	35	6.7	15.4	6.2	13.3	10	45.72	6 of 9
	34	1.8	8.9	1.5	6.6	8.3	66.33	6 of 10
	33	6.3	8.2	6	6.1	7.7	76.6	6 of 9
	32	3.6	4.4	3.5	2.2	7.7	77.31	6 of 8
	31	2.6	15.1	2.1	12.8	6.5	62.75	9 of 10
	30	4.8	11.9	4.4	9.7	14	30.81	5 of 7
	29	2.2	13	1.8	10.7	13.3	87.32	3 of 7
	28	incoh.						
	27	2.5	7.6	2.3	5.3	8.8	109.87	4 of 8
	26	incoh./N						
	25	4.2	−4.2	4.4	−6.4	18.3	26.12	4 of 9
	24	8.9	−0.4	9	−2.5	14.3	42.09	4 of 6
	23	32	7.8	31.6	6.8	0.4	82.43	6 of 7
Jim Camp Wash beds	22	2.8	1.5	2.8	−3.3	19.1	24.2	4 of 6

Continued

Stratigraphy	Site ID	Geo. Dec.	Geo. Inc.	Strat. Dec.	Strat. Inc.	α_{95}	k	N/N$_0$
	21	3.9	13	3.6	8.3	13.8	31.72	5 of 8
	20	14.5	2	14.5	−2.4	24.8	14.7	4 of 6
	19	incoh., 1 R, 1 N						
	18	8.4	7.3	8.2	2.7	22.5	12.54	5 of 6
	17	8	4.4	7.9	−0.2	5.9	170.97	5 of 7
	16	4.6	3.9	4.6	−0.8	23.3	29.15	3 of 6
	15	358.6	2.1	358.6	−2.8	9.3	98.05	4 of 8
	14	0.2	13.3	0	8.5	22.4	12.62	5 of 8
	13	16.4	11.4	16	7.1	11.7	44.07	5 of 9
	12	incoh.						
	11	incoh./R						
	10	incoh.						
	9	168	3.3	168	8.3	16.2	23.14	5 of 5
	8	incoh.						
	7	11.6	0.5	11.7	−4	11.5	44.87	5 of 9
	6	190.5	−4.7	190.4	−0.2	11.8	33.3	6 of 10
	5	2.6	3.3	2.6	−1.5	8.6	115.31	4 of 4
	4	179.5	−4.5	179.5	0.3	19.6	164.12	2 of 7
	3	186.7	11.9	187.2	16.5	12.2	57.98	4 of 4
	2	6.6	10.1	6.4	5.5	5.7	112.48	7 of 7
	1	6.8	−0.9	6.9	−5.5	7.9	58.75	7 of 7

APPENDIX II

Virtual geomagnetic pole (VGP) positions for all sites, including sites that were not included in grand mean direction calculations. Incoh. = incoherent/uninterpretable. Paleolat = paleolatitude.

Tepees

Stratigraphy	Site ID	VGP Lat	VGP Long	dp	dm	Paleolat	Paleolat error (+)	Paleolat error (−)
Blue Mesa Member								
Newspaper Rk. Ss	8	incoh.						
	7	incoh.						
	6	incoh.						
Lower Blue Mesa Mbr	5	63.1	36.6	11.5	21.8	11.6	24.9	0.9
	4	−52.5	246.1	5.5	10.9	2.6	8.2	−2.9
	3	incoh.						
	2	−45.4	244.1	15.1	29	9.6	27.8	−4.7
Mesa Redondo Member								
	1	−54.3	245.9	4.1	8.2	0.8	4.9	−3.4

Blue Mesa

Stratigraphy	Site ID	VGP Lat	VGP Long	*dp*	*dm*	Paleolat	Paleolat error (+)	Paleolat error (−)
Sonsela Member								
	15	incoh.						
	14	incoh.						
	13	incoh.						
Jasper Forest bed	12	incoh.						
	11	incoh.						
	10	incoh.						
	9	58.3	71.9	2.9	5.8	3.1	6.1	0.2
	8	59.2	71.9	5.6	11.2	4	9.8	−1.6
	7	−55.7	278.3	16.9	33.5	−3.7	13.7	−23.2
	6	56.4	56	6.8	13.5	2	8.9	−4.8
Camp Butte Ss	5	54.1	41.4	8.6	17.2	2.4	11.4	−6.3
Blue Mesa Member								
	4	−60.6	238.6	5	9.9	−5.9	−1	−11.1
	3	68.3	57.7	2.9	5.4	13.5	16.5	10.7
	2	−51.7	236.6	10.3	20.5	2.7	13.5	−7.7
Newspaper Rk equiv.	1	−45.7	231.3	9.2	17.9	7.7	17.7	−1.1

Milkshake Quarry

Stratigraphy	Site ID	VGP Lat	VGP Long	*dp*	*dm*	Paleolat	Paleolat error (+)	Paleolat error (−)
Sonsela Member								
Bowman Sandstone	7	incoh/N						
	6	−53.6	247.2	8.9	17.8	1.6	10.9	−7.4
	5	incoh/N						
	4	incoh/R						
	3	samples not analyzed						
	2	58.8	71	6.5	12.9	3.6	10.3	−2.8
Jim Camp Wash beds	1	58.4	71.9	8.4	16.8	3.3	12.1	−5.1

Peninsula/Martha's Butte

Sonsela Member

Stratigraphy	Site ID	VGP Lat	VGP Long	*dp*	*dm*	Paleolat	Paleolat error (+)	Paleolat error (−)
Martha's Butte beds	38	incoh.						
	37	incoh.						
	36	60.2	61.1	3.3	6.6	5.3	8.7	2
	35	61.4	57.3	5.2	10.2	6.7	12.2	1.7
	34	58.5	67.3	4.2	8.3	3.3	7.6	−0.9
	33	57.8	58.9	3.9	7.7	3.1	7	−0.8
	32	56.1	63.9	3.9	7.7	1.1	5	−2.8
	31	61.6	65.8	3.4	6.6	6.5	9.9	3.2
	30	59.8	61.4	7.2	14.2	4.9	12.4	−2.2
	29	60.6	66.5	6.8	13.5	5.4	12.6	−1.3
	28	incoh.						
	27	57.8	65.9	4.4	8.8	2.7	7.2	−1.8
	26	incoh.						
	25	51.8	63.1	9.2	18.4	−3.2	6	−13
	24	53	55.1	7.2	14.3	−1.3	6	−8.6
	23	47.1	20	0.2	0.4	3.4	3.6	3.2
Jim Camp Wash beds	22	53.5	65.5	9.6	19.1	−1.7	8.1	−11.6
	21	59.2	63.2	7	13.9	4.2	11.5	−2.8
	20	51.5	46.5	12.4	24.8	−1.2	11.6	−14.4
	19	incoh.						
	18	55.7	55.5	11.3	22.5	1.4	13.2	−10.2
	17	54.3	56.6	3	5.9	−0.1	2.9	−3.1
	16	54.5	62.2	11.7	23.3	−0.4	11.7	−12.6
	15	53.8	72.6	4.7	9.3	−1.4	3.3	−6.1
	14	59.5	70.2	11.4	22.6	4.3	16.7	−7.1
	13	55.4	41.2	5.9	11.8	3.6	9.7	−2.3
	12	incoh.						
	11	incoh.						
	10	incoh.						
	9	−49.4	168.8	8.2	16.3	4.2	12.8	−4
	8	incoh.						
	7	51.6	51.1	5.8	11.5	−2	3.8	−7.9
	6	−54	232.3	5.9	11.8	−0.1	5.9	−6.1
	5	54.4	65.7	4.3	8.6	−0.8	3.6	−5.1
	4	−55	251.1	9.8	19.6	0.2	10.3	−9.9
	3	−46.2	239.9	6.5	12.6	8.4	15.3	2.2
	2	57.4	58.3	2.9	5.7	2.8	5.7	−0.1
	1	51.9	59	4	7.9	−2.8	1.2	−6.8

REFERENCES

Anson, G.L., Kodama, K.P., 1987. Compaction-induced inclination shallowing of the post-depositional remanent magnetization in a synthetic sediment. Geophys. J. R. Astron. Soc. 88, 673–692.

Ash, S.R., 1987. Petrified Forest National Park, Arizona. In: Beus, S.S. (Ed.), CentennialField Guide, Rocky Mountain Section of Geological Society of America, vol. 2. University of Kansas Press, Lawrence, KS, pp. 405–410.

Atchley, S.C., Nordt, L.C., Dworkin, S.I., Ramezani, J., Parker, W.G., Ash, S.R., Bowring, S.A., 2013. A linkage among Pangean tectonism, cyclic alluviation, climate change, and biologic turnover in the Late Triassic: the record from the Chinle Formation, Southwestern United States. J. Sediment. Res. 83, 1147–1161.

Bazard, D.R., Butler, R.F., 1989. Paleomagnetism of the Chinle and Kayenta formations, Arizona and New Mexico; implications for North American Mesozoic apparent polar wander. Geol. Soc. Am. Abstr. Progr. 21, 55.

Bazard, D.R., Butler, R.F., 1991. Paleomagnetism of the Chinle and Kayenta formations, New Mexico and Arizona. J. Geophys. Res. 96, 9847–9871.

Besse, J., Courtillot, V., 2002. Apparent and true polar wander and the geometry of the geomagnetic field over the last 200 Myr. J. Geophys. Res. 107, B11.http://dx.doi.org/1029/2000JB000050.

Billingsley, G.H., 1985a. General stratigraphy of the Petrified Forest National Park, Arizona. In: Colbert, E.H., Johnson, R.R. (Eds.), In: The Petrified Forest Through the Ages: Museum of Northern Arizona Bulletin, vol. 54. pp. 3–8.

Billingsley, G.H., 1985b. Geologic Map of Petrified Forest National Park, Arizona (unpublished data). Petrified Forest Museum Association, Petrified Forest National Park, Arizona. scale 1:48,000.

Blakey, R.C., 1989. Triassic and Jurassic geology of the southern Colorado Plateau. In: Jenney, J.P., Reynolds, S.J. (Eds.), In: Geologic Evolution of Arizona: Arizona Geological Society Digest, vol. 17. pp. 369–396.

Blakey, R.C., Gubitosa, R., 1983. Late Triassic paleogeography and depositional history of the Chinle Formation, southeastern Utah and northern Arizona. In: Reynolds, M.W., Dolly, E.D. (Eds.), Mesozoic Paleogeography of the West-Central United States.pp. 57–76.

Cooley, M.E., 1957. Geology of the Chinle Formation in the Upper Little Colorado drainage Area, Arizona and New Mexico (M.S. thesis). University of Arizona, Tuscon, AZ. 317 p.

Deamer, G.A., Kodama, K.P., 1990. Compaction-induced inclination shallowing in synthetic and natural clay-rich sediments. J. Geophys. Res. 95, 4511–4529.

Dickinson, W.R., Gehrels, G.E., Stern, R.J., 2010. Late Triassic Texas uplift preceding Jurassic opening of the Gulf of Mexico: evidence from U-Pb ages of detrital zircons. Geosphere 6, 641–662.

Dubiel, R.F., 1987. Sedimentology of the Upper Triassic Chinle Formation, Southeastern Utah (Ph.D. dissertation). University of Colorado, Boulder, CO. 146 p.

Dubiel, R.F., 1989a. Depositional environments of the Upper Triassic Chinle Formation in the eastern San Juan Basin and vicinity. U.S. Geological Survey Bulletin, Albuquerque, NM. 1801B, 22 p.

Dubiel, R.F., 1989b. Depositional and climatic setting of the Upper Triassic Chinle Formation, Colorado Plateau. In: Lucas, S.G., Hunt, A.P. (Eds.), Dawn of the Age of the Dinosaurs in the American Southwest. pp. 171–187.

Dubiel, R.F., 1994. Triassic deposystems, paleogeography, and paleoclimate of the western interior. In: Caputo, M.V., Peterson, J.A., Franczyk, K.J. (Eds.), Mesozoic Systems of the Rocky Mountain Region. SEPM Rocky Mountain Section, Denver, CO, pp. 133–168.

Fisher, R.A., 1953. Dispersion on a sphere. Proc. R. Soc. Lond. A217, 295–305.

Golonka, J., 2007. Late Triassic and Early Jurassic paleogeography of the world. Palaeogeogr. Palaeoclimatol. Palaeoecol. 244, 297–307.

Heckert, A.B., Lucas, S.G., 1996. Stratigraphic description of the Tr-4 unconformity in west-central New Mexico and eastern Arizona. N. M. Geol. 18 (3), 61–70.

Heckert, A.B., Lucas, S.G., 2002. Revised upper Triassic stratigraphy of the Petrified Forest National Park, Arizona (USA). In: Heckert, A.B., Lucas, S.G. (Eds.), UpperTriassic Stratigraphy and Paleontology: New Mexico Museum of Natural History and Science Bulletin, vol. 21, pp. 1–23.

Heckert, A.B., Lucas, S.G., Dickinson, W.R., Mortensen, J.K., 2009. New ID-TIMS U-Pb ages for Chinle Group strata (Upper Triassic) in New Mexico and Arizona, correlation to the Newark Supergroup, and implications for the "long Norian". Geol. Soc. Am. Abstr. Progr. 41, 123.

Hounslow, M.W., Muttoni, G., 2010. The geomagnetic polarity timescale for the Triassic: linkage to stage boundary definitions. Geol. Soc. Lond., Spec. Publ. 334, 61–102.

Howell, E.R., Blakey, R.C., 2013. Sedimentological constraints on the evolution of the Cordilleran arc: New insights from the Sonsela Member, Upper Triassic Chinle Formation, Petrified Forest National Park (Arizona, USA). Geol. Soc. Am. Bull. 125, 1349–1368. http://dx.doi.org/10.1130/B30714.1.

Hunt, A.P., Lucas, S.G., 1993a. Triassic vertebrate paleontology and biochronology of New Mexico. New Mexico Museum of Natural History Bulletin, vol. 2, pp. 49–60.

Hunt, A.P., Lucas, S.G., 1993b. Stratigraphy and vertebrate paleontology of the Chinle Group (Upper Triassic), Chama Basin, north-central New Mexico. New Mexico Museum of Natural History Bulletin, vol. 2. pp. 61–69.

Hunt, A.P., Lucas, S.G., 1995. Two Late Triassic vertebrate faunas at Petrified Forest National Park. In: Santucci, V.L., McClelland, L., Santucci, V.L., McClelland, L. (Eds.), National Park Service Paleontological Research Technical Report NPS/NRPO/NRTR-95/16, pp. 89–93.

Irmis, R., Mundil, R., 2008. New age constraints from the Chinle Formation revise global comparisons of Late Triassic vertebrate assemblages. J. Vertebr. Paleontol. 28 (Suppl. 3), 95A.

Irmis, R.B., Nesbitt, S.J., Padian, K., Smith, N.D., Turner, A.H., Woody, D., Downs, A., 2007. A Late Triassic dinosauromorph assemblage from New Mexico and the rise of dinosaurs. Science 317, 358–361.

Irmis, R.B., Mundil, R., Martz, J.W., Parker, W.G., 2011. High-resolution U-Pb ages from the Upper Triassic Chinle Formation (New Mexico, USA) support a diachronous rise of dinosaurs. Earth Planet. Sci. Lett. 309, 258–267.

Johns, M.E., 1988. Architectural Element Analysis and Depositional History of the Upper Petrified Forest Member of the Chinle Formation, Petrified Forest National Park, Arizona. (M.S. thesis). Northern Arizona University, Flagstaff, AZ. 163 p.

Johnson, N.M., Opdyke, N.D., Lindsay, E.H., 1975. Magnetic polarity stratigraphy of Pliocene-Pleistocene terrestrial deposits and vertebrate faunas, San Pedro Valley, Arizona. Geol. Soc. Am. Bull. 86, 5–12.

Kent, D.V., Olsen, P.E., 1999. Astronomically tuned geomagnetic polarity time scale for the Late Triassic. J. Geophys. Res. 104, 12,831–12,841.

Kent, D.V., Olsen, P.E., 2000. Magnetic polarity stratigraphy and paleolatitude of the Triassic-Jurassic Blomidon Formation in the Fundy basin (Canada): implications for early Mesozoic tropical climate gradients. Earth Planet. Sci. Lett. 179, 311–324.

Kent, D.V., Irving, E., 2010. Influence of inclination error in sedimentary rocks on the Triassic and Jurassic apparent polar wander path for North America and implications for Cordilleran tectonics. J. Geophys. Res. 115, B10103. http://dx.doi.org/10.1029/2009JB007205.

Kent, D.V., Tauxe, L., 2005. Corrected Late Triassic latitudes for continents adjacent to the North Atlantic. Science 307, 240–244.

Kirschvink, J.L., 1980. The least squares line and plane and the analysis of paleomagnetic data. Geophys. J. R. Astron. Soc. 62, 699–718.

Litwin, R.J., 1986. The Palynostratigraphy and Age of the Chinle and Moenave Formations, Southwestern USA (Ph.D. dissertation). The Pennsylvania State University, College Park, MD. 265 p.

Larochelle, A., Currie, K.L., 1967. Paleomagnetic study of igneous rocks from the Manicouagan structure, Quebec. J. Geophys. Res. 72, 4163–4169.

Litwin, R.J., Traverse, A., Ash, S.R., 1991. Preliminary palynological zonation of the Chinle Formation, southwestern USA, and its correlation to the Newark Supergroup (eastern USA). Rev. Palaeobot. Palynol. 68, 269–287.

Long, R.A., Ballew, K.L., 1985. Aetosaur dermal armor from the Late Triassic of southwestern North America, with special reference to the Chinle Formation of Petrified Forest National Park. Mus. North. Ariz. Bull. 54, 45–68.

Lowrie, W., 1990. Identification of ferromagnetic minerals in a rock by coercivity and unblocking temperature properties. Geophys. Res. Lett. 17, 159–162.

Lucas, S.G., 1991. Sequence stratigraphic correlation of nonmarine Late Triassic biochronologies, western United States. Albertiana 4, 11–18.

Lucas, S.G., 1993. The Chinle Group: revised stratigraphy and biochronology of Upper Triassic nonmarine strata in the western United States. In: Morales, M. (Ed.), In: Aspects of Mesozoic Geology and Paleontology of the Colorado Plateau: Museum of Northern Arizona Bulletin, vol. 59, pp. 27–50.

Lucas, S.G., Hunt, A.P., 1992. Triassic stratigraphy and paleontology, Chama Basin and adjacent areas, north-central New Mexico. New Mexico Geological Society Guidebook, vol. 43. pp. 151-167.

Lucas, S.G., Zeigler, K.E., Heckert, A.B., Hunt, A.P., 2003. Upper Triassic stratigraphy and biostratigraphy, Chama Basin, north-central New Mexico. New Mexico Museum of Natural History and Science Bulletin, vol. 24, pp. 15–39.

Lucas, S.G., Zeigler, K.E., Heckert, A.B., Hunt, A.P., 2005. Review of Upper Triassic stratigraphy and biostratigraphy in the Chama Basin, northern New Mexico. New Mexico Geological Society Guidebook, vol. 56, pp. 170–181.

Martz, J.W., Parker, W.G., 2010. Revised lithostratigraphy of the Sonsela Member (Chinle Formation, Upper Triassic) in the southern part of Petrified Forest National Park, Arizona. PLoS ONE. 5.e9329. http://dx.doi.org/10.1371/journal.pone.0009329.

Martz, J.W., Parker, W.G., Skinner, L., Raucci, J.J., Umhoefer, P., Blakey, R.C., 2012. Geologic map of Petrified Forest National Park, Arizona. Arizona Geological Survey Contributed Map CM-12-A.

McFadden, P.L., McElhinny, M.W., 1990. Classification of the reversal test in paleomagnetism. Geophys. J. Int. 103, 725–729.

Molina-Garza, R.S., Geissman, J.W., Van der Voo, R., Lucas, S.G., Hayden, S.N., 1991. Paleomagnetism of the Moenkopi and Chinle formations in central New Mexico: implications for the North American apparent polar wander path and Triassic magnetostratigraphy. J. Geophys. Res. 96, 14239–14262.

Molina-Garza, R.S., Geissman, J.W., Lucas, S.G., 1993. Late Carnian-early Norian magnetostratigraphy from nonmarine strata, Chinle Group, New Mexico. New Mexico Museum of Natural History, Bulletin, vol. 3, pp. 345–352.

Molina-Garza, R.S., Geissman, J.W., Van der Voo, R., 1995. Paleomagnetism of the Dockum Group (Upper Triassic), northwest Texas: further evidence for the J-1 cusp in the North America apparent polar wander path and implications for rate of Triassic apparent polar wander and Colorado Plateau rotation. Tectonics 14, 979–993.

Molina-Garza, R.S., Geissman, J.W., Lucas, S.G., Van der Voo, R., 1996. Paleomagnetism and magnetostratigraphy of Triassic strata in the Sangre de Cristo Mountains and Tucumcari Basin, New Mexico, USA. Geophys. J. Int. 124, 935–953.

Molina-Garza, R.S., Acton, G.D., Geissman, J.W., 1998a. Carboniferous through Jurassic paleomagnetic data and their bearing on rotation of the Colorado Plateau. J. Geophys. Res. 103, 24,179–24,188.

Molina-Garza, R.S., Geissman, J.W., Gomez, A., Horton, B., 1998b. Paleomagnetic data from Triassic strata, Zuni uplift, New Mexico: further evidence of large-magnitude Triassic apparent polar wander of North America. J. Geophys. Res. 103, 24,189–24,200.

Molina-Garza, R.S., Geissman, J.W., Lucas, S.G., 2003. Paleomagnetism and magnetostratigraphy of the lower Glen Canyon and upper Chinle Groups, Jurassic-Triassic of northern Arizona and northeast Utah. J. Geophys. Res. 108, 24. http://dx.doi.org/10.1029/2002JB001909.

Murry, P.A., 1990. Stratigraphy of the Upper Triassic Petrified Forest Member (Chinle Formation) in Petrified Forest National Park, USA. J. Geol. 98, 780–789.

Nordt, L., Atchley, S., Dworkin, S., 2015. Collapse of the Late Triassic megamonsoon in western equatorial Pangea, present-day American Southwest. Geol. Soc. Am. Bull. 127 (11/12), 1798–1815.

Olsen, P.E., Kent, D.V., 1999. Long-period Milankovitch cycles from the Late Triassic and Early Jurassic of eastern North America and their implications for the calibration of the Early Mesozoic time-scale and the long-term behavior of the planets. Philos. Trans. R. Soc. Lond. 357, 1761–1786.

Olsen, P.E., Kent, D.V., Cornet, B., Witte, W.K., Schlische, R., 1996. High resolution stratigraphy of the Newark rift basin (Early Mesozoic, Eastern North America). Geol. Soc. Am. Bull. 108, 40–77.

Olsen, P.E., Koeberl, C., Huber, H., Montanaria, A., Fowell, S.J., Et-Touhami, M., Kent, D.V., 2002. Continental Triassic-Jurassic boundary in central Pangea: recent progress and discussion of an Ir anomaly. In: Koeberl, C., MacLeod, K.G. (Eds.), Catastrophic Events and Mass Extinctions: Impacts and Beyond. Geological Society of America, New York, NY, pp. 505–522.

Olsen, P.E., Kent, D.V., Whiteside, J.H., 2011. Implications for the Newark Supergroup-based astrochronology and geomagnetic polarity time scale (Newark-APTS) for the tempo and mode of the early diversification of the Dinosauria. Earth Environ. Sci. Trans. R. Soc. Edinb. 101, 201–229.

Parker, W.G., 2006. The stratigraphic distribution of major fossil localities in Petrified Forest National Park, Arizona. In: Parker, W.G., Ash, S.R., Irmis, R.B. (Eds.), A Century of Research at Petrified Forest National Park 1906-2006.In: Museum of Northern Arizona Bulletin, vol. 62. pp. 46–62.

Parker, W.G., Martz, J.W., 2011. The Late Triassic (Norian) Adamanian-Revueltian tetrapod faunal transition in the Chinle Formation of Petrified Forest National Park, Arizona. Earth Environ. Sci. Trans. R. Soc. Edinb. 101, 231–260. http://dx.doi.org/10.1017/S1755691011020020.

Peppe, D.J., Evans, D.A.D., Smirnov, A.V., 2009. Magnetostratigraphy of the Ludlow Member of the Fort Union Formation (Lower Paleocene) in the Williston Basin, North Dakota. Geol. Soc. Am. Bull. 121 (1–2), 65–79.

Ramezani, J., Bowring, S.A., Pringle, M.S., Winslow III, F.D., Rasbury, E.T., 2005. The Manicouagan impact melt rock: a proposed standard for the intercalibration of U-Pb and 40Ar/39Ar isotopic systems. Geochim. Cosmochim. Acta Suppl. 69, 321.

Ramezani, J., Hoke, G.D., Fastovsky, D.E., Bowring, S.A., Therrien, F., Dworkin, S.I., Atchley, S.C., Nordt, L.C., 2011. High-precision U-Pb zircon geochronology of the Late Triassic Chinle Formation, Petrified Forest National Park (Arizona, USA): temporal constraints on the early evolution of dinosaurs. Geol. Soc. Am. 123, 2142–2159.

Reeve, S.C., 1975. Paleomagnetic studies of sedimentary rocks of Cambrian and Triassic age (Ph.D. dissertation). University of Texas, Dallas,TX.

Reeve, S.C., Helsley, C.E., 1972. Magnetic reversal sequence of the upper portion of the Chinle Formation, Montoya, New Mexico. Geol. Soc. Am. Bull. 83, 3795–3812.

Riggs, N.R., Ash, S.R., Barth, A.P., Gehrels, G.E., Wooden, J.L., 2003. Isotopic age of the Black Forest Bed, Petrified Forest Member, Chinle Formation, Arizona: an example of dating a continental sandstone. Geol. Soc. Am. Bull. 115, 1315–1323.

Riggs, N.R., Oberling, Z.A., Howell, E.R., Parker, W.G., Barth, A.P., Cecil, M.R., Martz, J.W., 2016. Paleotopography and evolution of the Early Mesozoic Cordilleran margin as reflected in the detrital zircon signature of the basal Upper Triassic Chinle Formation, Colorado Plateau. Geosphere 12, 439–463.

Riggs, N.R., Reynolds, S.J., Lindner, P.J., Howell, E.R., Barth, A.P., Parker, W.G., et al., 2013. The early Mesozoic Cordilleran arc and late triassic paleotopography: the detrital record in upper triassic sedimentary successions on and off the Colorado Plateau. Geosphere 9 (3), 602–613.

Rinehart, L.F., Lucas, S.G., Heckert, A.B., Spielmann, J.A., Celeskey, M.D., 2009. The paleobiology of *Coelophysis bauri* (Cope) from the Upper Triassic (Apachean) Whitaker quarry, New Mexico, with detailed analysis of a single quarry block. New Mexico Museum of Natural History and Science Bulletin, vol. 45. 260 p.

Roadifer, J.E., 1966. Stratigraphy of the Petrified Forest National Park, Arizona (Ph.D. dissertation). University of Arizona, Tuscon, AZ. 152 p.

Robertson, W.A., 1967. Manicouagan, Quebec, paleomagnetic results. Can. J. Earth Sci. 4, 641–649.

Scotese, C.R., 1994. Late Triassic paleogeographic map. In: Klein, G.D. (Ed.), Pangea Paleoclimate, Tectonics, and Sedimentation during Accretion, Zenith, and Breakup of a Supercontinent. Geological Society of America Special Paper, vol. 288. p. 7.

Steiner, M.B., Lucas, S.G., 2000. Paleomagnetism of the Late Triassic Petrified Forest Formation, Chinle Group, western United States: Further evidence of "large" rotation of the Colorado Plateau. J. Geophys. Res. Solid Earth 105, 25,791–25,808.

Stewart, J.H., Poole, F.G., Wilson, R.F., 1972. Stratigraphy and origin of the Chinle Formation and related Upper Triassic strata of the Colorado Plateau region. U.S. Geological Survey Professional Paper 690, 336 p.

Tan, X., Kodama, K.P., Fang, D., 2002. Laboratory depositional and compaction-induced inclination errors carried by hematite and their implications in identifying inclination error of natural remanence in red beds. Geophys. J. Int. 151, 475–486.

Tan, X., Kodama, K.P., Gilder, S., Courtillot, V., 2007. Rock magnetic evidence for inclination shallowing in the Passaic Formation red beds from the Newark basin and a systematic bias of the Late Triassic apparent polar wander path for North America. Earth Planet. Sci. Lett. 354, 345–357.

Trendell, A.M., Atchley, S.C., Nordt, L.C., 2013. Facies analysis of probably large fluvial-fan depositional system: the Upper Triassic Chinle Formation at Petrified Forest National Park, Arizona, U.S.A. J. Sediment. Res. 83, 873–895. http://dx.doi.org/10.2110/jsr.2013.55.

Van der Voo, R., 1993. Paleomagnetism of the Atlantic, Tethys and Iapetus Oceans. Cambridge University Press, New York, NY. 411 p.

Weissmann, G.S., Hartley, A.J., Nichols, G., 2007. A new view of fluvial facies models—aggradational distributary networks. Geol. Soc. Am. Abstr. Progr. 39, 629.

Woody, D.T., 2006. Revised stratigraphy of the Lower Chinle Formation (Upper Triassic) of Petrified Forest National Park, Arizona. In: Parker, W.G., Ash, S.R., Irmis, R.B. (Eds.), A Century of Research at Petrified Forest National Park 1906-2006. Museum of Northern Arizona Bulletin, vol. 62. pp. 17–45.

Zeigler, K.E., 2008. Stratigraphy, paleomagnetism and magnetostratigraphy of the Upper Triassic Chinle Group, north-central New Mexico and preliminary magnetostratigraphy of the Lower Cretaceous Cedar Mountain Formation, eastern Utah (Ph.D. dissertation). University of New Mexico, Albuquerque, NM 224 p.

Zeigler, K.E., Lucas, S.G., Spielmann, J.A., Morgan, V., 2005. Vertebrate biostratigraphy of the Upper Triassic Salitral Formation, Chinle Group, north-central New Mexico. New Mexico Geological Society Guidebook, vol. 56, pp. 20–21.

Zeigler, K.E., Kelley, S.A., Geissman, J.W., 2008. Revisions to stratigraphic nomenclature of the Upper Triassic Chinle Group in New Mexico: new insights from geologic mapping, sedimentology, and magnetostratigraphic/paleomagnetic data. Rocky Mt. Geol. 43, 121–141.

Zeigler, K.E., Geissman, J.W., 2011. Magnetostratigraphy of the Upper Triassic Chinle Group of New Mexico: Implication for regional and global correlations among Upper Triassic sequences. Geosphere 7, 802–829.

Ziegler, A.M., Scotese, C.R., Barren, S.F., 1983. Mesozoic and Cenozoic paleogeographic maps. In: Sundermann, J., Brosche, P. (Eds.), Tidal Friction and the Earth's Rotation II. Springer-Verlag, Berlin, pp. 240–252.

Zijderveld, J.D.A., 1967. A.C. demagnetization of rocks: analysis of results. In: Collinson, D.W., Creer, K.M., Runcorn, S.K., Collinson, D.W., Creer, K.M., Runcorn, S.K., Collinson, D.W., Creer, K.M., Runcorn, S.K. (Eds.), Methods of Paleomagnetism. Elsevier, Amsterdam, pp. 254–286.

Kate Zeigler, PhD, CPG, is owner and senior geologist at Zeigler Geologic Consulting, LLC, where she works on a wide variety of projects, including geologic mapping, aquifer mapping, stratigraphic analyses, and paleontology resource management. She also continues to pursue the possibility of correlating Late Triassic strata in the American Southwest via magnetic polarity stratigraphy data in her spare time.

Magnetostratigraphy of the Upper Jurassic Morrison Formation at Dinosaur National Monument, Utah, and Prospects for Using Magnetostratigraphy as a Correlative Tool in the Morrison Formation

S.C.R. Maidment*, D. Balikova[†], A.R. Muxworthy[†]
*University of Brighton, Brighton, United Kingdom
[†]Imperial College London, London, United Kingdom

Contents

Terrestrial Depositional Systems
http://dx.doi.org/10.1016/B978-0-12-803243-5.00007-8

INTRODUCTION

The Upper Jurassic Morrison Formation of the Western Interior, USA, has been the subject of intensive research since the discovery of its diverse and well-preserved dinosaurian fauna in the latter part of the 19th century. Around 30 dinosaurian genera, including representatives of all major dinosaurian clades, are currently recognized from the Morrison Formation (Weishampel et al., 2004). Outcropping from Montana in the north to New Mexico in the south, the Morrison Formation covers over 12 degrees of latitude and, although the age of the formation has been difficult to definitively constrain, it appears to have been deposited over a time period of around 7 million years (Kowallis et al., 1998; Steiner, 1998; Litwin et al., 1998; Schudack et al., 1998; Trujillo and Kowallis, 2015). The formation comprises a series of fluvial, lacustrine, and floodplain sediments and, as is common in terrestrial environments, correlation on a region scale has proven incredibly difficult. Palynology (Litwin et al., 1998; Schudack et al., 1998), clay mineralogy (Peterson and Turner, 1998), paleopedology (Demko et al., 2004), and vertebrate biostratigraphy (Carpenter, 1998; Turner and Peterson, 1999; Bakker et al., 1990) have all been used in various attempts to correlate Morrison outcrops on the Colorado Plateau with outcrops in the Front Ranges of Colorado, the San Juan Basin of New Mexico and the Bighorn Basin of Wyoming. These methods, while apparently useful on a local scale, have not succeeded in correlating regionally and, to date, no robust long-range correlations for the Morrison Formation exist.

Poor correlation of the Morrison Formation on a regional scale has significantly hampered attempts to understand the dinosaurian fauna and its paleoecology. Stratigraphic relationships between dinosaur quarries remain unknown, meaning that the ages of dinosaurian taxa relative to each other are obscured. This has resulted in the dinosaurs of the Morrison being characterized as a single fauna (Foster, 2000), despite their broad geographic range and long temporal duration, a situation that appears extremely unlikely when compared with modern ecosystems. Regional correlation of Morrison outcrops would allow the ages of various dinosaur localities to be determined relative to each other. The Morrison Formation then has the potential to be an unparalleled case study of environmental and latitudinal biodiversity in the Late Jurassic world. Examination of latitudinal diversity patterns at a time in Earth's history when latitudinal climatic gradients are thought to have been reduced relative to today (Bender, 2013) is significant because it impacts our understanding of the causes of the modern latitudinal biodiversity gradient. The causative mechanisms behind this ubiquitous biodiversity pattern are currently unknown (Gaston, 2000; Willig et al., 2003; Jablonski et al., 2006; Mittelbach et al., 2007; Valentine et al., 2008), hampering attempts to predict biodiversity changes in response to modern day climate change.

In this study, we sampled two Morrison Formation sites at Dinosaur National Monument in the Uinta Basin, Utah, and analysed their magnetic polarity. The aims of this work are to (1) use magnetostratigraphy to correlate this well-dated, heavily studied area with previous magnetostratigraphic studies in Colorado, Wyoming, and New Mexico, and (2) to assess the potential for magnetostratigraphy to be used as a correlative tool across the Morrison Formation.

THE MORRISON FORMATION

The terrestrial sediments of the Morrison Formation were deposited in a broad, shallow, north-south trending basin subsequent to the regression of the Sundance Sea (Peterson, 1994). The Morrison depositional basin appears to have been continuously, although gently, subsiding throughout much of the Kimmeridgian and Tithonian (DeCelles and Currie, 1996), yet the causes of this subsidence, and the tectonic setting of the basin, remain the subject of some debate. Dynamic subsidence due to viscous coupling between the mantle wedge above a subducting oceanic plate along a continental margin to the west and the overlying North American craton has been invoked (Lawton, 1994). Most recent works, however, envisage the Morrison basin to be the back-bulge depozone of a retroarc foreland basin (DeCelles and Burden, 1992; Royse, 1993; DeCelles and Currie, 1996; Currie, 1998; Fuentes et al., 2009), with a contribution from dynamic subsidence (Roca and Nadon, 2007). The foreland basin itself has since been uplifted and eroded, a situation considered plausible as thrust fault restoration suggests up to 6 km of erosion may have occurred during the Early Cretaceous Sevier orogeny (Royse, 1993). The timing of development of Cordilleran orogenesis is controversial: while earlier workers considered the Early Cretaceous Sevier orogeny marked the onset of mountain-building in the Western Interior region (Heller et al., 1986; Yingling and Heller, 1992), there is an increasing body of evidence for a much earlier onset, perhaps as early as the Middle Jurassic (DeCelles and Burden, 1992; DeCelles and Currie, 1996; Cooley and Schmidt, 1998; Currie, 1998; Fuentes et al., 2009). This evidence includes regional metamorphism, and rapid exhumation and movement on thrusts (Fuentes et al., 2009), and it is this Upper Jurassic Cordilleran orogenic phase, which resulted in the flexural basin into which the Morrison deposits accumulated.

The Morrison Formation overlies the Middle Jurassic marine deposits of the Sundance Sea in the north, while to the south, it overlies sabkha and aeolian dune facies deposited on the Sea's arid shores. There is debatable evidence for the presence of an unconformity, the so-called "J5," in some areas at the base of the formation (Pipringos and O'Sullivan, 1978; Peterson and Turner, 1998; Anderson and Lucas, 1998; Bilbey, 1998; Turner and Peterson, 1999). Overlying the Morrison are the terrestrial foreland basin deposits of the Sevier orogeny (DeCelles and Currie, 1996; Currie, 1998). Although these deposits are known by different names in different geographic locations (Kootenai Formation in Montana; Cloverly and Lakota Formations in Wyoming; Cedar Mountain Formation in Utah; Burro Canyon Formation in Colorado; Lakota Formation in South Dakota), they comprise gray-purple mudstones and fluvial sandstones and possess widespread coarse conglomerates at their base (Pryor Conglomerate, Buckhorn Conglomerate, and Kootenai

Conglomerate). The base of the coarse conglomerates is usually considered an unconformity, the "K1" (Pipringos and O'Sullivan, 1978; Peterson and Turner, 1998; Turner and Peterson, 1999; Greenhalgh and Britt, 2007; Kirkland and Madsen, 2007), although some workers place the "K1" above the conglomerates and consider them to represent the last stages of Morrison deposition (Aubrey, 1998; Roca and Nadon, 2007). In areas where a coarse conglomerate is absent, the top of the Morrison Formation can be difficult to differentiate from overlying formations, but is often characterized by a change from reddish mudstones below to purple-gray mudstones above, a change that is clear from a distance but difficult to identify in individual sections (Ostrom, 1970; Greenhalgh and Britt, 2007; Kirkland and Madsen, 2007; Doelling and Kuehne, 2013; Sprinkel et al., 2012).

The Morrison Formation itself was deposited by rivers, in shallow lakes and on floodplains in a semiarid to seasonally wet environment (Ash and Tidwell, 1998; Demko and Parrish, 1998; Dunagan and Turner, 2004; Good, 2004; Hasiotis, 2004). Across its outcrop area, the formation has been divided into numerous members which in general can only be identified locally (Peterson, 1994). This is particularly true of the Morrison in southern Utah and New Mexico, where the formation hosts vast uranium–vanadium deposits and where it was mined for these minerals from the 1940s to the 1980s (Chenoweth, 1998). In this area, a proliferation of units, formal and informal, (e.g., the Poison Canyon Sandstone, the "K" shale, the Westwater Canyon) probably relate to facies, rather than distinct horizons (Turner-Peterson, 1986), or are synonymous with previously named members (Anderson and Lucas, 1998) and are thus not useful for correlation. Throughout the Colorado Plateau, however, the Morrison is readily divided into two mappable members: the Salt Wash Member below and the Brushy Basin Member above. The Salt Wash is dominated by the deposits of braided and anastomosing rivers with a source area to the southwest (Tyler and Ethridge, 1983; Turner-Peterson, 1986; Godin, 1991; Kjemperud et al., 2008; Owen et al., 2015), whereas the Brushy Basin is dominated by floodplain and lacustrine mudstones and sands deposited by low sinuousity, anastomosing rivers (Demko et al., 2004; Kirkland, 2006; Galli, 2014). To the east and north of the Colorado Plateau, further from the source area of the Salt Wash, the two members are no longer recognizable. The Tidwell Member is also recognized in parts of Utah and Colorado. It comprises red-brown mudstones, thin sands, and gypsum, and was deposited in sabkha and shallow marine environments (Anderson and Lucas, 1998) as the Sundance Sea retreated. Similarly, previous workers have recognized the shallow marine Windy Hill Member at the base of the Morrison Formation in northern Utah and Wyoming (Peterson, 1994; Peterson and Turner, 1998).

Various attempts have been made to identify correlatable horizons across the Morrison Formation. Peterson and Turner (1998) identified a horizon they termed the "clay change," purported to be a change from illitic, nonswelling clays below to smectitic, swelling clays above, and mappable in the field due to the tendency for swelling clays to create a distinct "popcorn" texture on unvegetated, weathered slopes. Peterson and Turner (1998) considered this a time-correlative horizon as it indicated a time when the ash influx into the Morrison basin suddenly increased. X-ray diffraction analysis of clay minerals in the Morrison Formation revealed, however, that there were both illitic and smectitic clays throughout the entirety of the sections sampled, and that no such distinct transition existed (Jennings and Hasiotis, 2006; Trujillo, 2006). The mappable "clay change" horizon was thus interpreted as a weathering product, perhaps dependent on slope angle and local environment. Mudstones in the upper part of the Morrison Formation in Utah and western Colorado do appear to contain more clays that weather to a "popcorn" texture than mudstones in the lower part of the formation, and thus the "clay change" might be useful for correlation at a local scale, particularly across the Colorado Plateau and in southernmost Wyoming (Jennings and Hasiotis, 2006; Kirkland, 2006; Galli, 2014). However, at most localities it is impossible to place the transition from nonswelling to swelling clays at a distinct horizon (Jennings and Hasiotis, 2006; Trujillo, 2006), and in any case, Turner and Peterson (1999) stated that the "clay change" was not identifiable further north than southern Wyoming, so it cannot be used for correlation on a regional scale.

Paleosols have also been used in attempts to correlate the Morrison. Demko et al. (2004) recognized very well-developed paleosols at the base of the Morrison Formation, in the lower part of the Brushy Basin Member, and at the top of the formation, and suggested that these were sequence-bounding unconformities, representing significant periods of time when the land surface was exposed. However, there are many paleosol horizons in the Morrison, and when logging sections it is difficult to distinguish those identified by Demko et al. (2004) as regional, as opposed to less long-lived, local events. Furthermore, the three unconformities identified by Demko et al. (2004) were not present in every section, confounding attempts to use them for correlation. Kirkland (2006) has hypothesized that the "clay change," the lower Brushy Basin paleosol, and a jump in magnetic pole positions from the Salt Wash to the Brushy Basin (Steiner, 1998) correspond with a regional unconformity within the Morrison Formation.

The biostratigraphy of a range of animal and plant fossils has been used in an attempt to both date and correlate the Morrison. Charophytes, ostracodes (Schudack et al., 1998), conchostrachans (Lucas and Kirkland, 1998), and molluscs (Evanoff et al., 1998) have been used but their lack of precise stratigraphic distribution in the Upper Jurassic of the Western Interior has proven confounding. Dinosaur biostratigraphy has also been attempted (Bakker et al., 1990; Carpenter, 1998; Turner and Peterson, 1999), both locally and regionally. On a local scale, the small number of specimens used makes the results questionable, although stratigraphic correlations between quarries are well supported. Regionally, the lack of independent means of stratigraphic location of dinosaur quarries makes attempts to identify faunal chrons inherently circular.

Thus, alternative methods are needed to correlate the Morrison across its outcrop area. Both sequence stratigraphy (Currie, 1997, 1998) and magnetostratigraphy (Van Fossen and

Kent, 1992; Steiner et al., 1994; Swierc and Johnson, 1996; Steiner, 1998) have been used with some success to correlate sections locally, and these methods therefore appear promising, although broader, regional studies are required.

PALEOMAGNETISM AND MAGNETOSTRATIGRAPHY

One of the main characteristics of the Earth's geomagnetic field is that it can reverse polarity. During normal polarity, the magnetic north pole is located in the geographic northern hemisphere with a positive inclination in the northern hemisphere and negative inclination in the southern hemisphere, but the field can reverse, switching the inclinations. The random occurrences of these reversals give measured polarity sequences a stratigraphic value: Although polarity reversals have a mean duration of ~300,000 years, this number varies from ~20,000 to tens of million years. The polarity reversals are considered to be globally synchronous (Langereis et al., 2010).

The signature of the ancient magnetic field is recorded in rocks during their formation. During rock-forming processes, the magnetic moments of magnetic minerals partially align with the prevailing direction of the magnetic field, and are "locked in" giving a paleomagnetic recording. This recording is referred to as natural remanent magnetization (NRM; Tauxe, 2010). There are several types of NRM; in sedimentary rocks, two dominate: detrital remanent magnetism (DRM) and chemical remanent magnetization (CRM). DRM is acquired during deposition and lithification of magnetic grains within sedimentary rocks. Upon compaction and dewatering, grains are mechanically fixated, preserving the direction of the ambient field (Butler, 1992; Langereis et al., 2010; Tauxe, 2010). CRM refers to the magnetization acquired via chemical changes that result in the formation or alteration of ferromagnetic minerals. A primary NRM can be overprinted by secondary magnetic components and the total NRM often consists of a vector sum of different magnetic elements. Overprinting can be achieved through weathering, thermochemical alteration and the acquisition of viscous remanent magnetization (VRM) via exposure to weak ambient magnetic fields after deposition. These secondary components are "softer" than the primary NRM and can be removed through various demagnetization techniques (see methodology; Butler, 1992; Langereis et al., 2010).

The geomagnetic polarity timescale (GPTS) is a compilation of the marine magnetic anomaly and magnetostratigraphic records, and is widely used for correlation of isochronous events globally (Langereis et al., 2010; Tauxe, 2010).

PREVIOUS MAGNETOSTRATIGRAPHIC STUDIES OF THE MORRISON FORMATION

A number of studies have attempted to use magnetostratigraphy to date and correlate the Morrison Formation. Steiner and Helsley (1975), Steiner et al. (1994), and Steiner (1998) investigated the magnetostratigraphy of five sections in Colorado and New Mexico, and were able to correlate between sections locally. The composite Morrison section produced from their results recorded 16 polarity reversals. The Morrison Formation in the area studied by Steiner and Helsley (1975), Steiner et al. (1994), and Steiner (1998) is readily divided into the lower, sandstone-dominated Salt Wash Member, and the upper, mudstone-dominated Brushy Basin Member. The Salt Wash Member was found to be deposited during time of equal lengths of normal and reversed polarity, while the Brushy Basin member displayed predominantly reversed polarity with thin normal intervals. The composite Morrison section was interpreted to represent magnetochrons 22–27 (CM22-27), corresponding with an age of 156.05–147.06 million years (Gee and Kent, 2007), and indicating deposition through much of the Kimmeridgian and Tithonian.

Swierc and Johnson (1996) examined the magnetostratigraphy of three sections of the Morrison Formation in the Bighorn Basin, Wyoming, near the historic Howe Dinosaur Quarry. Reporting preliminary results, the study identified nine magnetic reversals, which were correlated with CM22-19, corresponding with an age of 150.04–141.63 million years (Gee and Kent, 2007). However, the data presented are at odds with the interpretation: Data presented for the CW and SS sites (Swierc and Johnson, 1996: Figs. 4–6) indicate positive inclinations and thus normal polarity throughout the sampled sections, in contrast with the interpreted series of reversals, which seem to have been based on magnetic declination, rather than inclination. Taken at face value, the results of this study indicate a younger age for the Morrison in the Bighorn Basin than further south in Colorado and New Mexico, raising the possibility that Morrison Formation deposition continued into the Berriasian stage of the Lower Cretaceous. However, questions clearly remain regarding how interpretations were made.

These studies demonstrate that magnetostratigraphy has the potential to be used for correlative purposes between sections locally within the Morrison Formation. Herein, we apply similar techniques to sections in the vicinity of Dinosaur National Monument, further north than the studies of Steiner and colleagues (Steiner and Helsley, 1975; Steiner et al., 1994; Steiner, 1998) in Colorado and New Mexico, but further south than the Bighorn Basin study of Swierc and Johnson (1996).

METHODS
Sections Sampled

Two sections were logged and sampled in the vicinity of Dinosaur National Monument, Uintah Basin, Utah. The Dinosaur Quarry section (DQ) extended along a southerly trending drainage northwest of the quarry building (40°26.524′N, 109°18.205′W) to the quarry building access road, and then moved along strike to the trail to the east of the quarry building (40°26.402′N, 109°18.052′W), before moving back to the drainage, and finishing behind the quarry building. The US40 section was measured in a drainage that extended south from US Highway 40 (40°19.436′N, 109°15.783sW) about 32.5 km southeast of Vernal, and 14 km south–southeast of the DQ section (Fig. 1). Exposure at

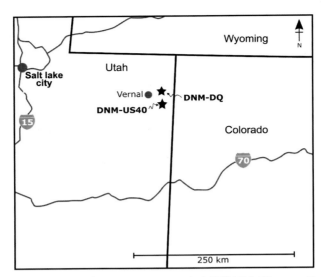

Fig. 1 Map showing the location of the sections sampled. *DNM-DQ*, Dinosaur Quarry section; *DNM-US40*, US40 section.

both locations was generally excellent. Sections were cored using a rock drill with a water-cooled diamond bit. Although continuous sampling through the sections was desired, a significant number of beds were poorly consolidated and therefore undrillable. As a result, well-consolidated sandstone beds were targeted, leaving our samples widely spaced (Fig. 2). Azimuth and plunge of drilling direction were recorded for each core, using a sun and/or magnetic compass and an inclinometer. The cores were then orientated by marking the drilling direction on them with a brass wire. This method can yield an error of ±3°, which is reduced by collecting a high number of samples (Tauxe, 2010); at least three cores were collected per bed where possible. At DQ, seven beds were sampled, including the quarry sandstone. A total of 45 cores were collected. Fifteen beds were sampled at US40, including in the underlying Stump Formation. A total of 66 cores were obtained. Large sampling gaps were present in unconsolidated sandstones and in thick mudstone sections.

Demagnetization Techniques

The collected cores were cut into 1-in. specimens, some of which were fragmented and reassembled using an adhesive. A JR5A spinner magnetometer with a noise level of $\sim 5 \times 10^{-10}$ A m^2 at Imperial College London was used to measure all of the specimens. As in previous studies (Van Fossen and Kent, 1992; Steiner et al., 1994), alternating field demagnetization was found to be ineffective in separating magnetic components and fully demagnetizing the specimens; thus thermal demagnetization was carried out in

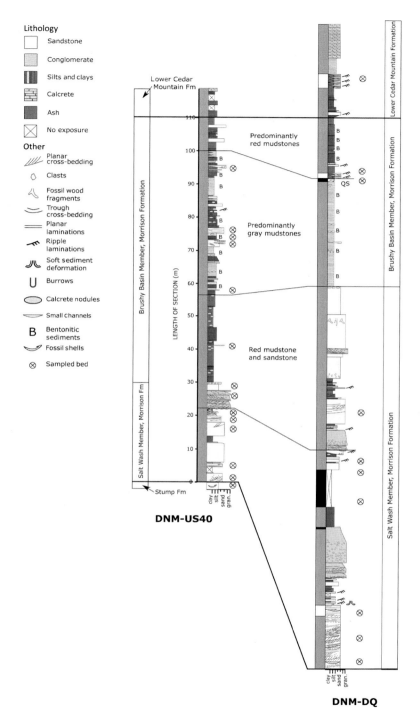

Fig. 2 Logged sections showing sedimentology, lithostratigraphy, sampled horizons, and horizons with accepted magnetostratigraphic results. *DNM-DQ*, Dinosaur Quarry section; *DNM-US40*, US40 section; *Fm*, Formation; *gran.*, granules; *QS*, quarry sand equivalent level.

controlled-field paleomagnetic ovens at Imperial College London. Specimens were heated in 10 steps between 100°C and 690°C. High temperatures of >600°C were chosen due to the likely presence of hematite (Van Fossen and Kent, 1992).

Data Analysis

Thermal demagnetization data were analyzed using Puffin Plot (Lurcock and Wilson, 2012). Specimens were corrected for tectonic tilt and regional modern geomagnetic field. Magnetization directions were identified by hand picking points with steady directions using Zijderveld plots (Zijderveld, 1967), followed by principal component analysis, which generates a best-fit line through single component of data (Kirschvink, 1980; Tauxe, 2010). Fisher statistics were employed to calculate the mean direction and a confidence limit of each individual bed, giving rise to either negative or positive inclinations (Butler, 1992). Virtual geomagnetic poles (VGPs) were calculated from the mean directions in order to assess the reliability of the paleomagnetic data obtained.

RESULTS
Sedimentology

The base of the Morrison Formation at DQ rests directly on top of a light gray, well-sorted sandstone with symmetrical ripples and disarticulated fossil shells, indicative of the marine Stump Formation (Sprinkel, 2000). The lower part of the section, corresponding to the Salt Wash Member, comprises 113 m of laterally discontinuous, often poorly consolidated, sandstones and clast-supported conglomerates, with occasional silty mudstone intervals. Strata corresponding to the Tidwell Member or Windy Hill Member were not observed. The sands of the Salt Wash are overlain by 52 m of monotonous, gray, bentonitic silty mudstones, at the top of which lies the laterally discontinuous quarry sandstone: a poorly sorted, coarse, litharenite. Overlying this are 18 m of variegated red and green silty, bentonitic mudstones, and thin sands. The Morrison–Cedar Mountain Formation boundary is identified in the area as the first appearance of nonbentonitic mudstones containing clasts and pebbles (Greenhalgh and Britt, 2007; Sprinkel et al., 2012), and this boundary is identified 18 m below a thick calcrete that was formerly considered to be the top of the Morrison Formation (e.g., Aubrey, 1998; Currie, 1998). The calcrete is now recognized as marking the boundary between the upper and lower parts of the Yellow Cat Member of the Cedar Mountain Formation (Doelling and Kuehne, 2013; Fig. 2).

The quarry sand is not present in the section we measured as it pinches out to the west of the quarry building; however, we sampled it along strike, and its stratigraphic level is indicated on the log. Black boxes indicate normal polarity sections; white boxes indicate reversed polarity. Gray indicates sections where there is no data or the data quality was not good enough to confirm the direction of magnetization.

The lower contact between the Morrison Formation and underlying Stump Formation is also present at the US40 section. The lower, sand-dominated Salt Wash Member is just 30 m thick at this locality and compromises coarse sandstones and fining-upwards, clast-supported, conglomerates interbedded with occasional silty mudstones. Again, the Tidwell Member was not identified. The Salt Wash is overlain by 26 m of predominantly reddish silty mudstones, followed by 43 m of greenish-gray bentonitic mudstones. The remaining 10.5 m of the section comprises red bentonitic silty mudstones and thin sands. The Morrison–Cedar Mountain Formation boundary is marked by a change from bentonitic red mudstone to poorly exposed purple silty mudstones containing pebbles and clasts, and occurs 7.5 m below the well-developed calcrete that marks the division between the lower and upper Yellow Cat Member of the Cedar Mountain Formation (Fig. 2).

Generally, the conglomerates and sands in the sections are laterally discontinuous, show basal scour, fine-upwards, and may contain trough cross bedding and ripple cross laminations on their upper surfaces. They are thus interpreted as fluvial channel sandstones. The mudstones are easily divisible into those that are predominantly reddish; contain thin, laterally discontinuous sandstone beds that may be ripple cross-laminated; and contain abundant calcrete nodules; and those that are predominantly gray or green-gray, bentonitic, and contain thin, laterally discontinuous greenish tuffaceous sandstones. The reddish silty mudstones are interpreted as overbank deposits of an oxidizing floodplain where the water table was low, while the gray–green mudstones are interpreted as overbank deposits of a reducing floodplain, with a high water table, and where ephemeral ponds and standing water were common (Dodson et al., 1980).

The DQ section is 183 m thick, while the US40 section is 109 m thick. However, stratigraphic correlation between the two sections is relatively straightforward (Fig. 2). The top of the Salt Wash Member at US40 is marked by a thick, well-consolidated, laterally continuous, clast-supported conglomerate, and a very similar conglomerate is observed in the lower part of the Salt Wash at DQ. This conglomerate is much more laterally continuous than other sandstone and conglomerate beds in the area, which pinch out along strike, and it appears to be the same horizon. A sequence of gray, ash-rich, silty mudstones overlain by variegated silty mudstones is present at both sites. If the gray mudstone interval is assumed to correlate, the top of the Salt Wash at DQ apparently correlates with the red mudstones at the base of the Brushy Basin member at US40, and the difference in thickness of the sections is attributable to greater accumulations of sand at DQ. This suggests that the top of the Salt Wash Member is correlative with the oxidizing floodplain deposits at the base of the Brushy Basin Member at US40, and underlines the point that the Salt Wash Member and Brushy Basin Member are lithostratigraphic units rather than time-correlative chronostratigraphic packages.

Paleomagnetic Signal and Magnetostratigraphy

In general, specimens were magnetically weak with low intensities of $<1 \times 10^{-3}$ A/m. During the demagnetization process, some specimens reached intensities lower than the magnetometer's sensitivity levels, in which case measuring was terminated. This resulted in noisy data, yielding high mean angular deviations (MAD) for individual specimens, and high α_{95} (95% confidence limit) values for the unit means. Although errors are relatively high and most samples did not fully demagnetize, the majority exhibited clear magnetization directions (e.g., Figs. 3 and 4). A high percentage of samples yielded steep and positive inclinations between 50 and 70 degrees (47%) and continuous unblocking

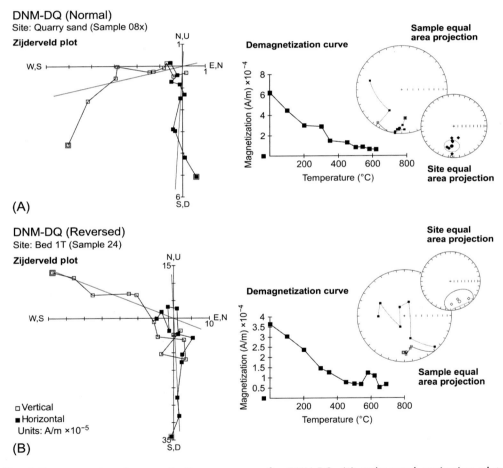

Fig. 3 Representative demagnetization responses for DNM-DQ: (1) orthogonal projection plots (Zijderveld plots), (2) demagnetization intensity plots, and (3) sample and site equal area projection plots for samples (A) 08× (normal), and (B) 24 (reversed). On the site means, α_{95} confidence limits are shown.

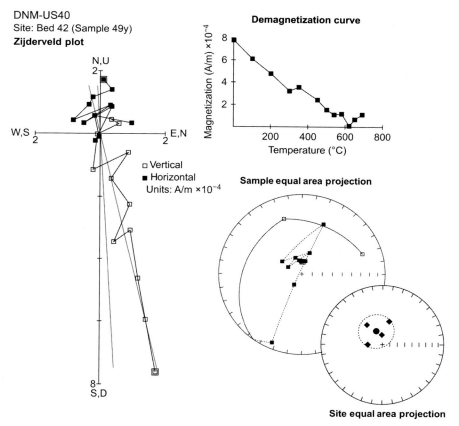

Fig. 4 Representative demagnetization response for DNM-US40 (Sample 49Y): (1) orthogonal projection plot (Zijderveld plot), (2) demagnetization intensity plot, and (3) sample and site equal area projection plots. On the site means, α_{95} confidence limits are shown.

spectra ranging from 580°C to 620°C (76%). High α_{95} values also accompany sites with three or fewer specimen points.

At DQ, 65% of specimens exhibited positive inclinations with either a south (52%) directed or north (48%) directed declination. The remainder of specimens exhibited positive inclination with south-directed declination. The main dinosaur-bearing bed (the quarry sand) yielded a positive inclination of 36 degrees and southerly declination of 186 degrees (Fig. 3). Just under a half of the specimens (43%) yielded undeterminable magnetization directions, exhibiting either an incoherent response to demagnetization or scattered results with no mean bed direction. Beds that exhibited reliable paleomagnetic data are shown in Fig. 2.

The majority of specimens (71%) from US40 yielded normal polarity with a steep inclination (Fig. 4). Declination directions varied from north to south. Four specimens produced data with unstable directions and were withdrawn from the final analysis. Overall, both inclination and declination directions obtained for this locality were very noisy, and as a result interpretation of magnetization directions was difficult.

DISCUSSION

We collected magnetostratigraphic samples using a rock core drill with a water-cooled bit (see "Methods"). The unconsolidated nature of many of the sandstones meant that drilling was extremely difficult, and we recovered less than one-third of the cores we drilled. The mudstones quickly disintegrated upon drilling because of the amount of water necessary to cool the drill bit. Finally, the magnetic weakness of the vast majority of samples we collected resulted in extremely noisy data that precluded easy interpretation. Our results are therefore of poor quality and somewhat unsatisfactory from a magnetostratigraphy point of view. Previous studies (Swierc and Johnson, 1996; Steiner et al., 1994) sampled mudstones by collecting hand samples; we also attempted to collect hand specimens of the mudstones in square centimeter boxes, but the mudstones were too friable. Trenching at least a meter into the formation to reach unweathered mudstones might result in the easier collection of hand specimens, but the time necessary to dig deep trenches through tens of meters of section at multiple locations precludes this as a method for widespread use in the Morrison Formation.

VGPs and Apparent Polar Wander Paths

In order to assess whether the magnetic data obtained from the two sites recorded the magnetic field of the Upper Jurassic or had been subsequently overprinted, we calculated VGPs and plotted them on the most recent apparent polar wander paths (APWP) for the North American craton (Torsvik et al., 2012). The calculated VGPs and associated errors for each locality are presented in Table 1. The VGP inclinations (latitude) were inverted in order to plot in the Southern Hemisphere, onto which APWP for the North American craton is projected (Torsvik et al., 2012). Since the original magnetization directions had large errors, the corresponding VGPs are also associated with high α_{95}. The Colorado Plateau is thought to have experienced a clockwise rotation relative to the stable

Table 1 Mean site virtual geomagnetic poles and their accompanying errors

Locality	Degree of rotation	VGP λ (Longitude)	VGP φ (Latitude)	α_{95}
DNM-DQ	5° Clockwise	300.3°	−70.3°	60.3°
DNM-DQ	10° Clockwise	310.1°	−67.3°	60.3°
DNM-US40	5° Clockwise	285.7°	−37.9°	30.4°
DNM-US40	10° Clockwise	291.3°	−35.8°	30.4°

VGPs were inverted to plot in the Southern Hemisphere. Rotations used a local Euler pole of 37°N and 103°W.

North American craton during the Late Cretaceous–early Paleogene Laramide orogeny and subsequently in the Cenozoic during opening of the Rio Grande Rift (Hamilton, 1981; Steiner, 1998), but the amount of rotation has been a matter of dispute, and estimates range from 0 to 11 degrees (Bryan and Gordon, 1990; Steiner, 1998). Bryan and Gordon (1990) argued for rotation of no more than 5 ± 2.4 degrees (95% confidence limit), by creating APWPs from numerous poles of a variety of ages from both the Colorado Plateau and elsewhere in the North American craton. However, Steiner (1998) argued that this method averaged data, potentially smoothing real differences between poles. By comparing paleopoles from the same formations occurring both on and off the Colorado Plateau, Steiner (1998) suggested rotation of 10 ± 3 degrees. Since the debate regarding the amount of rotation of the Colorado Plateau has not been satisfactorily resolved, the VGPs have been corrected for both 5 and 10 degree clockwise rotation (Table 1).

The corrected VGP for DQ (Fig. 5) falls close to the Late Jurassic paleopole, but taking into account both the VGP and APWP error margins, the magnetization age of the section could range anywhere between 260 Ma to the present day. The US40 VGP plots anomalously further north and outside the Late Jurassic paleopole error margin, instead overlapping the 260 and 310 Ma error ellipse. Additionally, mean site directions obtained at US40 are very scattered, yielding a mean suite inclination of almost

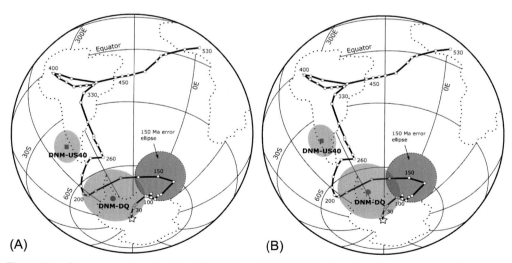

Fig. 5 Virtual geomagnetic poles for DNM-DQ and DNM-US40 plotted on an apparent polar wander path for the North American craton (after Torsvik et al., 2012). *Gray circle*, DNM-DQ; *Gray square*, DNM-US40. *Gray ellipses* around the VGPs are error ellipses. Numbers next to *white circles* are ages, in million years before present, for the location of paleopoles. (A), VGPs corrected for 5 degree rotation of the Colorado Plateau; (B), VGPs corrected for 10 degree rotation of the Colorado Plateau. Note that the error ellipses of the DNM-DQ VGP and the 150 Ma paleopole overlap, but the DNM-US40 VGP is located anomalously distant from the Morrison-age paleopoles.

90 degrees; this suggests random data distribution with no overriding signal, and indicates that no useful paleomagnetic information was preserved in samples from US40.

Lack of Magnetostratigraphic Resolution at US40

The lack of inclination variation at the US40 section is surprising, considering that there are abundant polarity reversals during the Upper Jurassic (Gee and Kent, 2007) and that polarity reversals were identified at the DQ section, just 14 km away. Large error margins on most data points and the anomalous location of the VGP derived from the section make interpretation of the causes of lack of inclination variation problematic; however, we consider it most likely that remagnetization of the section occurred during a time interval of normal polarity, which would have completely overprinted the primary magnetic signal.

Age of the Quarry Sand

At DQ, we identified a polarity reversal immediately above the quarry sand (Fig. 2). ^{40}Ar/^{39}Ar dating of an ash bed immediately underlying the quarry sand at Dinosaur National Monument yielded an age range of 150.91 ± 0.43 Ma (Trujillo and Kowallis, 2015), which corresponds with the lower part of CM23, and is predominantly of reversed polarity. However, a short interval of normal polarity, CM23n.2n, is dated at 150.9–150.93 Ma (Gee and Kent, 2007), lying within the error bars of the radiometric date (Fig. 6). It is therefore possible that the quarry sand was deposited within this 30,000-year interval of normal polarity. Because the radiometric date was obtained just below the quarry sand, however, it is also possible that the quarry sand was deposited during the overlying normal, CM23n.1n chron, between 150.69 and 150.04 Ma, and the reversal in the sands above corresponds to CM22Ar (150.04–149.72 Ma; Gee and Kent, 2007; Fig. 6). Nevertheless, using a combination of existing radiometric dating and our new magnetostratigraphic results, it is possible to constrain the age of the quarry sand to between 150.91 and 150.04 Ma. This result is consistent with previous work (Steiner, 1998; Steiner et al., 1994), which linked the upper Morrison Formation in Colorado and New Mexico to CM22.

Magnetic Reversal in the Lower Cedar Mountain Formation

The uppermost sandstone sampled at the DQ site lies above the Morrison–Cedar Mountain contact. The age of the Cedar Mountain Formation across its outcrop area has been difficult to constrain for many of the same reasons that the age of the Morrison Formation has historically been problematic: terrestrial deposits often lack zone fossils, and the formation is probably time-transgressive. Radiometric dating of the Yellow Cat Member, the lowest member of the Formation, indicates ages ranging anywhere from latest Berriasian to Albian (Kirkland, pers. comm. 2016). During the Cretaceous, there was

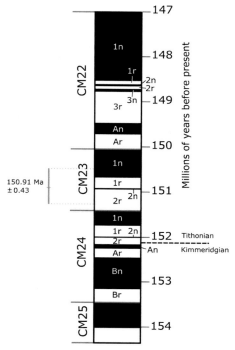

Fig. 6 Correlation of the geomagnetic polarity timescale of Gee and Kent (2007) with the radiometric date obtained from a mudstone underlying the quarry sand by Trujillo and Kowallis (2015). The quarry sand is normal polarity (*black*), while the interval immediately above it is reversed (*white*), and thus the quarry sand represents either CM 23n.2n or CM23n.1n.

a prolonged interval of normal polarity from 120.6 to 83.0 Ma (Tauxe, 2010). Reversed magnetization in the sands below the calcrete at DNM-DQ (Fig. 2) constrains the age of the lowermost Cedar Mountain Formation to older than the lowermost Aptian, prior to 120.6 Ma and the Cretaceous normal superchron, in the Uinta Mountains.

Comparison With Other Morrison Dinosaur Quarries

As of 2015, there are more than 275 published records of dinosaur quarries in the Morrison Formation (Foster, 2003; Paleobiology database (PBDB) www.paleobiodb.org), but very few have been dated, and a lack of long-range correlation across the Formation means that their relative stratigraphic positions remain unknown. Radiometric or magnetostratigraphic dates do exist, however, for the following quarries. The Valley of Death site at Garden Park, near Cañon City, Colorado: ^{40}Ar/^{39}Ar dating on sanidines gives an age of 152.29 ± 0.27 Ma (Trujillo and Kowallis, 2015); the Mygatt-Moore Quarry, west of Grand Junction, on the Utah–Colorado state line: ^{206}U/^{238}Pb on zircons gives an age of 152.18 ± 0.49 Ma (Trujillo et al., 2014); the Cleveland-Lloyd Dinosaur Quarry, on the northwest flank of the San Rafael Swell in Utah: $^{40}K/^{40}Ar$ dating of biotites above the quarry interval gives an age of 148.2–145.8 Ma (Bilbey, 1998); the Howe and Howe-Stephens quarries, in the Bighorn Basin, near Graybull, Wyoming: magnetostratigraphy gives a date of CM20 (Swierc and Johnson, 1996), which corresponds to 145.52–143.36 Ma (Gee and Kent, 2007). The range of different techniques used, however, causes problems with direct comparisons. Recent recalibration of legacy $^{40}Ar/^{39}Ar$ dates, including the Valley of Death, by Trujillo and Kowallis (2015), and the complete age report for the $^{206}U/^{238}Pb$ date at Mygatt-Moore (Trujillo et al., 2014) means that these two dates are directly comparable. However, the $^{40}K/^{40}Ar$ date at Cleveland-Lloyd is rather old, has large error estimates, and is not directly comparable to ages obtained using other systems. Furthermore, the magnetostratigraphic data presented for the Howe Quarry by Swierc and Johnson (1996) may have been incorrectly interpreted (see "Previous Magnetostratigraphic Studies of the Morrison Formation" discussed earlier).

The quarry sand at Dinosaur National Monument, dated herein as 150.91–150.04 Ma, is therefore only readily comparable with the Valley of Death quarry and the Mygatt-Moore Quarry. The Valley of Death quarry falls within the error bars of the Mygatt-Moore Quarry, so these appear to be roughly the same age. The quarry sand at Dinosaur National Monument is 1–2 million years younger.

Despite being roughly the same age, no taxa are shared between the Valley of Death quarry and Mygatt-Moore; in contrast, four genera are shared between Mygatt-Moore and the younger Dinosaur National Monument Quarry. The Valley of Death quarry shares a single taxon with Dinosaur National Monument (Table 2). The most likely explanation for the faunal differences between Valley of Death and Mygatt-Moore is that the former is undersampled relative to the latter, only two genera being known from the quarry. At generic level, there appears to be little faunal difference between the Mygatt-Moore Quarry and the Dinosaur National Monument Quarry, which is ~1–2 million years younger. Species-level taxonomy of many Morrison dinosaurs is problematic due to the often fragmentary and disarticulated nature of many of the specimens, and the inadequate nature of many historic holotypes, both of which preclude adequate comparisons. A lack of reliably dated dinosaur quarries and problematic species-level taxonomy means that, at present, it is not possible to examine changes in the Morrison Formation dinosaur fauna through time.

CONCLUSIONS

The unconsolidated nature of many Morrison Formation sandstones and the predominance of mudstone through much of the formation preclude the use of rock drills for collecting magnetostratigraphic samples, and future studies should focus on collecting hand specimens. Trenching at least 1 m into the formation is required to encounter fresh rock, and the time taken for such sampling may prove prohibitive to regional

Table 2 Dinosaur faunas from quarries with radiometric or magnetostratigraphic dates

Taxa	Valley of Death, near Cañon City, Colorado 152.29 ± 0.27 Ma[a]	Mygatt-Moore Quarry, near Fruita, Colorado 152.12 ± 0.49 Ma[b]	Dinosaur National Monument Quarry, Uintah Basin, Utah 150.91 – 150.04 Ma	Cleveland-Lloyd Dinosaur Quarry, San Rafael Swell, Utah 146.8 ± 1 Ma[c]	Howe Quarry, Bighorn Basin, Wyoming CM 20[d] (145.52 – 143.36 Ma)
Allosaurus		*	*	*	*
Apatosaurus		*	*	*	*
Barosaurus		*	*		*
Brachiosaurus					*
Camarasaurus		*	*	*	*
Camptosaurus				*	*
Ceratosaurus		*	*	*	
Coelurus	*			*	
Diplodocus		*	*		*
Dryosaurus			*		*
Kaatedocus					*
Marshosaurus				*	
Mymooropelta		*			
Othnielosaurus	*				
Stegosaurus			*	*	*
Stokesosaurus				*	*
Tanycolagreus				*	
Torvosaurus			*		
Uteodon			*		

Asterisks (*) indicate the presence of a taxon in the specified quarry.
Faunal data from the Paleobiology database (www.paleobiodb.org)
[a]Trujillo and Kowallis (2015).
[b]Trujillo et al. (2014).
[c]Bilbey (1998).
[d]Swierc and Johnson (1996).

magnetostratigraphic studies. Samples that were recovered were highly resistant to alternating field demagnetization, and the mineralogy often altered under thermal demagnetization, resulting in noisy data and large error margins.

Despite these problems, a polarity reversal was identified immediately above the quarry sand at Dinosaur National Monument, and recent radiometric dating of a horizon immediately below the quarry sand (Trujillo and Kowallis, 2015) allows us to determine that the polarity reversal is either within CM23 or is the boundary between CM23 and CM22. This constrains the age of the quarry sand to 150.91–150.04 Ma. The quarry sand is therefore younger than the Valley of Death site at Garden Park, and the Mygatt-Moore Quarry west of Grand Junction, Colorado. Due to a lack of reliably dated quarries, it is currently not possible to examine faunal change through time in the Morrison Formation. However, as we have demonstrated herein, magnetostratigraphy could help to constrain the ages of Morrison dinosaur quarries by reducing error bars on radiometric dates and allowing quarries that are not proximal to ash-bearing mudstones to be dated based on correlation with the GPTS.

ACKNOWLEDGMENTS

Sampling within Dinosaur National Monument was carried out under a National Park Service permit issued to SCRM. Thanks to Dan Chure and staff at Dinosaur National Monument, Justin Snyder (Utah Bureau of Land Management), Doug Sprinkel (Utah Geological Survey), and Jim Kirkland (Utah Geological Survey) for help with logistics and discussion of the Morrison Formation in Utah. Graham Nash, Sarah Dodd, Radchagrit Supakulopas, Jay Shah, and Thomas Berndt (Imperial College London) helped with sample preparation and analysis. Reviewers John Foster and Jim Kirkland, and editor Kate Zeigler, provided extremely helpful comments on an earlier draft of this manuscript.

REFERENCES

Anderson, O., Lucas, S.G., 1998. Redefinition of the Morrison Formation (Upper Jurassic) and related San Rafael Group strata, southwestern U.S.A. Mod. Geol. 22, 39–69.

Ash, S.R., Tidwell, W.D., 1998. Plant megafossils from the Brushy Basin Member of the Morrison Formation near Montezuma Creek trading post, southeastern Utah. Mod. Geol. 22, 321–340.

Aubrey, W.M., 1998. A newly discovered, widespread fluvial facies and unconformity marking the Upper Jurassic/Lower Cretaceous boundary, Colorado Plateau. Mod. Geol. 22, 209–233.

Bakker, R.T., Carpenter, K., Galton, P.M., Siegwarth, J.D., Filla, J., 1990. A new latest Jurassic vertebrate fauna, from the highest levels of the Morrison Formation at Como Bluff, Wyoming. Hunteria 2, 1–18.

Bender, M.L., 2013. Paleoclimate: Princeton Primers in Climate. Princeton University Press, Princeton, NJ.

Bilbey, S.A., 1998. Cleveland-Lloyd dinosaur quarry—age, stratigraphy and depositional environments. Mod. Geol. 22, 87–120.

Bryan, P., Gordon, R.G., 1990. Rotation of the Colorado Plateau: an updated analysis of paleomagnetic poles. Geophys. Res. Lett. 17, 1501–1504.

Butler, R.F., 1992. Paleomagnetism: Magnetic Domains to Geologic Terranes. Blackwell Scientific Publications, Oxford (2004 electronic edition).

Carpenter, K., 1998. Vertebrate biostratigraphy of the Morrison Formation near Cañon City, Colorado. Mod. Geol. 22, 407–426.

Chenoweth, W.L., 1998. Uranium mining in the Morrison Formation. Mod. Geol. 22, 427–440.

Cooley, J.T., Schmidt, J.G., 1998. An anastomosed fluvial system in the Morrison Formation (Upper Jurassic) of southwest Montana. Mod. Geol. 22, 171–208.

Currie, B.S., 1997. Sequence stratigraphy of nonmarine Jurassic-Cretaceous rocks, central Cordilleran foreland-basin system. Geol. Soc. Am. Bull. 109, 1206–1222.

Currie, B.S., 1998. Upper Jurassic—Lower Cretaceous Morrison and Cedar Mountain Formations, NE Utah—NW Colorado: relationships between nonmarine deposition and early Cordilleran foreland-basin development. J. Sediment. Res. 68, 632–652.

DeCelles, P.G., Burden, E.T., 1992. Non-marine sedimentation in the overfilled part of the Jurassic-Cretaceous Cordilleran foreland basin: Morrison and Cloverly Formations, central Wyoming, USA. Basin Res. 4, 291–313.

DeCelles, P.G., Currie, B.S., 1996. Long-term sediment accumulation in the Middle Jurassic-Early Eocene Cordilleran retroarc foreland-basin system. Geology 24, 591–594.

Demko, T.M., Parrish, J.T., 1998. Paleoclimatic setting of the Upper Jurassic Morrison Formation. Mod. Geol. 22, 283–296.

Demko, T.M., Currie, B.S., Nicoll, K.A., 2004. Regional paleoclimatic and stratigraphic implications of paleosols and fluvial/overbank architecture in the Morrison Formation (Upper Jurassic) Western Interior, USA. Sediment. Geol. 167, 115–135.

Dodson, P., Behrensmeyer, A.K., Bakker, R.T., McIntosh, J.S., 1980. Taphonomy and paleoecology of the dinosaur beds of the Jurassic Morrison Formation. Paleobiology 6, 208–232.

Doelling, H.H., Kuehne, P.A., 2013. Geologic Maps of the Klondike Bluffs, Mollie Hogans, and the Windows Section 7.5' Quadrangles, Grand County, Utah. Utah Geological Survey, Salt Lake City, UT. 31 pp. + 6 pl.

Dunagan, S.P., Turner, C.E., 2004. Regional paleohydrologic and paleoclimatic settings of wetland/lacustrine depositional systems in the Morrison Formation (Upper Jurassic), Western Interior, USA. Sediment. Geol. 167, 269–296.

Evanoff, E., Good, S.C., Hanley, J.H., 1998. An overview of the freshwater mollusks from the Morrison Formation (Upper Jurassic, Western Interior, USA). Mod. Geol. 22, 423–450.

Foster, J.R., 2000. Paleobiogeographic homogeneity of dinosaur faunas during the Late Jurassic in western North America. In: Lucas, S.G., Heckert, A.B. (Eds.), In: New Mexico Museum of Natural History and Science Bulletin, vol. 17. New Mexico Museum of Natural History and Science, Albuquerque, NM, pp. 47–50.

Foster, J.R., 2003. Paleoecological analysis of the vertebrate fauna of the Morrison Formation (Upper Jurassic). Rocky Mountain Region, U.S.A. New Mexio Musuem of Natural History and Science Bulletin, vol. 23. New Mexico Museum of Natural History and Science, Albuquerque, NM, pp. 1–85.

Fuentes, F., DeCelles, P.G., Gehrels, G.E., 2009. Jurassic onset of foreland basin deposition in northwestern Montana, USA: implications for along-strike synchroneity of Cordilleran orogenic activity. Geology 37, 379–382.

Galli, K.G., 2014. Fluvial architecture element analysis of the Brushy Basin Member, Morrison Formation, western Colorado, USA. Volumina Jurassica 12, 69–106.

Gaston, K.J., 2000. Global patterns in biodiversity. Nature 405, 220–227.

Gee, J.S., Kent, D.V., 2007. Source of oceanic magnetic anomalies and the geomagnetic polarity time scale. Treatise Geophys. 5, 455–507.

Godin, P.D., 1991. Fining-upward cycles in the sandy braided-river deposits of the Westwater Canyon Member (Upper Jurassic), Morrison Formation, New Mexico. Sediment. Geol. 70, 61–82.

Good, S.C., 2004. Paleoenvironmental and paleoclimatic significance of freshwater bivalves in the Upper Jurassic Morrison Formation, Western Interior, USA. Sediment. Geol. 167, 163–176.

Greenhalgh, B.W., Britt, B.B., 2007. Stratigraphy and sedimentology of the Morrison-Cedar Mountain Formation boundary, east-central Utah. In: Willis, G.C., Hylland, M.D., Clark, D.L., Chidsey Jr., T.C. (Eds.), Central Utah—Diverse Geology of a Dynamic Landscape. Utah Geological Association, Salt Lake City, UT, pp. 81–100. UGA Publication 36.

Hamilton, W., 1981. Plate tectonic mechanism of Laramide deformation. Contrib. Geol. 19, 87–92. University of Wyoming.

Hasiotis, S.T., 2004. Reconnaissance of Upper Jurassic Morrison Formation ichnofossils, Rocky Mountain Region, USA: paleoenvironmental, stratigraphic and paleoclimatic significance of terrestrial and freshwater ichnocoenoses. Sediment. Geol. 167, 177–268.

Heller, P.L., Bowdler, S.S., Chambers, H.P., Coogan, J.C., Hagen, E.S., Shuster, M.W., Winslow, N.S., Lawton, T.F., 1986. Time of initial thrusting in the Sevier orogenic belt, Idaho-Wyoming and Utah. Geology 14, 388–391.

Jablonski, D., Roy, K., Valentine, J.W., 2006. Out of the tropics: evolutionary dynamics of the latitudinal biodiversity gradient. Science 314, 102–106.

Jennings, D.S., Hasiotis, S.T., 2006. Paleoenvironmental and stratigraphic implications of authigenic clay distributions in Morrison Formation deposits, Bighorn Basin, Wyoming. In: Foster, J.R., Lucas, S.G. (Eds.), New Mexico Museum of Natural History and Science Bulletin, vol. 36. New Mexico Museum of Natural History and Science, Albuquerque, NM, pp. 25–34.

Kirkland, J.I., 2006. Fruita Paleontological Area (Upper Jurassic, Morrison Formation), western Colorado: an example of terrestrial taphofacies analysis. In: Foster, J.R., Lucas, S.G. (Eds.), New Mexico Museum of Natural History and Science Bulletin, vol. 36. New Mexico Museum of Natural History and Science, Albuquerque, NM, pp. 67–95.

Kirkland, J.I., Madsen, S.K., 2007. The Lower Cretaceous Cedar Mountain Formation, Eastern Utah: The View Up an Always Interesting Learning Curve (Field Trip Guide). UGA Publication 35.

Kirschvink, J.L., 1980. The least-squares line and plane and the analysis of paleomagnetic data. Geophys. J. Int. 62, 699–718.

Kjemperud, A.V., Schomacker, E.R., Cross, T.A., 2008. Architecture and stratigraphy of alluvial deposits, Morrison Formation (Upper Jurassic), Utah. AAPG Bull. 92, 1055–1076.

Kowallis, B.J., Christiansen, E.H., Deino, A.L., Peterson, F., Turner, C.E., Kunk, M.J., Obradovich, J.D., 1998. The age of the Morrison Formation. Mod. Geol. 22, 235–260.

Langereis, C.G., Krijgsman, W., Muttoni, G., Menning, M., 2010. Magnetostratigraphy—concepts, definitions, and applications. Newsl. Stratigr. 43, 207–233.

Lawton, T.F., 1994. Tectonic setting of Mesozoic sedimentary basins, Rocky Mountain Region, United States. In: Caputo, M.V., Peterson, J.A., Franczyk, K.J. (Eds.), Mesozoic Systems of the Rocky Mountains Region, USA. SEPM, Denver, CO, pp. 1–25.

Litwin, R.G., Turner, C.E., Peterson, F., 1998. Palynological evidence on the age of the Morrison Formation, Western Interior U.S. Mod. Geol. 22, 297–320.

Lucas, S.G., Kirkland, J.I., 1998. Preliminary report on Conchostraca from the Upper Jurassic Morrison Formation, western United States. Mod. Geol. 22, 415–422.

Lurcock, P.C., Wilson, G.S., 2012. PuffinPlot: a versatile, user-friendly program for paleomagnetic analysis. Geochem. Geophys. Geosyst. 13, 1–6.

Mittelbach, G.G., Schemske, D.W., Cornell, H.V., Allen, A.P., Brown, J.M., Bush, M.B., Harrison, S.P., Hurlbert, A.H., Knowlton, N., Lessios, H.A., McCain, C.M., McCune, A.R., McDade, L.A., McPeek, M.A., Near, T.J., Price, T.D., Ricklefs, R.E., Roy, K., Sax, D.F., Schluter, D., Sobel, J.M., Turelli, M., 2007. Evolution and the latitudinal diversity gradient: speciation, extinction and biogeography. Ecol. Lett. 10, 315–331.

Ostrom, J.H., 1970. Stratigraphy and paleontology of the Cloverly Formation (Lower Cretaceous) of the Bighorn Basin Area, Wyoming and Montana. Bull. Peabody Mus. Nat. Hist. 35, 1–234.

Owen, A., Nichols, G.J., Hartley, A.J., Weissmann, G.S., Scuderi, L.A., 2015. Quantification of a distributive fluvial system: the Salt Wash DFS of the Morrison Formation, SW U.S.A. J. Sediment. Res. 85, 544–561.

Peterson, F., 1994. Sand dunes, sabkhas, streams and shallow seas: Jurassic paleogeography in the southern part of the Western Interior Basin. In: Caputo, M.V., Peterson, J.A., Franczyk, K.J. (Eds.), Mesozoic Systems of the Rocky Mountains Region, USA. SEPM, Denver, CO, pp. 233–272.

Peterson, F., Turner, C.E., 1998. Stratigraphy of the Ralston Creek and Morrison Formations (Upper Jurassic) near Denver, Colorado. Mod. Geol. 22, 3–38.

Pipiringos, G.N., O'Sullivan, R.B., 1978. Principal Unconformities in Triassic and Jurassic Rocks, Western Interior United States—A Preliminary Report: US Geological Survey Professional Paper 1035A. United States Government Printing Office, Washington, DC. 26 pp.

Roca, X., Nadon, G.C., 2007. Tectonic control on the sequence stratigraphy of nonmarine retroarc foreland basin fills: insights from the Upper Jurassic of central Utah, USA. J. Sediment. Res. 77, 239–255.

Royse Jr., F., 1993. Case of the phantom foredeep: early Cretaceous in west-central Utah. Geology 21, 133–136.

Schudack, M.E., Turner, C.E., Peterson, F., 1998. Biostratigraphy, paleoecology and biogeography of char-ophytes and ostracodes from the Upper Jurassic Morrison Formation, Western Interior USA. Mod. Geol. 22, 379–414.

Sprinkel, D.A., 2000. Geologic guide along Flaming Gorge Reservoir, Flaming Gorge National Recreation Area, Utah-Wyoming. In: Anderson, P.B., Sprinkel, D.A. (Eds.), Geologic Road, Trail and Lakes Guides to Utah's Parks and Monuments. Utah Geological Association, Salt Lake City, UT, pp. 1–20.

Sprinkel, D.A., Madsen, S.K., Kirkland, J.I., Waanders, G.L., Hunt, G.J., 2012. Cedar Mountain and Dakota Formations Around Dinosaur National Monument—Evidence of the First Incursion of the Cretaceous Western Interior Seaway Into Utah: Utah Geological Survey Special Study 143. Utah Department of Natural Resources, Salt Lake City, UT. 21 pp.

Steiner, M.B., 1998. Age, correlation and tectonic implications of Morrison Formation paleomagnetic data, including rotation of the Colorado Plateau. Mod. Geol. 22, 261–281.

Steiner, M.B., Helsley, C.E., 1975. Reversal pattern and apparently polar wander for the Late Jurassic. Geol. Soc. Am. Bull. 86, 1537–1543.

Steiner, M.B., Lucas, S.G., Shoemaker, E.M., 1994. Correlation and age of the Upper Jurassic Morrison Formation from magnetostratigraphic analysis. In: Caputo, M.V., Peterson, J.A., Franczyk, K.J. (Eds.), Mesozoic Systems of the Rocky Mountains Region, USA. SEPM, Denver, CO, pp. 315–330.

Swierc, J.E., Johnson, G.D., 1996. A local chronostratigraphy for the Morrison Formation, northeastern Bighorn Basin, Wyoming. Wyoming Geological Association Guidebook, vol. 47. Wyoming Geological Association, Casper, WY, pp. 315–327.

Tauxe, L., 2010. Essentials of Paleomagnetism. University of California Press, Berkeley, CA.

Torsvik, T.H., Van der Voo, R., Preeden, U., Niocaill, C.M., Steinberger, B., Doubrovine, P.V., Van Hinsbergen, D.J.J., Domeier, M., Gaina, C., Tohver, E., Meert, J.G., McCausland, P.J., Cocks, L.R.M., 2012. Phanerozoic polar wander, paleogeography and dynamics. Earth-Sci. Rev. 114, 325–368.

Trujillo, K.C., 2006. Clay mineralogy of the Morrison Formation (Upper Jurassic –? Lower Cretaceous), and its use in long distance correlation and paleoenvironmental analyses. In: Foster, J.R., Lucas, S.G. (Eds.), New Mexico Museum of Natural History and Science Bulletin, vol. 36. New Mexico Museum of Natural History and Science, Albuquerque, NM, pp. 17–23.

Trujillo, K.C., Kowallis, B.J., 2015. Recalibrated legacy ^{40}Ar/^{39}Ar ages for the Upper Jurassic Morrison Formation, Western Interior, U.S.A. Geol. Intermountain West 2, 1–8.

Trujillo, K.C., Foster, J.R., Hunt-Foster, R.K., Chamberlain, K.R., 2014. A U/Pb age for the Mygatt-Moore quarry, Upper Jurassic Morrison Formation, Mesa County, Colorado. Volumina Jurassica 12, 107–114.

Turner, C.E., Peterson, F., 1999. Biostratigraphy of Dinosaurs in the Upper Jurassic Morrison Formation of the Western Interior, USA. Utah Geological Survey Miscellaneous Publications, Salt Lake City, UT. 99–1, 77–114.

Turner-Peterson, C.E., 1986. Fluvial sedimentology of a major uranium-bearing sandstone—a study of the Westwater Canyon Member of the Morrison Formation, San Juan Basin, New Mexico. In: Turner-Peterson, C.E., Santos, E.S., Fishman, N.S. (Eds.), AAPG Studies in Geology, vol. 22, pp. 47–76.

Tyler, N., Ethridge, F.G., 1983. Depositional setting of the Salt Wash Member of the Morrison Formation, southwest Colorado. J. Sediment. Petrol. 53, 67–82.

Valentine, J.W., Jablonski, D., Krug, A.Z., Roy, K., 2008. Incumbency, diversity and latitudinal gradients. Paleobiology 34, 169–178.

Van Fossen, M.C., Kent, D.V., 1992. Paleomagnetism of the Front Range (Colorado) Morrison Formation and an alternative model of Late Jurassic North American apparent polar wander. Geology 20, 223–226.

Weishampel, D.B., Barrett, P.M., Coria, R.A., Le Loeff, J., Xu, X., Zhao, X., Sahni, A., Gomani, E.M.P., Noto, C.R., 2004. Dinosaur distribution. In: Weishampel, D.B., Dodson, P., Osmólska, H. (Eds.), The Dinosauria, second ed. University of California Press, Berkeley, CA, pp. 517–606.

Willig, M.R., Kaufman, D.M., Stevens, R.D., 2003. Latitudinal gradients of diversity: pattern, process, scale and synthesis. Annu. Rev. Ecol. Evol. Syst. 34, 273–309.

Yingling, V.L., Heller, P.L., 1992. Timing and record of foreland sedimentation during the initiation of the Sevier orogenic belt in central Utah. Basin Res. 4, 279–290.

Zijderveld, J.D.A., 1967. A. C. demagnetization of rocks: analysis of results. In: Collinson, D., Creer, K., Runcorn, S. (Eds.), Methods in Paleomagnetism. Elsevier, Amsterdam, pp. 254–286.

Dr Susannah Maidment is a vertebrate paleontologist who uses geological techniques to examine paleobiological problems. Her research focuses on the ornithischian dinosaurs, particularly the stegosaurs.

Dominika Balikova completed a degree in Geological Sciences at Imperial College London, and this work formed part of her final year dissertation. She currently works as a research analyst for Clarksons.

Dr Adrian Muxworthy is a geophysicist who uses paleo- and rock magnetic techniques to solve Earth and planetary problems. His research is diverse, and includes Solar System formation, oil recovery, and geohazards.

Terrestrial Carbon Isotope Chemostratigraphy in the Yellow Cat Member of the Cedar Mountain Formation: Complications and Pitfalls

M.B. Suarez*, C.A. Suarez†, A.H. Al-Suwaidi‡, G. Hatzell†, J.I. Kirkland§, J. Salazar-Verdin*, G.A. Ludvigson¶, R.M. Joeckel**

*University of Texas San Antonio, San Antonio, TX, United States
†University of Arkansas Fayettville, Fayettville, AR, United States
‡Petroleum Institute University & Research Centre, Abu Dhabi, United Arabic Emirates
§Utah Geological Survey, Salt Lake City, UT, United States
¶Kansas Geological Survey, Lawrence, KS, United States
**University of Nebraska, Lincoln, NE, United States

Contents

Terrestrial Depositional Systems
http://dx.doi.org/10.1016/B978-0-12-803243-5.00008-X

INTRODUCTION

The original description of the Cedar Mountain Formation (CMF) by Stokes (1944) and revised by Young (1960), Sprinkel et al. (1999), and Kirkland (2005) as a relatively unfossiliferous shale and sandstone unit between the dinosaur-rich continental Middle to Upper Jurassic Morrison Formation and the coastal to marine Late Cretaceous Dakota Formation has since been revised to include a surprising number of new dinosaur taxa (Kirkland et al., 1997; Kirkland and Madsen, 2007). As paleontological interest in this relatively thin unit has increased, so too has the sedimentological and stratigraphic research, specifically related to understanding the timing and radiation of dinosaurian taxa recognized in the CMF and the paleoecology and paleogeography of the localities. While it is well known that an unconformity exists between the Late Jurassic (~147 Ma, Currie, 2002) and Early Cretaceous, the exact duration of this unconformity and the approximate age of the base of the CMF have been elusive.

The base of the CMF was initially described from the type section in Emery County, Utah, as the Cedar Mountain Shale, comprising drab shale between the Buckhorn Conglomerate and Dakota Formation (Stokes, 1944). Stokes (1952) revised the boundary between the Morrison Formation and the overlying Cedar Mountain strata to include the Buckhorn Conglomerate as the basal member of the renamed CMF, as it unconformably overlies the Jurassic Morrison Formation. Stokes (1952) also noted that the transition from the Morrison Formation to the CMF where the Buckhorn Conglomerate is not present may be marked as a change in color from highly variegated mudstones and claystones to more drably variegated mudstones and claystones, an increase in isolated carbonate nodules to coalesced carbonate nodules, the presence of polished chert pebbles, and the lack of dinosaur bones (Kirkland, 2005). This last characteristic has proven to be unrepresentative, given the large number of new dinosaurian and other taxa discovered in the CMF (cf. Kirkland et al., 1998; Kirkland, 2005; Thayn et al., 1985; Nelson and Crooks, 1987; Eaton and Cifelli, 2001; Kirkland and Madsen, 2007; Carpenter et al., 2008; McDonald et al., 2010).

The initial goal of this research was to construct a high-resolution organic carbon (OC) isotope ($\delta^{13}C_{org}$) record near the area of recent paleontological finds that can be correlated to chronologically well-constrained Early Cretaceous marine carbon isotope records. It was thought that such a correlation would help constrain the age of the lowest member of the CMF, the Yellow Cat Member (YCM), providing a better estimate for the unconformity between the Jurassic Morrison Formation and the Cretaceous CMF. However, in doing so, we encountered a number of difficulties. Here we outline the methods we use, demonstrate the limitations to the method and offer an explanation of the problems encountered, and provide solutions, as a guide to others who may wish to do similar work.

CONTINENTAL CARBON ISOTOPE CHEMOSTRATIGRAPHY

The Cretaceous Period is characterized by numerous instances of oceanic anoxic events, believed to be the result of perturbations to the global carbon cycle recorded as carbon isotope excursions (CIEs). These perturbations are often marked by excursions in the $\delta^{13}C$ of carbonate and OC in marine strata (Scholle and Arthur, 1980). The primary cause of the individual isotope excursions is the subject of significant research. Individual isotope excursions are commonly associated with Large Igneous Province activity, methane hydrate destabilization, and/or igneous intrusions in organic-rich shales, among other causes (cf. Hesselbo et al., 2000; Jenkyns, 2010).

The key principle behind carbon isotope chemostratigraphy is that global carbon cycle perturbations influence, and are reflected in, the carbon isotope composition of carbonate and OC in both marine and continental depositional systems. $\delta^{13}C$ is exchanged between surface ocean carbon pools and continental organic matter via a well-mixed atmosphere. As such any major perturbation in the global carbon pool should be recorded in the sedimentary record in both continental and marine depositional systems. Bulk organic matter C-isotope chemostratigraphy of continental deposits works if the organic matter sampled represents the isotopic composition of the surrounding plant material, and that plant material is influenced by the shifts in the isotopic composition of the atmosphere. Many studies in the Cretaceous and other time periods in which significant carbon cycle perturbations have occurred (e.g., Paleocene-Eocene Thermal Maximum (PETM), Valenginian-Barremian, Cenomanian-Turonian) have correlated the pattern of positive and negative CIEs associated with carbon cycle perturbations in continental sections to those of marine sections (cf. Gröcke et al., 1999; Ando et al., 2002; Magioncalda et al., 2004; Robinson and Hesselbo, 2004; Hesselbo et al., 2000; Whiteside et al., 2010; Tipple et al., 2011; Suarez et al., 2014), regardless of the cause of the C-cycle perturbation.

BACKGROUND GEOLOGY

The CMF is exposed in the San Rafael Swell, a large Laramide anticline in central Utah and on the Gunnison Plateau (Nelson and Crooks, 1987; Weiss and Roche, 1988), and represents continental sediments deposited in the distal foredeep to forebulge of the Sevier fold-thrust belt (Yingling and Heller, 1992; DeCelles et al., 1995;

DeCelles and Currie, 1996). It consists of five members: the Buckhorn Conglomerate, Yellow Cat, Poison Strip Sandstone, Ruby Ranch, and Mussentuchit, representing a wide range of depositional environments and sediments from lacustrine to volcanoclastic, palustrial, fluvial, alluvial plain, and paleosol deposits. Due to lateral variation in thickness of the members and local disconformities, not all members are present in any given locality (DeCelles and Currie, 1996; Currie, 1997; Kirkland et al., 1997, 1998; Kirkland and Madsen, 2007; Greenhalgh and Britt, 2007; Fig. 1).

Regionally, the presence of carbonate represents a large-scale change in climate for the area. Considering the proximity of the sites to each other (within 8 km), this change in climate likely affected all sites at approximately the same time and is used as a regional correlation.

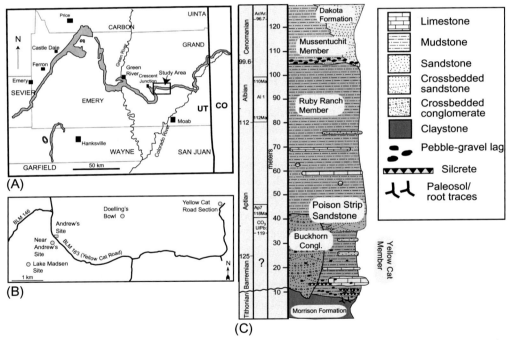

Fig. 1 Location and representative stratigraphic column of the Cedar Mountain Formation. (A) Outcrop belt *(gray)* of the Cedar Mountain Formation with the Yellow Cat Member study site area outlined by the black rectangle. (B) The location of the study sites *(gray circles)* relative to Yellow Cat Road. (C) Representative stratigraphic column of the Cedar Mountain Formation. The age of the base of the Yellow Cat Member is unknown and has been hypothesized to be as old as Berriasian and as young as lower Aptian (cf. Stokes, 1952; Kirkland et al., 1997, 1998; Kirkland and Madsen, 2007; Greenhalgh et al., 2006; Greenhalgh and Britt, 2007; Mori, 2009; Sames et al., 2010; Hunt et al., 2011; Sprinkel et al., 2012; Hendrix et al., 2015). Known ages are shown in the second age column based on geochronology (numerical dates) or C-isotope chemostratigraphy *(gray bars)*.

Age of the CMF

While the length of time represented by the unconformity at the top of the Morrison Formation as well as the age range of the CMF have continued to elude researchers, a number of studies have provided new chronological constraints for the unit. One of the most comprehensive studies was published by Ludvigson et al. (2010) and included detrital zircon dates, U-Pb carbonate dates, and carbonate carbon isotope stratigraphy. The study concluded that the Cedar Mountain strata sampled, represented by the uppermost YCM to the base of the Mussentuchit Member, are early Aptian to Albian. Detrital zircon data presented by Britt et al. (2007) suggest that the maximum depositional age of the YCM is 124 + 2/−2.8 Ma. This study as well as earlier work by Greenhalgh et al. (2006) and Greenhalgh and Britt (2007) suggests that YCM could be as old as Late Barremian (~125 Ma). A differing interpretation comes from Sames et al. (2010) on the basis of ostracode biostratigraphy and affinities to European ostracode taxa. This study suggests the YCM could be as old as Late Berriasian to Valanginian (~140–133 Ma) and has been recently supported by detrital zircons derived from paleosols (Hendrix et al., 2015). This disparity in ages is profound and requires further investigation. The upper age limit of the member is constrained by a U-Pb date from a uranium-enriched micritic limestone found near the boundary between the YCM and the overlying Poison Strip Sandstone, determined to be 119.4 ± 2.4 Ma (Ludvigson et al., 2010).

In addition to the U-Pb date, Ludvigson et al. (2010) identified two prominent positive CIEs that are correlative to CIEs in the upper Aptian (Ap7 and Ap10) and in the lower Albian (Al1) (Herrle et al., 2004). No carbon isotope record has previously been constructed through the YCM so we have hypothesized, based on the age constraints that have been detailed, that the prominent negative CIE in the earliest Aptian (C3, Bralower et al., 1999) should be present within the YCM. However, if the YCM is Berriasian to Valanginian, it is possible that there will be either a very small or no major CIE, with the exception of a minor positive excursion associated with the Valanginian "Weissert Event" (Erba et al., 2004).

LOCALITY DESCRIPTIONS
Near Andrew's Site

This location is named for a fossil locality found by the paleontologist for the St. George Dinosaur Discovery Site, Andrew R.C. Milner (McDonald et al., 2010). Andrew's Site is located close to Yellow Cat Road, approximately 9.5 km south of interstate I-70 in the upper part of the YCM, a few meters below the crest of a ridge that is capped by the Poison Strip Member. The type locality of the basal ankylosaur *Gastonia burgei* and the giant dromaeosaur *Utahraptor ostromaysorum* also occurs in the upper YCM,

1.4 km south–southwest of this location and a few meters lower stratigraphically (Kirkland et al., 1993, 1998). Samples for carbon isotope analyses were collected from a section 14.75 m thick, on the west side of Yellow Cat Road (Fig. 2A) at the type locality of *Nedcolbertia justinhoffmani* (Kirkland et al., 1998).

Fig. 2 Images of the measured and sampled sections. (A) Near Andrew's Site, (B) Doelling's Bowl, (C) Yellow Cat Road Section where the samples were taken above the first occurrence of carbonate because it was covered by overburden, and (D) Lake Madsen Section, where samples were taken from several short sections due to thick colluvial cover of Poison Strip Sandstone boulders. Note redoximorphic paleosols of the lower Yellow Cat Member. Inset of D shows the section from a different viewpoint to show the lacustrine units.

Here, the contact with the underlying Morrison Formation is often gradational and difficult to place but likely occurs at the base of a red sandy interval with floating chert pebbles averaging 5–10 mm in diameter. This red sandy interval occurs below distinctive silcretes associated with silicified roots that occur just below the start of the stratigraphic section sampled. These silcretes are present near the base of the CMF in other localities, including the Doelling's Bowl location (see the next section) as well as in localities in the San Rafael Swell (Greenhalgh and Britt, 2007; Kirkland and Madsen, 2007). However, these silcretes, to date, have not been identified directly in contact with classic Morrison Formation lithologies. The overlying lower 3.5 m of the Near Andrew's Site Section consists of red and green mottled mudstone with occasional small black silicified root traces and blocky peds. Colors become more gleyed and less brightly variegated up-section, shifting to primarily weak red to light purple with green and yellow mottled mudstone before a sharp contact with a 75-cm-thick nodular limestone bed (calcrete) that is also color mottled. Above the limestone bed, the section consists of dark reddish gray calcareous mudstone with green mottles and occasional soil carbonates. Laterally, the small theropod *Nedcolbertia* was recovered from this horizon in a pale greenish gray to olive gray mudstone that contains floating crystals of barite, lungfish tooth plates, other fish teeth, hybodont shark material, and small crocodilian teeth (Kirkland et al., 1998). The first sandstone bed occurs at about 11.25 m above the Morrison Formation contact. Laterally, this horizon contains dinosaur tracks (Lockley et al., 1999; Kirkland and Madsen, 2007) that can be traced to the east between the upper and lower bone horizons at Andrew's Site (McDonald et al., 2010). The Poison Strip Sandstone at this section is somewhat thin (less than 5 m), but thickens rapidly both to the west and the east of the road forming the crest of prominent escarpments.

Doelling's Bowl Section

The Doelling's Bowl area is located ~3 km east-northeast of the Near Andrew's Site location, and here the YCM is much thicker than Near Andrew's Site (about 30.25 m thick). This area has been the site of numerous new Cretaceous dinosaur discoveries in the lower YCM (Kirkland and Madsen, 2007; Kirkland et al., 2012). The base of the measured section begins about 120 cm above what is confidently identified as the Morrison Formation (Fig. 2B). The lowest two meters consist of partially covered sandy units (sandy mudstone to pebble conglomerate). Between 5 and 13 m above the base of the section, the lithology consists of reddish gray to dark reddish gray noncalcareous mudstones and sandstones with green mottles, Fe-oxide nodules and cements, small root traces, and at least two thin discontinuous silcrete horizons at 6.5 and 8.4 m above the base of the section. These silcretes are interpreted to be correlative with silcretes that are present at the base of the main fossil-producing strata in Doelling's Bowl. Some of these siliceous beds form low root mounds with sinuous and tapering root structures or meter-scale laminated mounds that may represent spring seeps. An approximately

1- or 2-m-thick nodular limestone (calcrete) separates the lowermost 14 m of stacked mature paleosols of the lower YCM from the uppermost ~15 m of the section. Also above the calcrete in Doelling's Bowl are ganoid fish scales and juvenile sauropod bones. The overlying upper YCM Section consists of dark reddish gray calcareous mudstone with green mottles and thin beds of sandstone, limestone, carbonate nodules, slickensided mudstones, ostracodes, and root traces. The first major sandstone beds of the Poison Strip Member occur at 30.25 m above the basal contact of the YCM with the Morrison Formation and are 7.25 m thick, comprising cross-bedded coarse sand and pebble conglomeratic facies that are locally ridge forming.

Yellow Cat Road Section

The Yellow Cat Road Section is located east of the Doelling's Bowl Section by approximately 2.5 km (Figs. 1 and 2) and is approximately 16 m thick (Fig. 2C). The section was sampled below a steep face of the Poison Strip Sandstone. The lowermost mudstones are grayish-red in color and fissile and are overlain by red mudstones with large slickensides and small carbonate nodules that are between about 1 and 2.5 m above the base of the section. A thin fine-grained sandstone at 2.8 m separates the lower red mudstone from a pale red mudstone with large slickensides and small carbonate nodules and filled vertical cracks. A second thin, fine-grained sandstone at about 7.5 m above the base of the measured section occurs at the base of a silty mudstone unit with abundant small carbonate nodules, large drab redox patches or mottles, and large cylindroid carbonate nodules as well as small vertical trains of carbonate nodules. This unit is about 5 m thick. A third small, fine-grained channel sand separates this unit from a light red and light greenish gray silty mudstone with large slickensides, large palustrine carbonate nodules, and large redox mottling. This unit occurs between about 12.5 and 14 m above the base of the section and is bounded at its upper contact by the Poison Strip Sandstone.

Lake Madsen Section

The "Lake Madsen Discovery Section" (here referred to as the Lake Madsen Section) is located 1 km southwest of the Near Andrew's Site Section. The YCM is stratigraphically much thinner than the Doelling's Bowl Section (Fig. 2D). The lower part of the section is dominated by redoximorphic paleosols of highly mottled, poorly sorted, red, purple, and green mudstones with floating granule-sized pebbles throughout. There are abundant root traces and metal oxide nodules suggesting highly weathered soils. No carbonate was observed in the lower 5 m of the section. At approximately 5.25 m, small, light tan-colored carbonate nodules appear within much paler purple, pink, and green mudstones. The size of carbonate nodules increases up-section. At 6.75 m, the first notable carbonate-bearing unit occurs as a tan sandy limestone. At 7 m, dense well-cemented carbonate similar to a "calcrete" occurs with domains of red and orange sandstone surrounded by micritic carbonate. At approximately 7.75 m, there is a sharp transition to dark green, laminated, very fine-grained sandstone and mudstone interpreted to represent a lacustrine setting. This appears to be the deepest water facies of an extensive lake or wetlands and was found to contain the only pollen known to be preserved in the YCM, as well as abundant ostracodes, conchostracans, fish, and fossil material pertaining to lungfish and hybodont sharks. This unit alternates between dark green-gray and brown laminated mudstone and sandstone that contains occasional thin lenses of ostracod grain stones (Sames et al., 2010). At approximately 16.25 m above the base of the section, the lacustrine mudstone is interbedded with the first sandstones of the overlying Poison Strip Sandstone.

METHODS

Samples from the Near Andrews Site, Doelling's Bowl, and Lake Madsen Sections were collected every 25 cm. Samples for the Yellow Cat Road Section were collected at random intervals for making thin sections since the original goal of sampling at this locality was to assess paleosols in the section and not to conduct a high-resolution carbon isotope study. Hand samples from each section were crushed using mortar and pestle to produce approximately 1–2 g of fine powder. These samples were decarbonated with 3 M hydrochloric acid (HCl) at 60°C for approximately 4–8 h. The samples were then rinsed to neutrality using deionized water and dried for at least 24 h at 50°C in an oven. Samples were analyzed at the University of Texas at San Antonio (UTSA) for the Near Andrew's Site and Doelling's Bowl Section, the Keck Paleoenvironmental and Environmental Stable Isotope Lab (KPESIL) at the University of Kansas for the Yellow Cat Road Section, and at the University of Arkansas Stable Isotope Laboratory (UASIL) for the Lake Madsen Section.

At UTSA, approximately 80 mg from each sample were analyzed by combustion on a Costech 4010 Elemental Analyzer at 1000°C coupled to a Thermo Finnigan Delta PlusXP isotope ratio mass spectrometer. OC isotope values ($\delta^{13}C_{org}$) are reported relative to Vienna Pee Dee Belemenite (VPDB). Precision was monitored by daily analyses of National Institute of Standards and Technology (NIST) standard 1547, peach leaves with an average value of −26.2‰ and $\sigma = 0.2$. At UASIL, depending on the amount of OC present in samples, approximately 3–18 mg were weighed into tin capsules and pressed closed. Based on preliminary samples, we know that organic C preserved within strata from the CMF is low (<1%). The tin capsules were then combusted in a Thermo Finnigan NC2500 Elemental Analyzer attached to a Thermo Finnigan Delta Plus IRMS at the UASIL to determine. Instrument stability and precision was monitored via analysis of UASIL lab standards Black Weeks 036 with an average value = −25.09‰ and $\sigma = 0.676$, White River trout with an average value $= -26.63‰$ and $\sigma = 0.118$, and corn maize with an average value $= -11.32‰$ and $\sigma = 0.059$, to within 0.8‰ of actual values. Samples at the KPESIL were analyzed on a Costech 4010 Elemental Analyzer connected to a Thermo MAT 253 IRMS. Montana Soil (NIST Ref. Mat. 2711) and Peach Leaves (NIST Ref. Mat. 1547) standards were used to determine precision at KPESIL, and results compiled over 2 years show the precision to be better than $\pm0.22‰$.

RESULTS

Fig. 3 shows the results of the carbon isotope analyses relative to the stratigraphic location of the samples. To reduce noise of the individual values, and to show the overall trend, the carbon isotope curves shown are 3-point running averages. Similar moving averages have been applied to other chemostratigraphic records (cf. Bains et al., 2003; Herrle et al., 2004). CIEs are labeled for descriptive purposes for any of the sections that contain notable CIEs. All results are found in Appendix I.

Near Andrew's Site

The $\delta^{13}C$ of bulk OC in this section fluctuates between $-25‰$ and $-28‰$ with five pronounced cycles of fluctuation and do not appear to correspond with any lithological change. The oscillatory pattern of the $\delta^{13}C_{org}$ values does not allow us to identify any specific positive or negative carbon isotope trends and likely represents cycles in lake productivity or periodic changes in detrital organic matter input.

Doelling's Bowl Section

The $\delta^{13}C_{org}$ values in the first ~5 m of the Doelling's Bowl Section are relatively invariant with a slight decrease (isotope segment a, Fig. 3B). A first positive step in $\delta^{13}C$ isotope values (segment b, Fig. 3B) occurs just above 6 m from the base of the section with values decreasing to $-27.6‰$ (segment c, Fig. 3B). The $\delta^{13}C_{org}$ values then remain low for approximately 10 m (segment d, Fig. 3B), before increasing in $\delta^{13}C$ to $-26.5‰$ at 16 m above the base (segment e, Fig. 3B). Between 16 and 23 m, the $\delta^{13}C_{org}$ values fluctuate from $-28.5‰$ to $-27‰$. Above 23 m, the $\delta^{13}C$ values show a more protracted decrease, then increase (segments f and g, Fig. 3B), before returning to a more oscillatory pattern.

Yellow Cat Road Section

In the Yellow Cat Road Section, $\delta^{13}C_{org}$ values show several CIEs but overall display a negative carbon isotope trend. Values for samples from 0 to 2.5 m decrease to approximately $-27.5‰$ (segment a, Fig. 3C), and values then increase overall from 2.5 m to the

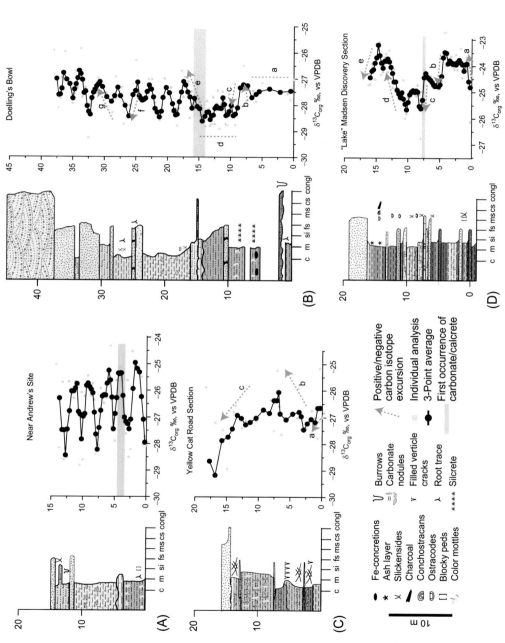

Fig. 3 Measured stratigraphic columns and C-isotope chemostratigraphic profiles for each of the four sites analyzed in the Yellow Cat Member. The gray bar in the stratigraphic curves represents the first occurrence of carbonate or the characteristic "calcrete" that Kirkland and Madsen (2007) and Suarez et al. (2014) consider the boundary between the lower and upper Yellow Cat Member. The Yellow Cat Road Section was sampled above the first occurrence of carbonate. Regionally the presence of carbonate represents a large-scale change in climate for the area. Considering the proximity of the sites to each other (within 8 km), this change in climate likely affected all sites at approximately the same time and is used as a regional correlation. Carbon isotope excursions and segments are indicated by the *letters* and are described in the text. The carbon isotope curves from the four localities show significant scatter and no obvious, reproducible isotope trends. Explanations for this disparity are explored in the text.

most enriched values ($\sim$−25.2‰) at $\sim$7 m (segment b, Fig. 3C). Above this, the δ^{13}C curve for the remaining uppermost section decreases to a low of −29.5‰ (segment c, Fig. 3C).

Lake Madsen Section

For samples from the Lake Madsen Section, the δ^{13}C of bulk OC ranged from −22.72 to −26.76‰ with a mean value of −24.49‰ (Fig. 3C). From 0 to 0.75 m, there is a sharp increase in values followed by a more gradual increase to just below 5 m (segment a, Fig. 3D). There is a sharp two-step negative trend in the δ^{13}C values at about 5 m and then again at 7.75–8 m with a minimum of −26.76‰ (segments b and c, Fig. 3D). The δ^{13}C values oscillate slightly before increasing to a maximum of −22.95‰ at 14.50 m (segment d, Fig. 3D) and then decrease toward the top of the section at 16.25 m (segment e, Fig. 3D).

DEPOSITIONAL ENVIRONMENT INTERPRETATION

Environmental interpretation of the contact between the Morrison Formation and CMF where the YCM crops out is complex. By its nature, the contact represents a hiatus and as a result the depositional environment interpretation is obscure. Here we recognize a lower depositional unit that is characterized by noncalcareous, reddish mudstones and sandstones that contain yellow, green, and purple mottles present at the base of the Lake Madsen Section, the Doelling's Bowl Section, and the Near Andrew's Site Section, and the presence of Fe-oxide nodules, and silcretes at the Doelling's Bowl Section and the Near Andrew's Site Section. This depositional unit is composed primarily of reworked material from the underlying Morrison Formation (Hunt et al., 2011; Sprinkel et al., 2012). Patchy color mottles and Fe-oxide nodules suggest alternation between seasonally water-saturated conditions and seasonally well-drained drier conditions. Environmental conditions likely transitioned to more arid conditions, as represented by the calcrete that is well developed at the Near Andrew's Site Section, Doelling's Bowl Section, and Lake Madsen Section. Overlying this regional calcrete, the YCM includes calcareous drab shales and mudstones with fossils that indicate an overall lacustrine to palustrine facies that extends up-section to the base of the Poison Strip Member, but are replaced by carbonate-bearing paleosols up-section and laterally to the east and west.

CARBON ISOTOPE CORRELATIONS IN THE YCM

As a means of providing a first-order correlation or starting point, we use the first occurrence of carbonate as a stratigraphic datum with which to compare these sections, specifically the meter-scale calcrete that has previously been associated with the Morrison-Cedar Mountain boundary (Kirkland et al., 1997, 1998; Aubrey, 1998; Stikes, 2006; Greenhalgh and Britt, 2007). This horizon is used as an independent datum (gray shading in Figs. 3 and 4), as it occurs in all sections with the exception of the Yellow Cat Road Section. The majority of the Yellow Cat Road Section contains carbonate nodules, and is likely above the calcrete.

Although we prefer not to correlate these sections based on lithology alone, due to the variability of facies in continental depositional environments, the presence of carbonate at the same interval within each of the sections suggests a large-scale climatic shift to drier environments (Birkeland, 1999), and this feature has been used as a regional marker bed by many others (Aubrey, 1998; Kirkland and Madsen, 2007; Greenhalgh and Britt, 2007; Suarez et al., 2014). Previous research has suggested the calcrete is the result of a rain shadow initiated by the uplift of the Sevier Orogeny (Suarez et al., 2014; Hatzell, 2015). This climatic shift likely influenced all the sections we analyzed at approximately the same time, and therefore may represent an effectively isochronous horizon for the study area. Using this change in facies as a stratigraphic datum, we can produce more confident correlations with what would otherwise be almost uncorrelatable data.

In addition to providing a means of correlation, the presence of the calcrete also implies that a fairly significant unconformity exists between the lower and upper portions of the YCM. The calcrete occurs as a horizon of coalesced nodules and in some places in the CMF is more than a meter thick and exhibits characteristics of stage III (continuous layer often overlying a nodular horizon) to stage VI (cracking, recementation, autobrecciation, some laminar and pisolitic carbonate) morphologies in carbonate horizon development (Gile et al., 1966). The maturity of carbonate horizon development suggests the horizon may encompass tens of thousands to potentially millions of years of time (Gile et al., 1966; Retallack, 1997). In fact, Greenhalgh and Britt (2007) suggest the calcrete represents a "multimillion"-year depositional hiatus. We also note that dinosaur species found below the calcrete have thus far not been found in the YCM above the calcrete. This apparent biostratigraphic zonation has been documented for four dinosaur groups: iguanodontians, polacanthid ankylosaurs, basal macronarian sauropods convergent on titanosaurs, and dromaeosaurs (Kirkland et al., 2012). As dinosaur genera turn over about every 1–5 million years (Wang and Dodson, 2006; Wills et al., 2008), the time represented by this unconformity must represent at least a few millions of years.

Despite being aided by lithostratigraphic correlation via the calcrete and a tenuous dinosaur fauna biostratigraphic zonation, the lack of consistency between carbon isotope curves means there is no unified interpretation for all sections in the sample area with which we are completely confident. As a result, we present two hypotheses for correlating these units (Fig. 4A and B).

Hypothesis 1

The increase and decrease in $\delta^{13}C$ values composed of isotope segments a and b (Fig. 4A) in the Lake Madsen Section and the segments b and c (Fig. 4A) in the Doelling's Bowl Section are correlated to each other as well as the lowest increase and decrease in values at

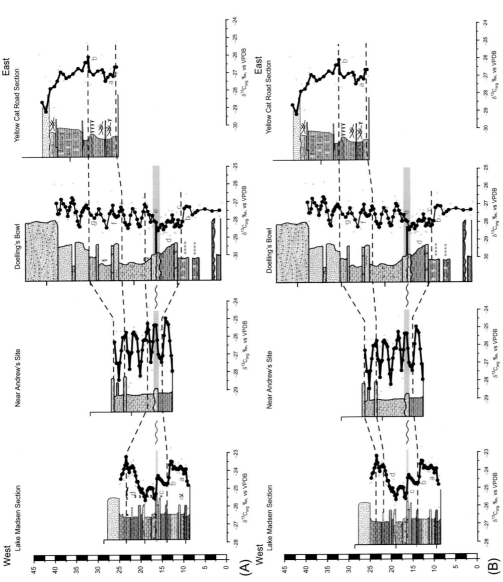

Fig. 4 Two potential Yellow Cat Member correlations. (A) Hypothesis 1 correlation suggests the Lake Madsen Section negative to positive trend bounded by the decreasing trend in d and increasing trend in correlates to the same trend bounded by the decreasing $\delta^{13}C$ trend c and increasing trend e in the Doelling's Bowl Section. The overlying "saddle" bounded by f and g in the Doelling's Bowl Section is subsequently correlated to a similar "saddle" bounded by a and b in the Yellow Cat Road Section. (B) Correlation for hypothesis 2, which suggests that the negative to positive trend bounded by the b+c and d in the Lake Madsen Section correlates to the "saddle" bounded by c and e in Doelling's Bowl. The overlying correlations include a small decrease near the apex of d in the Lake Madsen Section correlated to the "saddle" bounded by f and g in the Doelling's Bowl Section and the saddle bounded by a and b in the Yellow Cat Road Section. The protracted decreasing limb of the Yellow Cat Road Section is then correlated to the decreasing trend in $\delta^{13}C$ of the Lake Madsen Section.

the Near Andrew's Site Section (Fig. 4A). The next part of the YCM up-section that can potentially be correlated includes the transition to more calcareous horizons. In the Lake Madsen Section, this encompasses segments c and d (Fig. 4A) and in the Doelling's Bowl Section, segments d and e (Fig. 4A). Lastly, isotope segments f and g (Fig. 4A) in the Doelling's Bowl Section are correlated to isotope segments a and b (Fig. 4A) in the Yellow Cat Road Section. This decrease and increase in $\delta^{13}C$ values is correlated to the uppermost decrease and increase in $\delta^{13}C$ in the oscillatory pattern observed in the Near Andrew's Site Section, which is also supported by the lithostratigraphic correlation between the two sections.

Hypothesis 2

Like Hypothesis 1, the increase and decrease in $\delta^{13}C$ represented by isotope segments a and b in the Lake Madsen Section and b and c (Fig. 4B) in the Doelling's Bowl Section are correlated. The overlying correlation differs from hypothesis 1 in that isotope segments c and d of the Lake Madsen Section are correlated with a thicker portion of the Doelling's Bowl Section: isotope segments d, e, and the overlying oscillatory pattern just before isotope segment f (Fig. 4B). This correlation maintains a more consistent stratigraphic thickness between Lake Madsen, Near Andrew's Site, and Doelling's Bowl Sections than hypothesis 1. The overlying unit correlations are based on the correlation between Yellow Cat Road Section a and b segments with the f and g segments of the Doelling's Bowl Section. Like hypothesis 1, hypothesis 2 suggests a correlation between the uppermost decrease and increase in $\delta^{13}C$ at Near Andrew's Site. However, hypothesis 2 differs from hypothesis 1 in that it also encompasses a small decrease and increase in $\delta^{13}C$ that occurs within the overall increasing trend of isotope segment d (Fig. 4B) in the Lake Madsen Section.

DISCUSSION

Data from the YCM are an example of major challenges to using C-isotope chemostratigraphy in continental settings, especially with limited independent chronostratigraphic constraints (i.e., ash, pollen, or other microfossils). Additionally, correlation is difficult in these localities, which may be influenced by local processes that could affect sedimentary OC.

As is evident in all sections discussed here, and in previous studies, low OC preservation is a major challenge to ecological and chemostratigraphic studies (Brodie et al., 2011a,b; Bataille et al., 2013). When sections have low OC preservation, they can become easily contaminated by recycled or exogenous carbon or biased by the type of organic matter preserved. Also, samples with low OC are more likely to be altered by preanalysis sample preparation (Brodie et al., 2011a,b).

Brodie et al. (2011a,b) suggests that the decarbonation method can greatly alter the $\delta^{13}C_{org}$ composition of sedimentary organic matter. Three main methods exist for decarbonation: (1) the in-capsule method, (2) the fumigation method, and (3) the rinse method. The in-capsule method involves adding powdered sample to the silver or tin capsule used for analysis, then dropping small amounts (microliters) of acid (typically HCl) at varying molarity until the reaction has come to completion. Samples are then dried, and then wrapped again in capsules for analysis via combustion elemental analyzer. However, this method can be imprecise because instances of incomplete dissolution of carbonate have been documented (Brodie et al., 2011a,b), especially if the carbonate present is a less soluble form of carbonate such as dolomite or siderite. Another problem is that the dissolution of acid-soluble organic matter when using higher molarity (>3 M) acid can bias the final $\delta^{13}C_{org}$.

The fumigation method involves exposing the samples to acid fumes under vacuum to dissolve the inorganic carbon within the powdered sample. A small amount of sample, typically 90–500 μg, is added to silver capsules in a vacuum desiccator. The samples are placed in a tray above ~50 mL of an acid (typically HCl), and a small amount of water (~50 μL) is added to the sample. This allows acid fumes to dissolve into the water creating an acid solution that dissolves carbonate. Samples are fumigated in the capsules for ~6 h, then dried and wrapped shut for analysis. Although this method has been used with some success for modern soil studies, Brodie et al. (2011b) found significant variability that obscured the true value of the organic matter. For samples with high amounts of inorganic carbon (carbonate), the fumigation method tends to result in incomplete decarbonation and greater $\delta^{13}C$ values relative to the true value of OC.

The method used here, the rinse method, involves adding 30 mL of HCl to a specific amount of crushed sample (~1 g). The molarity of acid used varies between labs but usually ranges from 0.5 to 6 M. We used 3 M HCl for all of the samples presented here. Some methods also heat the samples to anywhere from 50°C to 60°C in order to dissolve any sulfur-rich molecules (e.g., pyrite) or accelerate the reaction if less soluble carbonates are suspected to be present, such as dolomite and siderite. Acid is either replaced at consistent intervals or acid pH is tested regularly to test the completion of the reaction. After no reaction is observed in the solution and the pH remains acidic, samples are centrifuged and the remaining solution is decanted off. Samples are then rinsed with deionized water, shaken, and decanted approximately three times or until the sample reaches neutrality. The number of times it takes a sample to reach neutrality will vary significantly depending on the clay content of the samples. Samples containing hydrophilic clays may take significantly longer to neutralize. It is critical that samples reach neutrality in order to avoid harsh acidic chemicals being inadvertently added to the mass spectrometer. Centrifugation at >3000 rpm is suggested to alleviate the chance for loss of fine particles of organic matter, a major source of bias observed by Brodie et al. (2011b). The rinse method is the preferred method suggested here. We also suggest using at least 3 M HCl because several authors have indicated that using HCl as low as 0.1 to 0.5 M results in incomplete reaction of the carbonate (Midwood and Boutton, 1998; Brodie et al., 2011a,b and references therein). However, using acid at much higher molarity (e.g., 6 M, or the "flash acid" method) can result in dissolution of more soluble organic matter such as carbohydrates, carboxyl groups, aliphatics, and ligins (Midwood and Boutton, 1998).

Preservation of organic remnants can bias the $\delta^{13}C$ value and is a major limitation, if not the major limitation, to C-isotope chemostratigraphy in continental settings, especially in fluvial and overbank sediments in well-drained oxygenated soils. In a detailed study by Bataille et al. (2013) of the floodplain sediments spanning the PETM, the type of organic matter was a strong control of the chemostratigraphic profile generated. In fact, in one section the type of organic matter preserved was composed of recycled and ^{13}C-enriched organic matter and overwhelmed the dispersed organic matter (bulk organic matter), thereby partially obscuring the "typical" PETM C-isotope curve. In Bataille et al. (2013), the section with the lowest amount of preserved organic matter had a greater likelihood of preserving recalcitrant organic matter such as vitrinite and other organic molecules that were likely recycled from "non–PETM sediments." In sediments with low soil moisture where the degree of organic matter oxidation is high, the proportion of recycled organic matter is probably higher as it is more likely to be preserved than indigenous organic matter, resulting in a more enriched $\delta^{13}C$ value. Thus, poorly drained soils with greater potential for preservation of organic matter are preferred and result in C-isotopic compositions that are more reminiscent of plant matter.

The nature of continental sedimentation is much more dynamic than marine depositional systems. Migration of rivers across the landscape, erosion, differences in geomorphology, and myriad other processes result in well-drained vs poorly drained soils, and the diachronous nature of floodplain sediments can result in variable depositional rates and unconformities of unknown amounts of time. In this case, the time-transgressive nature of the floodplain is likely an even greater factor as the base of the CMF sits unconformably on the Jurassic Morrison Formation. As described in Greenhalgh and Britt (2007), there was probably low aggradation space for deposition in the lower parts of the CMF, resulting in weathering and modification of the underlying Morrison Formation via pedogenesis. The variable nature of depositional rates and presence of unconformities of variable durations can result in carbon isotope curves that have missing CIEs, are condensed, expanded, or vary significantly from one locality to another making the sections difficult to correlate. One way to potentially alleviate the problem of expanded or condensed isotope records is to convert the vertical thickness of the section to time. This can be done using a variety of methods such as lithologic depositional rate, astronomically tuned depositional rates, and high-precision geochronology. For example, Whiteside et al. (2010) converted their strata to an age model that was then compared to an astronomically calibrated geomagnetic polarity time scale generated by Kent and Olsen (1999) for the Newark/Hartford basin to place their chemostratigraphic profile relative to time rather than vertical thickness. In continental settings, however, where sedimentation is discontinuous, and biological processes add to the complexity of the sedimentation (resulting in vertical mixing of sediment), constructing an astronomically calibrated time scale becomes almost impossible.

Despite the many challenges to using chemostratigraphy in continental settings, the preservation of $\delta^{13}C_{org}$ can result in chemostratigraphic curves that can be correlated to global records (cf. Jahren et al., 2001; Ando et al., 2002; Robinson and Hesselbo, 2004; Suarez et al., 2013) when a high enough proportion of organic matter is preserved, or if steps are taken to resolve the origin and thermal maturity of low-OC samples (Robinson and Hesselbo, 2004; Bataille et al., 2013). One must recognize the limitations of the technique, identify when the technique should be called into question, and carefully choose strata that are appropriate for C-isotope chemostratigraphic study. Here are some suggestions for successful implementation of this stratigraphic method.

SUGGESTIONS FOR CHEMOSTRATIGRAPHIC STUDIES

For successful implementation of carbon isotope chemostratigraphy in continental depositional systems, we recommend the following tactics, constraints, and methodologies:

1. Longer chemostratigraphic profiles that can be statistically correlated
2. Higher-resolution sampling
3. Adding further geochronologic or other stratigraphic constraints
4. Horizon or stratigraphic datum selection for correlation
5. Isolating OC and sample preparation

Long Chemostratigraphic Profiles and Statistical Correlation

Until successful statistical methods of correlation are developed, such as statistical cross correlation that can account for unconformities and changes in depositional rate (Waterman and Raymond, 1987), C-isotope chemostratigraphy is very much "wiggle-matching." Although statistical correlation methods do exist and have been discussed by various authors (e.g., Waterman and Raymond, 1987; Sageman and Hollander, 1999; Lisiecki and Lisiecki, 2002), they are rarely applied to C-isotope chemostratigraphy. These methods need to become more commonplace so that we can place some degree of statistical confidence on correlations. The necessary methods required for more accurate chemostratigraphic correlation may only need slight modification to programs or code that already exists in other geoscience disciplines (e.g., geophysics, dendrochronology, sclerochronology).

When given the option to sample a short stratigraphic interval or a more expanded stratigraphic interval, we recommend the latter. Longer, more expanded sections tend to encompass a longer period of time. In continental sections where chronostratigraphic constraints are limited, longer sections allow improved correlation as they offer a more extended carbon isotope profile that is more likely to capture the CIE of interest and also allow the recognition of unconformities, as gaps in the carbon isotope profile may be more visible. That being said, when sampling and measuring section, careful observation of potential long-term unconformities should be noted. Observations that may indicate unconformities are present in a section can include deposition of course-grained clastics such as cobble-rich, fluvial deposits, especially those that show erosive lower boundaries, and extensive paleosol development.

High-Resolution Sampling

Higher-resolution sampling could also help resolve problematic correlations between and among carbon isotope curves. In the current study, the samples were collected every 0.25 m wherever possible. In reexploring these sections, it may be useful to sample at a higher resolution of 0.10 m, for example, which may help resolve trends that occur in stratigraphically condensed levels. For conducting preliminary studies to understand the general carbon isotope trends through a thick succession, 1-m intervals may be sufficient. Subsequently, 0.05–0.5-m sampling resolution may be employed, once one has some constraint on the section. Ultimately, regardless of sampling resolution or time constraints in the field, it is important to sample systematically instead of randomly or based on lithological changes.

Geochronologic Constraints

Time constraints are of utmost importance when it comes to documenting chemostratigraphic profiles. Some amount of absolute age control is needed in continental sections to limit the range of carbon isotope curves to which to correlate local sections. Radiometric ages from ash beds near the top and bottom of the section would provide the most ideal situation for chronostratigraphic control; however, this is not common in continental depositional environments. Biostratigraphy, palynology, or magnetostratigraphy can also be useful, although biostratigraphy in continental sections is generally not resolved at a high enough resolution to allow us to constrain any given taxon or assemblage of animals to a specific stage boundary. Biostratigraphic data can provide constraints to an epoch and in some cases, the first appearance of dinosaur families may be associated with a specific stage. Pollen may provide some chronostratigraphic control where certain pollens have a limited range in the geological record (cf. Bebout, 1981; Goodwin et al., 1999; Markevitch, 1994; Doyle and Robbins, 1977). However, the preservation of pollen and taphonomic bias can present significant problems with the use of palynostratigraphy. Magnetostratigraphic data is also at risk of depositional bias, including erosional events removing the record of chrons that would otherwise be recorded in the strata. Additionally, sedimentary rocks carry a magnetic remanence via either a detrital signal, susceptible to inclination-shallowing effects, and/or a chemical signal, which can incorporate a significant time lag (see Zeigler and Kodama, Chapter 5, this volume).

When volcanic horizons are absent, stratigraphers may turn to detrital zircon analyses for determining maximum depositional age, which may help to constrain the age of deposition of portions of stratigraphic sections (cf. Greenhalgh and Britt, 2007; Dickinson and Gehrels, 2009). High-precision U/Pb dates of carbonates have also provided useful age constraints for stratigraphic sections where appropriate carbonate beds are available (cf. Rasbury et al., 1997; Wang et al., 1998; Rasbury et al., 2000).

Horizon Correlation Selection

Despite variability in deposition in continental depositional environments, stratigraphic workers should also take into consideration lithostratigraphic correlation. In this study, a horizon of carbonate nodules or indurated calcrete that occurs within the YCM has been identified as a useful correlation horizon or datum because it is present and identifiable in all sections, although the horizon differs in thickness between sections. Here, the assumption is made that this horizon is likely of similar age because it probably represents a regional change in climate that would affect the deposition and soil development of the entire Yellow Cat depositional system. For example, within the Ruby Ranch Member, a regionally extensive calcrete is also present and has also been used to help correlate different localities in the Price River area in the northern San Rafael Swell (Ludvigson et al., 2015). Identifying possible or obviously correlative surfaces ultimately allows us to independently correlate carbon isotope curves, and may help constrain the "wiggle matching" between local and regional sections.

Isolating OC

The idea behind bulk organic matter C-isotope chemostratigraphy is that the organic matter sampled represents the isotopic composition of the surrounding plant material. That plant material is influenced by shifts in the isotopic composition of the atmosphere, which is sensitive to global carbon cycle perturbations. Since carbon isotope chemostratigraphy depends on preserved OC, the more organic rich the sediment the greater likelihood that the $\delta^{13}C$ profile generated will reflect that of plant material. For sediments that are low in organic matter, there is a higher likelihood of contamination by recycled material or during the preparation process and/or bias by ^{13}C-enriched recalcitrant organic matter such as vitrinite.

Sections that are deposited in overall wet conditions with OC values ~1%–3% or more are preferred. If sections do not meet those criteria, chemostratigraphic study can still be possible by also analyzing pedogenic carbonates, careful sample preparation, and detailed characterization of the organic matter preserved. For lacustrine deposits, care needs to be taken that changes in lake productivity are not biasing the $\delta^{13}C$ profile.

For example, as lake productivity increases, the amount of ^{13}C-depleted material buried increases, thereby making surface waters enriched in ^{13}C. Carbonate production will therefore be enriched in ^{13}C relative to OC. Developing a carbonate $\delta^{13}C_{carbonate}$ curve will assist in identifying these instances. A negative correlation between $\delta^{13}C_{org}$ and $\delta^{13}C_{carbonate}$, as well as a negative correlation between $\delta^{13}C_{org}$ and OC, is suggestive of lake productivity controlling the $\delta^{13}C$ profile rather than atmospheric $\delta^{13}C$ perturbations.

In low-OC settings, organic matter contamination and preservation bias can obscure the indigenous nature of the $\delta^{13}C$ chemostratigraphic curve. In this setting, it may be worth separating the organic matter fraction and identifying the OC present via petrographic analysis, which will also allow for evaluation of vitrinite reflectance and thermal maturity. Bataille et al. (2013) and references therein give a detailed description of methods to separate and assess disparate organic matter preserved in deposits low in total OC.

Careful sample preparation is important to the integrity of $\delta^{13}C$ chemostratigraphic profiles. We prefer the rinse method for sample decarbonation and typically use 3 M HCl, and test the pH before rinsing and after rinsing to be sure all acid is removed from the supernatant. This is especially important in highly smectitic deposits as the polar nature of clay particles causes acid molecules to "stick" to the clay particles. Another crucial step is that all samples are centrifuged at ≥ 3000 rpm prior to decanting the acid or rinse solution. The fine fraction of organic matter poured off may contribute to biases in the organic matter preserved as has been observed in other studies (Brodie et al., 2011a, b; Bataille et al., 2013). When samples are crushed before decarbonation and after decarbonation, care must be taken to clean mortar and pestle between samples. Potential contaminants such as lab wipes may leave residual fibers; thus mortar and pestles should be thoroughly cleaned with lab-grade (organic free) soap, organic solvent such as methanol or acetone, and dusted with compressed air and carefully checked between samples for contamination. It is also advisable to clean the pestle and mortar as usual and then place in an ultrasonic water bath for 15 min to remove any particles that may be trapped in any fractures or pits in the pestle and mortar. In samples with less than 0.5% OC, cross-contamination from pestle and mortars that are not cleaned frequently or adequately could have a serious effect on the sample. Finally, in low-OC settings it may be worth taking the time to separate the types of organic matter (Bataille et al., 2013).

CORRELATION TO GLOBAL C-ISOTOPE CURVES

Despite the problematic nature of the isotope curves we have generated from the YCM, we can attempt to provide a tentative correlation to global C-isotope records (cf. Gröcke et al., 1999; Jahren et al., 2001; Robinson and Hesselbo, 2004). We emphasize that this is a tentative correlation, and the many problems with the poor reproducibility between sections is a major caveat to this correlation. We can attempt two correlations based on the two hypothetical correlations presented previously (Fig. 4). Depending on which regional hypothesis is used, we correlate the C3 negative excursion documented by many researchers, including Jahren et al. (2001), in at least one of the Yellow Cat Sections (Fig. 5) for each hypothesis.

For correlation hypothesis 1, the Lake Madsen Section sediments are older than the C3 early Aptian negative CIE as is the majority of Doelling's Bowl as well as the section between segments a and b in the Yellow Cat Road Section (Fig. 5A). If the correlation hypothesis 2 is chosen, then the majority of the Lake Madsen Section, Doelling's Bowl, and the segments bounded by a and b in the Yellow Cat Road Section (Fig. 5B) are Barremian in age (or at least pre-Aptian). The section from the apex of d in the Lake Madsen Section, the section after the increasing $\delta^{13}C$ trend in g of the Doelling's Bowl Section, and the decreasing trend after b in the Yellow Cat Road Section are early Aptian in age and correlate to the C3 negative CIE (Fig. 5B).

CONCLUSION

Several $\delta^{13}C$ organic chemostratigraphic curves were generated from the Cretaceous CMF with the purpose of correlating the dinosaur fauna contained at each site, as well as to better constrain the age of the YCM. Results produced chemostratigraphic records that have poor reproducibility between sections. This is likely due to the low amount of OC preserved in the majority of the samples. This highlights the challenges of correlating continental depositional section to global C–isotope chemostratigraphic curves that generally are referenced to marine strata. Potential error in continental sedimentary records include preservation bias, preservation of local seasonal excursions rather than global C–isotope perturbations, contamination from reworked organic matter, contamination during sample collection and analysis, and degradation of organic matter during preanalysis preparation. In order to avoid such downfalls, strata selected for C-isotopic study should be from more organic rich strata, preferably 1%–3% OC. If this cannot be achieved, the type of organic matter should be characterized, careful sample preparation should be conducted, and additional proxies such as pedogenic carbonates should also be analyzed for comparison. To alleviate confusion due to expanded or condensed sections, samples should be plotted relative to time if possible, and evidence of major unconformities should be identified as these boundaries can contain missing important CIEs. This requires consideration of lithology and sedimentology of units that are being correlated. If possible, stratigraphically long sections should be analyzed so that there is a higher likelihood that a number of correlatable CIEs are preserved. In all cases, it is of utmost importance to make every attempt to obtain numerical age constraints.

In some settings, C–isotope chemostratigraphic analysis alone is unadvisable; however, given the caveats noted, C–isotope chemostratigraphy is still a powerful tool for

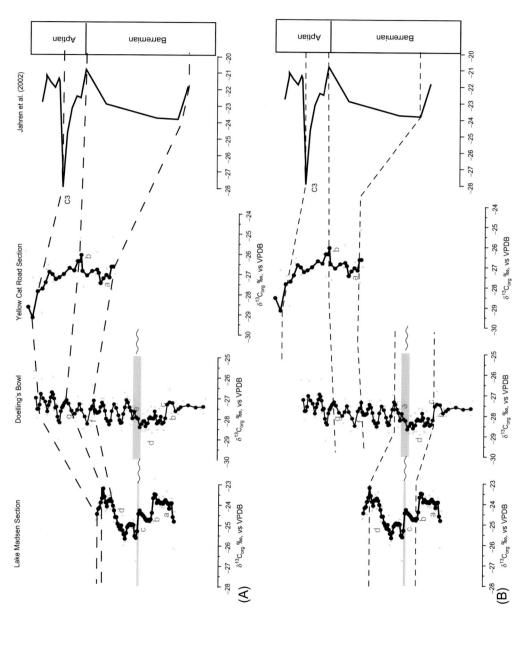

Fig. 5 Two potential global correlations. In both hypothesized correlations, the decreasing δ¹³C trend near the top of the Yellow Cat Road Section is correlated with the C3 negative carbon isotope excursion in the early Aptian identified in the data from Jahren et al. (2001). (A) Global correlation if regional correlation hypothesis 1 is chosen, and (B) global correlation if regional correlation hypothesis 2 is chosen.

correlating global C-cycle perturbations to the continental record, constraining age, and correlating sections across depositional basins. To that end we tentatively hypothesize that the carbon isotope record constructed from the YCM correlates to carbon isotope curves that characterize the Barremian to early Aptian.

ACKNOWLEDGMENTS

Funding for this project was provided by a Geological Society of America Student Research Grant to M. Suarez, startup funds from the University of Texas at San Antonio to M. Suarez, and startup funds from the University of Arkansas to C. Suarez. Additional logistical support in the field was provided by Don DeBlieux, Scott Madsen, and Gary Hunt of the Utah Geological Survey. Jim Kirkland reviewed an earlier version of this manuscript. We also thank Greg Cane from the Keck Paleoenvironmental and Environmental Stable Isotope Lab at the University of Kansas and Erik Pollock and Lindsey Conaway from the University of Arkansas Stable Isotope Lab.

APPENDIX I

Sedimentary organic carbon $\delta^{13}C$ values.

Location	Sample name	Meters	$\delta^{13}C$ ‰ vs VPDB	%OC
Lake Madsen	LMDS–contact	0.00	−24.3	0.06
	LMDS-0	0.25	−26.1	0.06
	LMDS-1	0.50	−24.0	0.04
	LMDS-2	0.75	−23.7	0.03
	LMDS-3	1.00	−24.1	0.02
	LMDS-4	1.25	−24.0	0.02
	LMDS-5	1.50	−24.4	0.02
	LMDS-6	1.75	−23.3	0.03
	LMDS-7	2.00	−23.8	0.02
	LMDS-8	2.25	−24.1	0.02
	LMDS-9	2.50	−23.6	0.02
	LMDS-10	2.75	−24.1	0.02
	LMDS-11	3.00	−24.0	0.02
	LMDS-12	3.25	−23.6	0.02
	LMDS-13	3.50	−23.9	0.02
	LMDS-14	3.75	−23.8	0.02
	LMDS-15	4.00	−24.0	0.02
	LMDS-16	4.25	−22.7	0.02
	LMDS-17	4.50	−23.7	0.03
	LMDS-18	4.75	−24.6	0.03
	LMDS-19	5.00	−24.9	0.03
	LMDS-20	5.25	−24.7	0.02
	LMDS-21	5.50	−24.8	0.03
	LMDS-22	5.75	−24.7	0.02
	LMDS-23	6.00	−24.8	0.02
	LMDS-24	6.25	−24.7	0.03
	LMDS-25	6.50	−24.4	0.03

Location	Sample name	Meters	δ¹³C ‰ vs VPDB	%OC
	LMDS-26	6.75	−24.6	0.03
	LMDS-27	7.00	−24.5	0.1
	LMDS-28	7.25	−24.2	0.08
	LMDS-29	7.50	−24.1	0.04
	LMDS-30	7.75	−25.0	
	LMDS-31	8.00	−26.8	
	LMDS-32	8.25	−25.1	
	LMDS-33	8.50	−24.7	
	LMDS-34	8.75	−25.1	
	LMDS-35	9.00	−25.0	
	LMDS-36	9.25	−24.9	
	LMDS-37	9.50	−25.0	
	LMDS-38	9.75	−25.2	
	LMDS-39	10.00	−25.2	
	LMDS-40	10.25	−25.8	
	LMDS-41	10.50	−26.0	
	LMDS-42	10.75	−24.5	
	LMDS-43	11.00	−25.3	
	LMDS-44	11.25	−25.3	0.02
	LMDS-45	11.50	−25.1	0.03
	LMDS-46	11.75	−24.8	0.04
	LMDS-47	12.00	−24.9	0.05
	LMDS-48	12.25	−24.1	0.05
	LMDS-49	12.50	−24.8	0.05
	LMDS-50	12.75	−23.9	0.07
	LMDS-51	13.00	−23.4	0.11
	LMDS-52	13.25	−24.2	0.03
	LMDS-53	13.50	−23.5	0.05
	LMDS-54	13.75	−23.7	0.08
	LMDS-55	14.00	−24.2	0.06
	LMDS-56	14.25	−24.3	0.07
	LMDS-57	14.50	−23.1	0.07
	LMDS-58	14.75	−23.4	0.18
	LMDS-59	15.00	−23.2	0.20
	LMDS-60	15.25	−24.5	0.07
	LMDS-61	15.50	−24.0	0.07
	LMDS-62	15.75	−23.9	0.07
	LMDS-63	16.00	−24.7	0.06
	LMDS-64	16.25	−24.7	0.08
Near Andrew's Site	NAS-0	0.00	−28.1	0.05
	NAS-01	0.25	−27.7	0.02
	NAS-02	0.50	−28.0	0.03
	NAS-03	0.75	−25.5	0.01
	NAS-03 (org repeat)		−26.30	0.02

Continued

Location	Sample name	Meters	δ¹³C ‰ vs VPDB	%OC
	NAS–04	1.00	−25.0	0.01
	NAS–04		−25.30	0.01
	NAS–05	1.25	−25.2	0.01
	NAS–06	1.50	−25.3	0.01
	NAS–07	1.75	−25.1	0.01
	NAS–08	2.00	−25.84	0.01
	NAS–09	2.25	−24.5	0.01
	NAS–10	2.50	−28.2	0.04
	NAS–11	2.75	−28.5	0.03
	NAS–11 (org repeat)		−28.6	0.04
	NAS–12	3.00	−25.6	0.01
	NAS–12 (org repeat)		−28.0	0.04
	NAS–13	3.25	−27.6	0.03
	NAS–14	3.50	−28.3	0.06
	NAS–15	3.75	−25.1	0.03
	NAS–16	4.00	−25.2	0.06
	NAS–17	4.25	−25.7	0.03
	NAS–17R	4.25	−25.1	0.06
	NAS–18	4.50	−25.3	0.02
	NAS–19	4.75	−25.6	0.03
	NAS–20	5.00	−27.9	0.06
	NAS–21	5.25	−27.1	0.06
	NAS–22	5.50	−27.4	0.08
	NAS–23	5.75	−24.3	0.01
	NAS–24	6.00	−25.2	0.03
	NAS–24 (org repeat)		−25.5	0.03
	NAS–25	6.25	−26.3	0.03
	NAS–26	6.50	−25.8	0.03
	NAS–27	6.75	−26.5	0.03
	NAS–28	7.00	−26.3	0.03
	NAS–29	7.25	−25.9	0.02
	NAS–30	7.50	−28.1	0.04
	NAS–31	7.75	−28.2	0.04
	NAS–32	8.00	−28.4	0.06
	NAS–33	8.25	−26.4	0.04
	NAS–34	8.50	−25.8	0.04
	NAS–35	8.75	−26.2	0.10
	NAS–36	9.00	−26.0	0.03
	NAS–37	9.25	−25.5	0.02
	NAS–38	9.50	−25.8	0.02
	NAS–39	9.75	−26.2	0.03
	NAS–40	10.00	−28.6	0.05
	NAS–41	10.25	−26.1	0.02
	NAS–42	10.50	−28.8	0.05
	NAS–43	10.75	−25.8	0.02
	NAS–44	11.00	−25.8	0.02

Location	Sample name	Meters	δ¹³C ‰ vs VPDB	%OC
	NAS–45	11.25	−25.7	0.01
	NAS–46	11.50	−26.3	0.02
	NAS–47	11.75	−26.1	0.02
	NAS–49	12.25	−25.7	0.02
	NAS–50	12.50	−28.5	0.03
	NAS–51	12.75	−28.2	0.02
	NAS–52	13.00	−28.6	0.04
	NAS–53	13.25	−25.7	0.01
	NAS–55	13.75	−28.2	0.02
	NAS–57	14.25	−25.0	0.04
Doelling's Bowl	DB–0	0.00	−27.0	0.04
	DB–1	0.25	−27.80	0.02
	DB–2	0.50	−27.58	0.02
	DB–4	5.00	−27.1	0.04
	DB–5	5.50	−27.44	0.03
	DB–6	6.00	−27.86	0.04
	DB–7	6.50	−27.46	0.02
	DB–8	7.50	−27.82	0.04
	DB–9	8.00	−26.51	0.02
	DB–10	8.50	−27.4	0.05
	DB–11	9.00	−28.2	0.04
	DB–12	9.25	−28.8	0.07
	DB–13	9.50	−28.1	0.05
	DB–15	10.00	−28.4	0.05
	DB–16	10.25	−27.6	0.03
	DB–17	10.50	−28.8	0.03
	DB–18	10.75	−28.7	0.04
	DB–19	11.00	−27.7	0.03
	DB–20	11.25	−27.5	0.02
	DB–21	11.50	−28.0	0.04
	DB–22	11.75	−28.9	0.06
	DB–23	12.00	−27.5	0.03
	DB–24	12.25	−27.9	0.03
	DB–25	12.50	−28.7	0.04
	DB–25R	12.50	−28.7	0.04
	DB–26	12.75	−27.7	0.04
	DB–27	13.00	−28.7	0.06
	DB–28	13.25	−28.8	0.06
	DB–29	13.50	−28.1	0.04
	DB–30	13.75	−27.8	0.05
	DB–31	14.00	−28.8	0.05
	DB–33	14.50	−28.6	0.05
	DB–35	15.00	−27.8	0.03
	DB–35 r	15.00	−29.0	0.03
	DB–36	15.25	−27.4	0.06
	DB–38	15.75	−28.4	0.04
	DB–41	16.25	−27.5	0.03

Continued

Location	Sample name	Meters	$\delta^{13}C$ ‰ vs VPDB	%OC
	DB–42	16.50	−28.5	0.05
	DB–44	17.00	−27.1	0.06
	DB–45	17.25	−26.94	0.02
	DB–46	17.50	−27.7	0.03
	DB–47	17.75	−26.71	0.02
	DB–49	18.25	−28.14	0.03
	DB–50	18.50	−28.8	0.05
	DB–51	18.75	−28.1	0.03
	DB–52	19.00	−28.2	0.03
	DB–53	19.00	−28.3	0.05
	DB–54	19.25	−27.3	0.05
	DB–58	20.50	−27.01	0.02
	DB–60	21.00	−28.5	0.03
	DB–61	21.25	−28.58	0.04
	DB–63	21.75	−28.33	0.05
	DB–64	22.00	−27.92	0.06
	DB–65	22.25	−28.16	0.07
	DB–66	22.75	−27.03	0.04
	DB–67	23.00	−27.03	0.02
	DB–68	23.25	−27.7	0.06
	DB–69	23.50	−27.8	0.03
	DB–70	23.75	−27.8	0.03
	DB–71	24.00	−27.9	0.03
	DB–72	24.25	−27.9	0.13
	DB–73	24.50	−27.6	0.12
	DB–74	24.75	−27.5	0.02
	DB–75	25.00	−26.3	0.02
	DB–77	25.50	−28.54	0.04
	DB–079	26	−28.02	0.03
	DB–83	27.00	−28.7	0.03
	DB–084	27.25	−27.33	0.09
	DB–088	28.25	−26.93	0.05
	DB–089	28.25	−27.77	0.03
	DB–90	28.75	−28.3	0.03
	DB–094	29.75	−27.55	0.04
	DB–097	30.5	−27.26	0.04
	DB–98	30.75	−26.9	0.02
	DB–99	31.00	−26.98	0.02
	DB–100	31.25	−27.5	0.01
	DB–101	31.50	−27.4	0.03
	DB–102	31.75	−27.4	0.01
	DB–103	32.00	−27.8	0.03
	DB–104	32.25	−28.0	0.04
	DB–105	32.50	−29.3	0.01
	DB–106	32.75	−27.5	0.03
	DB–107	33.00	−27.3	0.02

Location	Sample name	Meters	δ¹³C ‰ vs VPDB	%OC
	DB-108	33.25	−27.6	0.02
	DB-109	33.50	−27.7	0.01
	DB-110	33.75	−25.86	0.04
	DB-111	34	−26.90	0.05
	DB-112	34.25	−27.35	0.05
	DB-113	34.5	−26.72	0.02
	DB-114	34.75	−27.59	0.04
	DB-115	35	−27.94	0.03
	DB-116	35.25	−27.7	0.03
	DB-120E	36.25	−25.97	0.03
	DB-121	36.5	−27.43	0.04
	DB-122	36.75	−26.98	0.02
	DB-123	37	−27.75	0.03
	DB-124	37.25	−28.11	0.12
	DB-125	37.5	−27.02	0.03
	DB-126	38	−25.92	0.01
Yellow Cat Road	PSS-6	18.34	−25.9	0.03
	PSS-5	18.04	−32.9	0.04
	PSS-4	16.84	−27.2	0.03
	PSS-3	15.64	−27.5	0.04
	PSS-2	14.94	−28.9	0.05
	PSS-1	13.91	−26.8	0.05
	MI-III-20 m down frm top	13.64	−26.6	0.48
	MI-III-120 cm down frm top	12.64	−27.4	0.22
	MI-III-180 cm down frm top	12.04	−27.1	0.22
	MI-III-220 cm down frm top	11.64	−27.3	0.2
	MI-III-340 cm down	10.44	−27.1	0.14
	MI-III-220 cm up frm base	9.29	−26.7	0.18
	MI-III-100 cm up frm base	8.09	−26.4	0.19
	MI-III-50 cm up frm base	7.59	−27.5	0.08
	MI-III-top of bed	7.09	−25.2	0.04
	MI-III-bottom of bed	6.79	−26.4	0.04
	MI-II-top of unit	6.78	−26.6	0.06
	MI-II-70 cm down frm top	6.39	−27.7	0.07
	MI-II-180 cm up frm base	3.91	−27.0	0.08
	MI-II-160 cm up frm base	3.71	−25.9	0.06
	MI-II-120 cm up frm base	3.31	−27.4	0.06
	MI-II-80 cm up frm base	2.91	−27.5	0.06
	MI-II-40 cm up frm base	2.51	−27.4	0.1
	MI-I-50 cm down frm top	1.61	−26.7	0.06
	MI-I-100 cm down frm top	1.11	−27.0	0.08
	MI-I-150 cm down frm top	0.61	−27.8	0.15
	MI-I-170 cm down frm top	0.41	−25.2	0.12
	MI-I-210 cm down frm top	0.00	−27.0	0.12

REFERENCES

Ando, A., Kakegawa, T., Takashima, R., Saito, T., 2002. New perspective on Aptian carbon isotope stratigraphy: data from $\delta^{13}C$ records of terrestrial organic matter. Geology 30, 227–230. http://dx.doi.org/10.1130/0091-7613(2002)030<0227:NPOACI>2.0.CO.2.

Aubrey, W.M., 1998. A newly discovered, widespread fluvial facies and unconformity marking the Upper Jurassic/Lower Cretaceous boundary, Colorado Plateau. In: Carpenter, K., Chure, D., Kirkland, J.I. (Eds.), The Upper Jurassic Morrison Formation—An Interdisciplinary Study, Part I: Modern Geology. In: 22, Taylor & Francis, Routledge, pp. 209–233.

Bains, S., Norris, R.D., Corfield, R.M., Bowen, G.J., Gingerich, P.D., Koch, P.L., 2003. Marine-terrestrial linkages at the Paleocene-Eocene boundary. Geol. Soc. Am. Spec. Pap. 369, 1–9.

Bataille, C.P., Mastalerz, M., Tipple, B.J., Bowen, G.J., 2013. Influence of provenance and preservation on the carbon isotope variations of dispersed organic matter in ancient floodplain sediments. Geochem. Geophys. Geosyst. 14, 4874–4891. http://dx.doi.org/10.1002/ggge.20294.

Bebout, J.W., 1981. An informal palynologic zonation for the Cretaceous system of the United States mid-Atlantic (Baltimore canyon area) outer continental shelf. Palynology 5, 159–194. http://dx.doi.org/10.1080/01916122.1981.9989224.

Birkeland, P.W., 1999. Soils and Geomorphology, third ed. Oxford University Press, Oxford. 372 pp.

Bralower, T.J., CoBabe, E., Clement, B., Sliter, W.V., Osburn, C.L., Longoria, J., 1999. The record of global change in Mid-Cretaceous (Barremian-Albian) sections from the Sierra Madre, northeastern Mexico. J. Foraminifer. Res. 29, 418–437.

Britt, B.B., Burton, D., Greenhalgh, B., Kowallis, B., Christiansen, E., Chure, D.J., 2007. Detrital zircon ages for the basal Cedar Mountain Formation (Early Cretaceous) near Moab, and Dinosaur National Monument, Utah. In: Geological Society of America Abstracts With Programs, p. 16.

Brodie, C.R., Casford, J.S.L., Lloyd, J.M., Leng, M.J., Heaton, T.H.E., Kendrick, C.P., Yongqiang, Z., 2011a. Evidence for bias in C/N, $\delta^{13}C$ and $\delta^{15}N$ values of bulk organic matter, and on environmental interpretation, from a lake sedimentary sequence by pre-analysis acid treatment methods. Quat. Sci. Rev. 30, 3076–3087. http://dx.doi.org/10.1016/j.quascirev.2011.07.003.

Brodie, C.R., Leng, M.J., Casford, J.S.L., Kendrick, C.P., Lloyd, J.M., Yongqiang, Z., Bird, M.I., 2011b. Evidence for bias in C and N concentrations and $\delta^{13}C$ composition of terrestrial and aquatic organic materials due to pre-analysis acid preparation methods. Chem. Geol. 282, 67–83. http://dx.doi.org/10.1016/j.chemgeo.2011.01.007.

Carpenter, K., Bartlett, J., Bird, J., Barrick, R., 2008. Ankylosaurs from the Price River Quarries, Cedar Mountain Formation (Lower Cretaceous), east-central Utah. J. Vertebr. Paleontol. 28, 1089–1101. http://dx.doi.org/10.1671/0272-4634-28.4.1089.

Currie, B.S., 1997. Sequence stratigraphy of non-marine Jurassic-Cretaceous rocks, central Cordilleran foreland-basin system. Geol. Soc. Am. Bull. 109, 1206–1222. http://dx.doi.org/10.1130/0016-7606(1997)109<1206:SSONJC>2.3.CO;2.

Currie, B.S., 2002. Structural configuration of the Early Cretaceous Cordilleran foreland-basin system and Sevier thrust belt, Utah and Colorado. J. Geol. 110, 697–718. http://dx.doi.org/10.1086/342626.

DeCelles, P.G., Currie, B.S., 1996. Long-term sediment accumulation in the Middle Jurassic-early Eocene Cordilleran retroarc foreland-basin system. Geology 24, 591–594. http://dx.doi.org/10.1130/0091-7613(1996)024<0591:LTSAIT>2.3.CO;2.

DeCelles, P.G., Lawton, T.F., Mitra, G., 1995. Thrust timing, growth of structural culminations, and synorogenic sedimentation in the type Sevier orogenic belt, western United States. Geology 23, 699–702. http://dx.doi.org/10.1130/0091-7613(1995)023<0699:TTGOSC>2.3.CO;2.

Dickinson, W.R., Gehrels, G.E., 2009. U–Pb ages of detrital zircons in Jurassic eolian and associated sandstones of the Colorado Plateau: evidence for transcontinental dispersal and intraregional recycling of sediment. Geol. Soc. Am. Bull. 121, 408–433.

Doyle, J.A., Robbins, E.I., 1977. Angiosperm pollen zonation of the continental cretaceous of the Atlantic coastal plain and its application to deep wells in the Salisbury embayment. Palynology 1, 41–78. http://dx.doi.org/10.1080/01916122.1977.9989150.

Eaton, J.G., Cifelli, R.L., 2001. Multituberculate mammals from near the Early-Late Cretaceous boundary, Cedar Mountain Formation, Utah. Acta Palaeontol. Pol. 46, 453–518.

Erba, E., Bartolini, A., Larson, R.L., 2004. Valanginian Weissert oceanic anoxic event. Geology 32, 149–152. http://dx.doi.org/10.1130/G20008.1.

Gile, L.H., Peterson, F.F., Grossman, R.B., 1966. Morphological and genetic sequences of carbonate accumulaion in desert soils. Soil Sci. 101, 347–360.

Goodwin, M.B., Clemens, W.A., Hutchison, J.H., Wood, C.B., Zavada, M.S., Kemp, A., Duffin, C.J., Schaff, C.R., 1999. Mesozoic continental vertebrates with associated palynostratigraphic dates from the northwestern Ethiopian plateau. J. Vertebr. Paleontol. 19, 728–741. http://dx.doi.org/10.1080/02724634.1999.10011185.

Greenhalgh, B.W., Britt, B.B., 2007. Stratigraphy and sedimentology of the Morrison-Cedar Mountain Formation boundary, east-central Utah. In: Willis, G.C., Hylland, M.D., Clark, D.L., Chidsey Jr., T.C. (Eds.), Central Utah: Diverse Geology of a Dynamic Landscape. Utah Geological Association, Salt Lake City, UT, pp. 81–100. Utah Geological Association Pub. 36.

Greenhalgh, B.W., Britt, B.B., Kowallis, B.J., 2006. New U-Pb age control for the lower Cedar Mountain Formation and an evaluation of the Morrison Formation/Cedar Mountain Formation boundary, Utah. In: Geological Society of America Abstracts With Programs.

Gröcke, D.R., Hesselbo, S.P., Jenkyns, H.C., 1999. Carbon-isotope composition of Lower Cretaceous fossil wood: ocean-atmosphere chemistry and relation to sea-level change. Geology 27, 155–158. http://dx.doi.org/10.1130/0091-7613(1999)027<0155:CICOLC>2.3.CO;2.

Hatzell, G.A., 2015. Paleoclimate Implications from Stable Isotope Analysis of Sedimentary Organic Carbon and Vertebrate Fossils from the Cedar Mountain Formation, UT, USA. University of Arkansas, Fayetteville, AR.

Hendrix, B., Moeller, A., Ludvigson, G.A., Joeckel, R.M., Kirkland, J.I., 2015. A new approach to date paleosols in terrestrial strata: a case study using u-pb zircon ages for the yellow cat member of the Cedar Mountain Formation of eastern Utah. Geol. Soc. Am. Abs. Prog. 47 (7). https://gsa.confex.com/gsa/2015AM/webprogram/Paper269057.html.

Herrle, J.O., Kößler, P., Friedrich, O., Erlenkeuser, H., Hemleben, C., 2004. High-resolution carbon isotope records of the Aptian to Lower Albian from SE France and the Mazagan Plateau (DSDP Site 545): a stratigraphic tool for paleoceanographic and paleobiologic reconstruction. Earth Planet. Sci. Lett. 218, 149–161. http://dx.doi.org/10.1016/S0012-821X(03)00646-0.

Hesselbo, S.P., Gröcke, D.R., Jenkyns, H.C., Bjerrum, C.J., Farrimond, P., Morgans Bell, H.S., Green, O.R., 2000. Massive dissociation of gas hydrate during a Jurassic oceanic anoxic event. Nature 406, 392–395. http://dx.doi.org/10.1038/35019044.

Hunt, G.J., Lawton, T.F., Kirkland, J.I., 2011. Detrital zircon U-Pb geochronological provenance of Lower Cretaceous strata, foreland basin, Utah. In: Sprinkel, D.A., Yonkee, W.A., Chidsey Jr., T.C. (Eds.), Sevier thrust belt: northern and central Utah and adjacent areas. pp. 193–211. Utah Geological Association Publication 40.

Jahren, A.H., Arens, N.C., Sarmiento, G., Guerrero, J., Amundson, R., 2001. Terrestrial record of methane hydrate dissociation in the Early Cretaceous. Geology 29, 159–162. http://dx.doi.org/10.1130/0091-7613(2001)029<0159:TROMHD>2.0.CO;2.

Jenkyns, H., 2010. Geochemistry of ocean anoxic events. Geochem. Geophys. Geosyst. 11, Q03004. http://dx.doi.org/10.1029/2009GC002788.

Kent, D.V., Olsen, P.E., 1999. Astronomically tuned geomagnetic polarity timescale for the Late Triassic. J. Geophys. Res. Solid Earth 104, 12831–12841.

Kirkland, J.I., 2005. Utah's newly recognized dinosaur record from the Early Cretaceous Cedar Mountain Formation. Utah Geol. Surv. Surv. Notes 37, 1–5.

Kirkland, J.I., Madsen, S.K., 2007. The Lower Cretaceous Cedar Mountain Formation, eastern Utah the view up an always interesting learning curve. In: Lund, W.R. (Ed.), Field Guide to Geological Excursions in Southern Utah, Geological Society of America Rocky Mountain Section 2007 Annual Meeting, Grand Junction Geological Society. Utah Geological Association Publication 35:1-108 CD-ROM.

Kirkland, J.I., Burge, D., Gaston, R., 1993. A large dromaeosaurid (Theropoda) from the Lower Cretaceous of Eastern Utah. Hunteria. 2. no. 10, 16 p.

Kirkland, J.I., Britt, B., Burge, D.L., Carpenter, K., Cifelli, R., DeCourten, F., Eaton, J., Hasiotis, S., Lawton, T., 1997. Lower to middle Cretaceous dinosaur faunas of the central Colorado Plateau: a key to understanding 35 million years of tectonics, sedimentology, evolution, and biogeography. Brigh. Young Univ. Geol. Stud. 42, 69–104.

Kirkland, J.I., Britt, B.B., Whittle, C.H., Madsen, S.K., Burge, D.L., 1998. A small coelurosaurian theropod from the Yellow Cat Member of the Cedar Mountain Formation (Lower Cretaceous, Barremian) of eastern Utah. Mus. Nat. Hist. Sci. Bull. 14, 239–248.

Kirkland, J.I., DeBlieux, D., Madsen, S.K., Hunt, G.J., 2012. New dinosaurs from the base of the Cretaceous in eastern Utah suggest that the "so-called" basal Cretaceous calcrete in the Yellow Cat Member of the Cedar Mountain Formation, while not marking the Jurassic-Cretaceous unconformity represents evolutionary time. J. Vertebr. Paleontol. 18 (Suppl.), 121–122. October 12, Abstracts and Program.

Lisiecki, L.E., Lisiecki, P.A., 2002. Application of dynamic programming to the correlation of paleoclimate records. Paleoceanography 17, 1049. http://dx.doi.org/10.1029/2001PA000733.

Lockley, M., Kirkland, J.I., DeCourten, F., Britt, B.B., Hasiotis, S., 1999. Dinosaur tracks from the Cedar Mountain Formation of eastern Utah: a preliminary report. In: Gillette, D. (Ed.), Vertebrate Paleontology in Utah, Utah Geological Survey. Utah Geological Survey, Salt Lake City, UT, pp. 253–257. Utah Geological Survey, Misc. Pub. 99-1.

Ludvigson, G.A., Joeckel, R.M., Gonzalez, L.A., Gulbranson, E.L., Rasbury, E.T., Hunt, G.J., Kirkland, J.I., Madsen, S., 2010. Correlation of Aptian-Albian carbon isotope excursions in continental strata of the Cretaceous foreland basin, eastern Utah, USA. J. Sediment. Res. 80, 955–974.

Ludvigson, G.A., Joeckel, R.M., Murphy, L.R., Stockli, D.F., González, L.A., Suarez, C.A., Kirkland, J.I., Al-Suwaidi, A., 2015. The emerging terrestrial record of Aptian-Albian global change. Cretac. Res. 56, 1–24. http://dx.doi.org/10.1016/j.cretres.2014.11.008.

Magioncalda, R., Dupuis, C., Smith, T., Steurbaut, E., Gingerich, P.D., 2004. Paleocene-Eocene carbon isotope excursion in organic carbon and pedogenic carbonate: direct comparison in a continental stratigraphic section. Geology 32, 553–556. http://dx.doi.org/10.1130/G20476.1.

Markevitch, V.S., 1994. Palynological zonation of the continental Cretaceous and lower Tertiary of eastern Russia. Cretac. Res. 15, 165–177. http://dx.doi.org/10.1006/cres.1994.1008.

McDonald, A.T., Kirkland, J.I., DeBlieux, D.D., Madsen, S.K., Cavin, J., Milner, A.R.C., Panzarin, L., 2010. New basal iguanodonts from the Cedar Mountain formation of Utah and the evolution of thumb-spiked dinosaurs. PLoS ONE. 5, e14075. http://dx.doi.org/10.1371/journal.pone.0014075.

Midwood, A.J., Boutton, T.W., 1998. Soil carbonate decomposition by acid has little effect on $\delta^{13}C$ of organic matter. Soil Biol. Biochem. 30, 1301–1307.

Mori, H., 2009. Dinosaurian Faunas of the Cedar Mountain Formation With Detrital Zircon Ages for Three Stratigraphic Sections and the Relationship Between the Degree of Abrasion and U-Pb LA-ICP-MS Ages of Detrital Zircons. Department of Geolological Sciences, Brigham Young University, Provo, UT. Unpublished M.S. Thesis, 102 p.

Nelson, M.E., Crooks, D.M., 1987. Stratigraphy and paleontology of the cedar mountain formation (lower cretaceous), eastern emery county, Utah. In: Paleontology and Geology of the Dinosaur Triangle. Grand Junction Geological Society, Grand Junction, CO, pp. 55–63.

Rasbury, E.T., Hanson, G.N., Meyers, W.J., Saller, A.H., 1997. Dating of the time of sedimentation using U-Pb ages for paleosol calcite. Geochim. Cosmochim. Acta 61, 1525–1529. http://dx.doi.org/10.1016/S0016-7037(97)00043-4.

Rasbury, E.T., Meyers, W.J., Hanson, G.N., Goldstein, R.H., Saller, A.H., 2000. Relationship of Uranium to petrography of caliche paleosols with application to precisely dating the time of dedimentation. J. Sediment. Res. 70, 604–618. http://dx.doi.org/10.1306/2DC4092B-0E47-11D7-8643000102C1865D.

Retallack, G.J., 1997. A Colour Guide to Paleosols. Wiley, West Sussex. 175p.

Robinson, S.A., Hesselbo, S.P., 2004. Fossil-wood carbon-isotope stratigraphy of the non-marine Wealden Group (Lower Cretaceous, southern England). J. Geol. Soc. 161, 133–145. http://dx.doi.org/10.1144/0016-764903-004.

Sageman, B.B., Hollander, D.J., 1999. Cross correlation of paleoecological and geochemical proxies: a holistic approach to the study of past global change. Geol. Soc. Am. Spec. Pap., 332, 365–384.

Sames, B., Cifelli, R.L., Schudack, M.E., 2010. The nonmarine Lower Cretaceous of the North American Western Interior foreland basin: new biostratigraphic results from ostracod correlations and early mammals, and their implications for paleontology and geology of the basin—an overview. Earth-Sci. Rev. 101, 207–224.

Scholle, P.A., Arthur, M.A., 1980. Carbon isotope fluctuations in Cretaceous pelagic limestones: potential stratigraphic and petroleum exploration tool. AAPG Bull. 64, 67–87.

Sprinkel, D.A., Weiss, M.P., Fleming, R.W., Waanders, G.L., 1999. Redefining the Lower Cretaceous stratigraphy within the central Utah foreland basin. Utah Geol. Surv. Spec. Stud. 97. 21 p.

Sprinkel, D.A., Madsen, S.K., Kirkland, J.I., Waanders, G.L., Hunt, G.J., 2012. Cedar Mountain and Dakota Formations around Dinosaur National Monument—Evidence of the First Incursion of the Cretaceous Western Interior Seaway into Utah. Utah Geological Survey, Salt Lake City, UT. Utah Geological Survey Special Study 143, 21 p, 6 appendices.

Stikes, M.W., 2006. Fluvial Facies and Architecture of the Poison Strip Sandstone, Lower Cretaceous Cedar Mountain Formation, Grand County, Utah. Utah Geological Survey, Salt Lake City, UT. Miscellaneous Publication 06-2, 84 p., CD-ROM.

Stokes, W.L., 1944. Morrison formation and related deposits in and adjacent to the Colorado Plateau. Geol. Soc. Am. Bull. 55, 951–992. http://dx.doi.org/10.1130/GSAB-55-951.

Stokes, W.L., 1952. Lower Cretaceous in Colorado Plateau. AAPG Bull. 36, 1766–1776.

Suarez, M.B., Ludvigson, G.A., González, L.A., Al-Suwaidi, A.H., You, H.-L., 2013. Stable isotope chemostratigraphy in lacustrine strata of the Xiagou Formation, Gansu Province, NW China. Geol. Soc. Lond. Spec. Publ. 382, 143–155.

Suarez, C.A., González, L.A., Ludvigson, G.A., Kirkland, J.I., Cifelli, R.L., Kohn, M.J., 2014. Multi-Taxa isotopic investigation of Paleohydrology in the Lower Cretaceous Cedar Mountain Formation, Eastern Utah, U.S.A.: deciphering effects of the Nevadaplano Plateau on regional climate. J. Sediment. Res. 84, 975–987.

Thayn, G.F., Tidwell, W.D., Stokes, W.L., 1985. Flora of the Lower Cretaceous Cedar Mountain Formation of Utah and Colorado. Part III: Icacinoxylon pittiense n. sp. Am. J. Bot. 72, 175–180. http://dx.doi.org/10.2307/2443544.

Tipple, B.J., Pagani, M., Krishnan, S., Dirghangi, S.S., Galeotti, S., Agnini, C., Giusberti, L., Rio, D., 2011. Coupled high-resolution marine and terrestrial records of carbon and hydrologic cycles variations during the Paleocene-Eocene Thermal Maximum (PETM). Earth Planet. Sci. Lett. 311, 82–92.

Wang, S.C., Dodson, P., 2006. Estimating the diversity of dinosaurs. Proc. Natl. Acad. Sci. U.S.A. 103 (37), 13601–13605. http://dx.doi.org/10.1073/pnas.0606028103.

Wang, Z.S., Rasbury, E.T., Hanson, G.N., Meyers, W.J., 1998. Using the U–Pb system of calcretes to date the time of sedimentation of clastic sedimentary rocks. Geochim. Cosmochim. Acta 62, 2823–2835. http://dx.doi.org/10.1016/S0016-7037(98)00201-4.

Waterman, M.S., Raymond Jr., R., 1987. The match game: new stratigraphic correlation algorithms. Math. Geol. 19, 109–127. http://dx.doi.org/10.1007/BF00898191.

Weiss, M.P., Roche, M.G., 1988. The Cedar Mountain Formation (Lower Cretaceous) in the Gunnison Plateau, central Utah. Geol. Soc. Am. Mem. 171, 557–570. http://dx.doi.org/10.1130/MEM171-p557.

Whiteside, J.H., Olsen, P.E., Eglinton, T., Brookfield, M.E., Sambrotto, R.N., 2010. Compound-specific carbon isotopes from Earth's largest flood basalt eruptions directly linked to the end-Triassic mass extinction. Proc. Natl. Acad. Sci. U.S.A. 107, 6721–6725. http://dx.doi.org/10.1073/pnas.1001706107.

Wills, M.A., Barrett, P.M., Heathcote, J.F., 2008. The modified gap excess ratio (GER*) and the stratigraphic congruence of dinosaur phylogenies. Syst. Biol. 57 (6), 891–904.

Yingling, V.L., Heller, P.L., 1992. Timing and record of foreland sedimentation during the initiation of the Sevier orogenic belt in central Utah. Basin Res. 4, 279–290. http://dx.doi.org/10.1111/j.1365-2117.1992.tb00049.x.

Young, R.G., 1960. Dakota group of Colorado Plateau. AAPG Bull. 44, 156–194.

M.B. Suarez is an assistant professor in Geological Sciences her hometown of San Antonio, at the University of Texas at San Antonio. Dr. Suarez received a PhD in Geology from the University of Kansas in 2009. Her research focuses on stable isotope geochemistry and terrestrial paleoclimate and depositional environments, especially of Greenhouse climates such as the Cretaceous.

INDEX

Note: Page numbers followed by *f* indicate figures, and *t* indicate tables.